Challenges and Opportunities in the Global Clean Energy Transition

Challenges and Opportunities in the Global Clean Energy Transition

Guest Editors

Nikolay Hinov
Georgi Todorov

MDPI

Basel • Beijing • Wuhan • Barcelona • Belgrade • Novi Sad • Cluj • Manchester

Guest Editors

Nikolay Hinov
Department of Computer Systems
Technical University of Sofia
Sofia
Bulgaria

Georgi Todorov
Department of Manufacturing Technology and Systems
Technical University of Sofia
Sofia
Bulgaria

Editorial Office
MDPI AG
Grosspeteranlage 5
4052 Basel, Switzerland

This is a reprint of the Special Issue, published open access by the journal *Energies* (ISSN 1996-1073), freely accessible at: https://www.mdpi.com/journal/energies/special_issues/9E21J9170O.

For citation purposes, cite each article independently as indicated on the article page online and as indicated below:

Lastname, A.A.; Lastname, B.B. Article Title. *Journal Name* **Year**, *Volume Number*, Page Range.

ISBN 978-3-7258-7382-1 (Hbk)
ISBN 978-3-7258-7383-8 (PDF)
https://doi.org/10.3390/books978-3-7258-7383-8

Contents

About the Editors vii

Preface ix

Piotr Olczak and Dominika Matuszewska
Analysis of Implementing Hydrogen Storage for Surplus Energy from PV Systems in Polish Households
Reprinted from: *Energies* **2025**, *18*, 3674, https://doi.org/10.3390/en18143674 1

Wenyu Wang, Runfan Zhu, Siyu Liu, Yong He, Wubin Weng, Shixing Wang, et al.
Investigation of Flame Structure and PAHs' Evolution in a Swirl-Stabilized Spray Flame at Elevated Pressure
Reprinted from: *Energies* **2025**, *18*, 3923, https://doi.org/10.3390/en18153923 19

Nikolay Hinov
Hybrid Small Modular Reactor—Renewable Systems for Smart Cities: A Simulation-Based Assessment for Clean and Resilient Urban Energy Transitions
Reprinted from: *Energies* **2025**, *18*, 3993, https://doi.org/10.3390/en18153993 35

Qingsheng Liu, Xuanhong Ye, Hai Huang, Junjie Cheng, Kai Meng, Qinglong Yu, et al.
Parametric Evaluation of Coolant Channels for Proton-Exchange Membrane Fuel Cell Based on Multi-Pass Serpentine Flow Field
Reprinted from: *Energies* **2025**, *18*, 4264, https://doi.org/10.3390/en18164264 62

Jiawei Fu, Xinchun Liu, Yuqing Kong, Ruyu Zhao, Yinyong Sun and Ahmed S. Abou-Elyazed
Enhancing the Adsorption Performance of HKUST-1 by Adding NH_4F During Room-Temperature Synthesis for Desulfurization of Fuel Oil
Reprinted from: *Energies* **2025**, *18*, 5344, https://doi.org/10.3390/en18205344 91

Chengyao Song, Amin Ghanizadeh and Christopher R. Clarkson
The Impact of Brine Saturation and Distribution on Lean Gas Huff-n-Puff EOR Performance of Tight Oil Reservoirs: Examples from the Montney Formation (Canada)
Reprinted from: *Energies* **2025**, *18*, 5537, https://doi.org/10.3390/en18205537 107

Maksymilian Mądziel
Phase-Specific Mixture of Experts Architecture for Real-Time NO_x Prediction in Diesel Vehicles: Advancing Euro 7 Compliance
Reprinted from: *Energies* **2025**, *18*, 5853, https://doi.org/10.3390/en18215853 123

Thu-Thao Le and Wen-Min Lu
Charting the Landscape of Data Envelopment Analysis in Renewable Energy and Carbon Emission Efficiency
Reprinted from: *Energies* **2025**, *18*, 6147, https://doi.org/10.3390/en18236147 153

Silvia Baeva, Ivelina Hinova and Plamen Stanchev
Analysis and Evaluation of the Operating Profile of a DC Inverter in a PV Plant
Reprinted from: *Energies* **2025**, *18*, 6306, https://doi.org/10.3390/en18236306 174

Nikolay Hinov
Solid-State Transformers in the Global Clean Energy Transition: Decarbonization Impact and Lifecycle Performance
Reprinted from: *Energies* **2026**, *19*, 558, https://doi.org/10.3390/en19020558 211

About the Editors

Nikolay Hinov

Nikolay Hinov is a Professor in the Department of Computer Systems at the Technical University of Sofia, Bulgaria. He graduated with a Master's degree in Electronics and Microelectronics at the Technical University of Sofia in 1995 and defended his PhD dissertation in 1998. In 2024, he also defended his dissertation for the degree of Doctor of Science, dedicated to the model-based design of power electronic devices. His professional career is inextricably linked to the Technical University of Sofia, where he held various academic positions—from senior assistant to professor. Over the years, he has held leadership positions, including Deputy Dean and Head of the Department of Power Electronics and Secretary of the Bulgaria Section CAS/SSC Chapter of IEEE. His scientific interests include industrial and power electronics, artificial intelligence systems, electronic converters, autonomous and electric vehicles, smart grids and cities, as well as modelling and optimization of mechatronic systems. He has over 350 scientific publications indexed in Scopus and Web of Science, over 1000 peer-reviewed reviews in IF and/or SJR journals, and has participated in more than 50 research projects, including being the leader of 10 of them. He holds 10 patents and 5 utility models, and under his supervision, seven doctoral students have successfully defended their dissertations. He has initiated, organized, and edited more than 10 Special Issues in various IF journals. He is the author of textbooks, laboratory manuals, and one monograph indexed in Scopus. He is a member of the editorial boards of three international scientific journals with an impact factor. He has received a number of awards, including the title of Senior Member of IEEE, the Gold Badge, and the Gold Medal "Prof. Assen Zlatarov" of the Federation of Scientific and Technical Unions in Bulgaria, an award for the best project of the National Science Fund (Bulgaria) in the field of Technical Sciences, received in 2024.

Georgi Todorov

Georgi Todorov is a researcher and professor at the Technical University of Sofia, Faculty of Industrial Technology, where he has served as Dean since 2019. Since 2024, he has also been Director of the National Centre of Excellence "Mechatronics and Clean Technologies" at the Technical University of Sofia. He is also a Corresponding Member of the Bulgarian Academy of Sciences. He obtained his BSc degree in 1986, MSc in 1988, PhD in 1998, and DSc in 2019 from the Technical University of Sofia. His research interests include 3D technologies, virtual engineering, rapid prototyping, new product development, medical implants, advanced manufacturing technologies, and energy storage systems. His work focuses on matrix approaches for integrating mechanical, electronic, sensor, and software systems to achieve cost-effective and reliable engineering solutions. He is the founder of a laboratory in implantology, a research group in 3D technologies and virtual engineering at the Technical University of Sofia, and has supervised 24 successful PhD students. Prof. Todorov has served as principal investigator of more than 20 international industrial projects and has participated in projects under the EU Framework Programmes FP7, Horizon 2020, and Horizon Europe. He has been a visiting professor at Harbin University of Science and Technology (China) and has completed research specializations in Japan and the United Kingdom in the field of CAD/CAM systems. He is the author of 243 scientific publications (96 indexed in Scopus/WoS, h-index 9) and co-author of 86 patents, including 26 international patents registered in Europe, the USA, China, Japan, and other countries. He has received numerous national awards for scientific achievements and innovation, including the "Pythagoras" National Award for Applied Sciences and the "Inventor of the Year" Award from the Patent Office of the Republic of Bulgaria.

Preface

This Reprint presents a collection of scientific contributions addressing current challenges and emerging opportunities in the global transition toward cleaner and more sustainable energy systems. The scope of this Reprint includes research related to energy harvesting, energy conversion and storage, green and intelligent transport, energy-efficient manufacturing, and digital technologies supporting sustainable industry and society.

The aim of this Reprint is to highlight interdisciplinary approaches that combine advances in energy engineering, electronics, mechatronics, intelligent sensing, modelling and control, and digital technologies. Such integration is essential for developing innovative technological solutions that improve energy efficiency, reduce environmental impact, and support the transformation of modern industrial and societal systems.

The scientific directions represented in this Reprint are closely related to the research priorities of the National Centre of Excellence in Mechatronics and Clean Technologies at the Technical University of Sofia, located in Campus Studentski Grad, Bulgaria. The centre serves as a national research infrastructure supporting advanced studies in mechatronics, clean technologies, intelligent systems, virtual engineering, bio-mechatronics, and sustainable technological development.

This Reprint is intended for researchers, engineers, and professionals working in the fields of energy systems, clean technologies, mechatronics, and sustainable engineering. By presenting recent research results and practical engineering solutions, this Reprint contributes to the broader discussion on technological innovation and interdisciplinary collaboration required to support the transition toward cleaner, smarter, and more efficient energy and industrial systems.

Nikolay Hinov and Georgi Todorov
Guest Editors

energies

Article

Analysis of Implementing Hydrogen Storage for Surplus Energy from PV Systems in Polish Households

Piotr Olczak [1,*] and Dominika Matuszewska [2]

[1] Mineral and Energy Economy Research Institute, Polish Academy of Sciences, 7A Wybickiego St., 31-261 Cracow, Poland
[2] Faculty of Energy and Fuels, AGH University of Kraków, 30 Mickiewicza Ave., 30-059 Cracow, Poland; dommat@agh.edu.pl
* Correspondence: olczak@min-pan.krakow.pl

Abstract

One of the methods for mitigating the duck curve phenomenon in photovoltaic (PV) energy systems is storing surplus energy in the form of hydrogen. However, there is a lack of studies focused on residential PV systems that assess the impact of hydrogen storage on the reduction of energy flow imbalance to and from the national grid. This study presents an analysis of hydrogen energy storage based on real-world data from a household PV installation. Using simulation methods grounded in actual electricity consumption and hourly PV production data, the research identified the storage requirements, including the required operating hours and the capacity of the hydrogen tank. The analysis was based on a 1 kW electrolyzer and a fuel cell, representing the smallest and most basic commercially available units, and included a sensitivity analysis. At the household level—represented by a single-family home with an annual energy consumption and PV production of approximately 4–5 MWh over a two-year period—hydrogen storage enabled the production of 49.8 kg and 44.6 kg of hydrogen in the first and second years, respectively. This corresponded to the use of 3303 kWh of PV-generated electricity and an increase in self-consumption from 30% to 64%. Hydrogen storage helped to smooth out peak energy flows from the PV system, decreasing the imbalance from 5.73 kWh to 4.42 kWh. However, while it greatly improves self-consumption, its capacity to mitigate power flow imbalance further is constrained; substantial improvements would necessitate a much larger electrolyzer proportional in size to the PV system's output.

Keywords: hydrogen; long-term household energy storage; photovoltaic; hybrid photovoltaic; hydrogen

1. Introduction

1.1. Polish Energy Mix and Challenges

The growing share of renewable energy sources in the energy mix in Poland brings new challenges. Storing energy on a daily basis using batteries is not always a viable solution, as increasing storage capacity may lead to underutilization of the energy generated during daylight hours due to insufficient nighttime demand. Consequently, the capacity is not fully utilized the next day. This occurs because daily household energy consumption, especially in summer, is often less than the amount of energy stored (for increased storage capacity), resulting in storage not being emptied overnight. This indicates that the assumed storage capacity is insufficiently large compared to actual consumption. One potential solution is energy storage using hydrogen.

Both rising electricity prices and their variability on a daily and seasonal scale, as well as the need to introduce an increasing volume of energy from renewable sources, create favorable conditions for the development and application of energy storage systems, including the aforementioned hydrogen storage systems. The energy transformation in Poland forces the use of increasingly new solutions in this area [1,2]. An analysis of the use of various types of renewable energy in Polish households was made by Wróblewski and Niekurzak [3]. They accounted for the impact of using various types of renewable energy sources (RES) combined with financing methods on reducing CO_2 emissions, specifically in the context of replacing old heating systems with new ones, while considering electricity generation from photovoltaic systems.

In Poland, the volume of energy produced from photovoltaics and wind is growing every year, despite the presence of many barriers to the development of these technologies and the seasonality of high energy production from these sources. Their seasonal excess affects both the national grid and the local scale and leads to various related problems. Locally, the problems are mainly related to photovoltaic micro-installations, because over 99% of micro-installations are photovoltaic systems [4]. The installed capacity of renewable energy sources has been increasing for several years, resulting in steadily rising energy production from these sources. From 2023, a temporary excess of energy that can be produced (on a daily scale) has led to the need to curtail some sources (mainly PV in 2024). At the same time, it is noticeable that wind energy predominantly generates in the winter period and photovoltaics in the spring and summer periods. Additionally, the scale of installed photovoltaic capacity is significant, mostly prosumer installations, of which there are currently about 1.5 million [4], while the installed wind power capacity consists only of large turbines. Support programs for smaller wind turbines on a household scale are only just emerging, but this is associated with high uncertainty in the annual productivity of these sources in various locations in the country, as shown, for example, by Kryzia et al. [5], but the local location system is also important.

The benefits and opportunities of energy storage at the level of households, enterprises, and local communities in Poland have also been analyzed [6,7]. From an economic point of view, without more than a 50% subsidy for energy storage, this type of installation is not profitable. To avoid the duck curve effect, energy storage facilities should be properly managed.

Currently, energy storage in battery storage and pumped storage power plants is a solution on a daily scale, as Olczak showed [8,9], but it is not a preferred solution on a household scale, because these households have a much higher energy surplus around noon than the maximum hourly energy demand at night. This trend intensifies with the increasing average power of photovoltaic micro-installations among prosumers: in 2021, this value was below 7 kWp, while in 2024, it was above 8 kWp [4]. Work is also being carried out (in the field of, among others, energy management systems) to increase the self-consumption rate in this type of household. This is, of course, purposeful and very necessary, but it is not possible to effectively utilize the additional several dozen kWh of energy on a household scale every day. There are simply no such energy needs [7,10], and even if there were, the size of the selected power of the PV installation would be worth considering. The introduction of dynamic tariffs may allow for an increase in energy consumption (around noon hours) also among those who do not use photovoltaics, which in the future will enable more efficient integration of energy into the grid by prosumers or photovoltaic farms. The occurrence of periods of even several days in a row with surpluses of energy from photovoltaics (around noon) contributes to the need for long-term storage of electricity for households. As Olczak showed, such storage would have to have large capacities, i.e., of the order of several megawatt hours for a household [11].

Therefore, partial storage of excess energy from photovoltaics into hydrogen is considered a viable option.

1.2. Literature Review

Kupecki et al. proposed a 10 kW power-to-gas system along with an assessment and analysis of the system's effectiveness for stabilizing operation, among others, of the national grid (in the case of both large-scale power and a large number of micro-installations) [12]. A similar analysis for a 5 kWp photovoltaic home installation (a household in Switzerland) and hydrogen production, as well as hydrogen storage installation, was conducted by Hollmuller before 2000 [13]. However, it concerned a different purpose and was based on different technologies from over 20 years ago, and these have developed significantly in recent years [14–16].

Since then, several studies have utilized the HOMER software to explore hydrogen's potential for energy storage in specialized applications. For example, Kalamaras et al. [17] used HOMER to simulate and calculate the feasibility of employing hydrogen for energy storage in specialized home installations, such as off-grid systems. As a result, they identified a configuration, including a 3.35 kW PV module, a 4.2 kW electrolyzer, a 40 kg hydrogen tank, and a 2.9 kW fuel cell, as enough combination to fulfill electrical and thermal load demand for an off-grid Greek island. Similarly, Lacko et al. [18] employed HOMER to analyze the use of hydrogen in a stand-alone renewable combined heat and power system. They included three scenarios of electricity and heat production using combinations of RES and fossil fuel to supply a reference household. Analyses were based on actual measured data and simulation. Kapen et al. [19], also using HOMER, evaluated the feasibility of integrating a hydrogen system to relieve the power system in Cameroon. They analyzed energy load and production, including a hydrogen tank (5000 kg) for three types of household electricity demand profiles. Attemene et al. [20] conducted optimization studies focused on system configuration, considering the total cost, including capital, maintenance, and replacement costs. Additionally, they analyzed energy reliability based on the Loss of Power Supply Probability (LPSP). Their research employed methods such as the Non-Dominated Sorting Genetic Algorithm II (NSGA-II) and included sensitivity analysis based on wind speed profiles and subsystem costs [20].

Budak and Devrim analyzed the system for a household. The methanol and water electrolyzers were designed for a 1.2 kW PEMFC hydrogen demand, which met the household's energy requirements. The analysis shows that the use of the methanol electrolyzer can be effective [21]. Hashimoto and Kawahara optimally selected the fuel cell (FC) and photovoltaic system for a single-family house [22]. Hassan carried out an optimization of the size of household devices in the field of hydrogen and renewable energy production [23]. Torok et al. developed a prototype solution for producing hydrogen for households [24].

The aging effects on the modeling and operation of a photovoltaic system with hydrogen storage over 10 years of operation were analyzed by Ceran et al. The analysis was based on experimental data, including photovoltaic modules, fuel cells, electrolyzers, hydrogen storage, and hydrogen backup systems [25]. The results showed that after 10 years of system operation, the electricity consumption per unit of hydrogen by the electrolyzer increases by at least 4.6%, while the hydrogen consumption per unit of electricity by the fuel cell increases by over 13%. Additionally, the energy efficiency of photovoltaic panels decreases. These factors, combined with the variability in production and energy demand caused by factors such as weather conditions, present challenges for selecting appropriate infrastructure for such systems.

Abdolmaleki and Berardi [26] studied hybrid renewable energy-based systems incorporating PV, fuel cell, and hydrogen. Among others, PV/hydrogen was compared using the HOMER software, available for major North American cities. They also investigated a hydrogen-fueled system for zero-energy buildings based on dynamic simulation with TRNSYS software [27]. Jahanbin et al. [28] analyzed integrating hybrid-battery hydrogen energy storage for academic buildings. As a result of integrating PV with gaseous hydrogen and metal hydride storage, the annual renewable factor increased to 22.6% (for PV alone, 14.4%).

Shao et al. [29] analyzed the use of a hydrogen-centered multi-energy microgrid for rural areas based on household application. They also conducted a risk analysis of using hydrogen in microgrids for rural areas. Zhuk et al. [30] investigated a fully automated hydrogen generator for domestic and mobile applications, and estimated the hydrogen cost at 5–8 USD/kg H_2. Nastasi et al. [31] compared hydrogen and battery solutions for storing solar energy applications in five buildings. The optimal storage time (based on CAPEX and OPEX) was calculated as weekly or monthly storage, depending on the electricity demand profile.

The objective of this study is to determine the appropriate size of a hydrogen storage tank for a household-scale system, given specified power capacities of the fuel cell and electrolyzer, and to address the associated energy challenges. Numerous studies have explored the optimal sizing of energy storage systems, often from an economic perspective (e.g., utilizing tools such as HOMER software) or by analyzing technical requirements for storing energy from renewable sources. However, these studies typically focus on determining the appropriate capacity of hydrogen storage infrastructure and its potential impact on the national energy grid. The latter is achieved by quantitatively estimating the scope of solving the "duck curve" phenomenon by storing energy in hydrogen, which remains a complex and multifaceted challenge [8,32].

Section 2 provides an overview of the installation used as the basis for the calculations and details the methodologies employed in this study. Section 3 presents the results of the calculations, followed by a discussion of these findings in Section 4. Finally, Section 5 concludes the article with key takeaways and insights.

2. Materials and Methods

2.1. Research Object

A household facility located in Małopolska with energy consumption corresponding to 4–5 MWh per year was selected for the analysis, rather than households located in blocks of flats, as only those with slightly increased electricity demand consider such solutions. The selected installation (capacity PPV = 5.04 kWp) has already been analyzed in several articles regarding both photovoltaics [33] and energy storage, but never in terms of hydrogen production. Examples of energy consumption as well as energy production from PV are shown in Figure A1.

Approximately 850 h with energy consumption exceeding 1 kWh were recorded annually. For PV, 1780 and 1830 h in particular years with energy consumption exceeding 1 kWh were observed, and approximately 3500 h with energy production above 0.1 kWh.

Surprisingly, for the installation studied, an 8–10 kWh energy storage system proved as effective at storing annual energy as a system ten times larger [11]. Nonetheless, to achieve relative energy self-sufficiency, a substantially larger storage capacity of around 2 MWh is necessary, and even then, performance will fluctuate with yearly weather patterns. The EUR 1 million [11] cost of this storage facility is exorbitant compared to average annual household energy expenses of under EUR 2000. Therefore, the advantages of energy storage do not directly lead to significant savings or reducing energy costs.

2.2. Hydrogen Installation

The above-mentioned issues related to battery storage suggest exploring solutions in other technologies, hence, the proposal to store household energy in hydrogen—Figure 1, and the energy calculation algorithm—Figure 2.

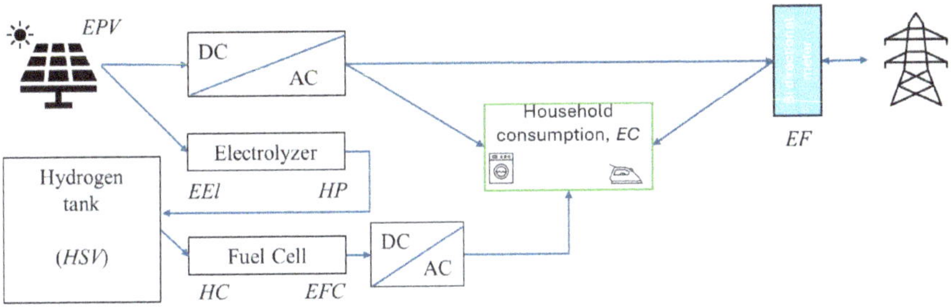

Figure 1. Proposed installation scheme with hydrogen storage. Symbols are explained in the equations below.

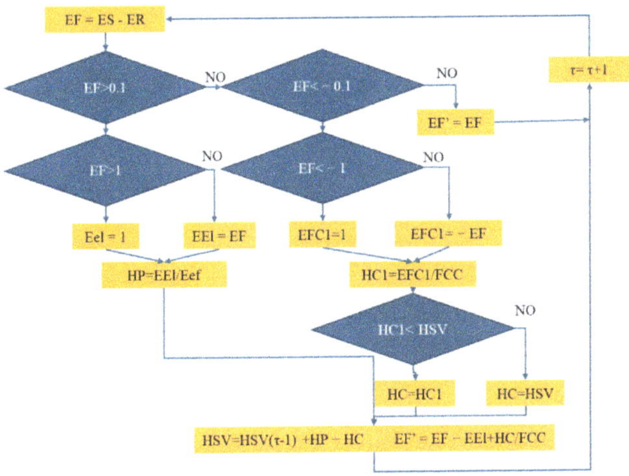

Figure 2. Simplified calculation algorithm (without the influence of storage capacity)—please see also equations in Appendix A.

Hourly mass value of hydrogen production in the electrolyzer (*HP*):

$$HP(\tau) = \frac{EEl(\tau)}{Eef} \tag{1}$$

Eef—electrolyzer energy consumption (for the production of 1 kg of H_2): 50 [34], kWh/kg H_2

EEl—please see Appendix A, kWh

Energy flow values after hydrogen system is installed (*EF'*):

$$EF'(\tau) = EF(\tau) - EEl(\tau) + \frac{HC(\tau)}{FCC} \tag{2}$$

The algorithm for calculating *HSV* and *EF'* is shown in the figure below. The input capacity of the electrolyzer and the output capacity of the fuel cell were assumed to be *PH* = 1 kW.

2.3. Impact on Duck Curve Calculation Method

Additionally, the impact of energy storage in hydrogen on the unevenness of energy flow from installation to the external grid (*EF*) was demonstrated, expressed using the *VDC* unevenness index, as shown by [32], for installations without storage:

$$VDC(day) = max(EF(day)) - min(EF(day)) \quad (3)$$

However, with a hydrogen storage system, the calculation is based on the following formula:

$$VDC\prime(day) = max(EF\prime(day)) - min(EF\prime(day)) \quad (4)$$

where:

VDC—maximum daily difference in energy flow without storage, kWh
VDC′—maximum daily difference in energy flow with hydrogen storage, kWh
EF or *EF′*—energy flow without or with hydrogen storage, kWh

2.4. Self-Consumption Ratio and Self-Sufficiency Ratio

Hourly self-consumption energy with a hydrogen storage system was calculated by:

$$ESC'(\tau) = EPV(\tau) - ES'(\tau, HSC0) \quad (5)$$

where:

ESC′—self-consumed energy with hydrogen installation, kWh
EPV—hourly PV energy production, kWh
ES′—hourly energy sent to the grid from installation with hydrogen installation—Appendix A, kWh
HSC0—hydrogen storage useful capacity—Appendix A, kg

3. Results

The results in the form of the amount of hydrogen produced in the electrolysis process and the amount of hydrogen consumed in the fuel cell, presented alongside the energy flow to and from the grid, are shown in Figure 3.

On the first day (1 June 2020), the maximum hourly amount of hydrogen production was 28.6 g H_2 (from 10 am to 5 pm), and the maximum hourly amount of hydrogen consumption was 54.9 g H_2. During each day shown in Figure 3, the cumulative production of hydrogen reached nearly 0.25 kg H_2/day. On the second day, the maximum value of *EF* was 3.76 kWh, and the minimum was −0.39 kWh (*VDC* = 4.15 kWh); with the hydrogen system, the values were 2.76 kWh and −0.20 kWh (*VDC′* = 2.96 kWh). The excess energy from PV transferred to the electrolysis process was 9.4 kWh on the first day, 8.3 kWh on the second day, and 8.0 kWh on the third day.

However, the total value of energy delivered by the fuel cell was 2 kWh on the first day (practical operation began in the afternoon, as the hydrogen tank was first filled at 8 a.m.), 2.1 kWh on the second day, and 2.8 kWh on the third day.

The total electricity directed to the electrolysis process was 1742 kWh in the first year and 1561 kWh in the second year—a total of 3303 kWh, which accounts for approximately 33% of the energy produced by the PV installation. The total electricity produced by the fuel cell was 546 kWh in the first year and 490 kWh in the second year.

The total hydrogen produced in the first year was 49.8 kg and in the second year, 44.6 kg, with the same amount of hydrogen consumed. The amount of energy allocated to the electrolysis process in relation to the total PV production is shown in Figure 4.

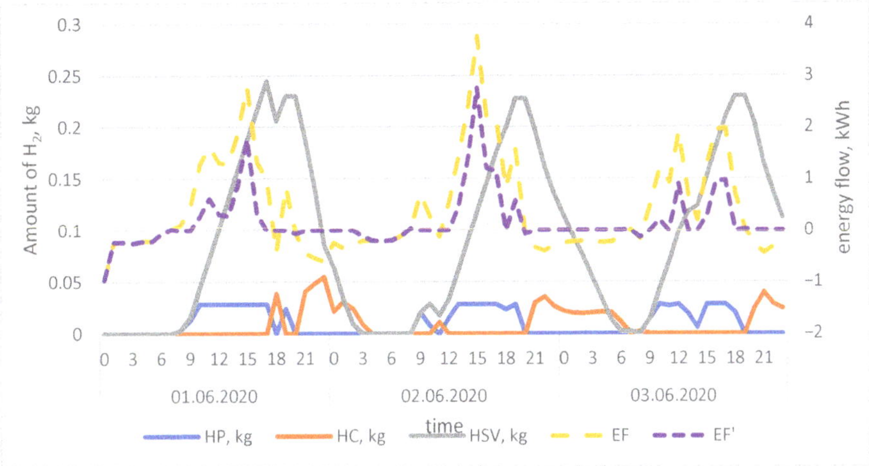

Figure 3. Quantitative (mass) hydrogen production and consumption, state of charge of the hydrogen storage, and energy flows in the installation—example for the period from 1 June 2020 to 3 June 2020. *HP*—hourly hydrogen production, *HC*—hourly hydrogen consumption, *HSV*—amount of hydrogen stored, *EF*—energy flow in the analyzed installation without hydrogen system (right-scale), *EF'*—energy flow with the hydrogen system (right-scale).

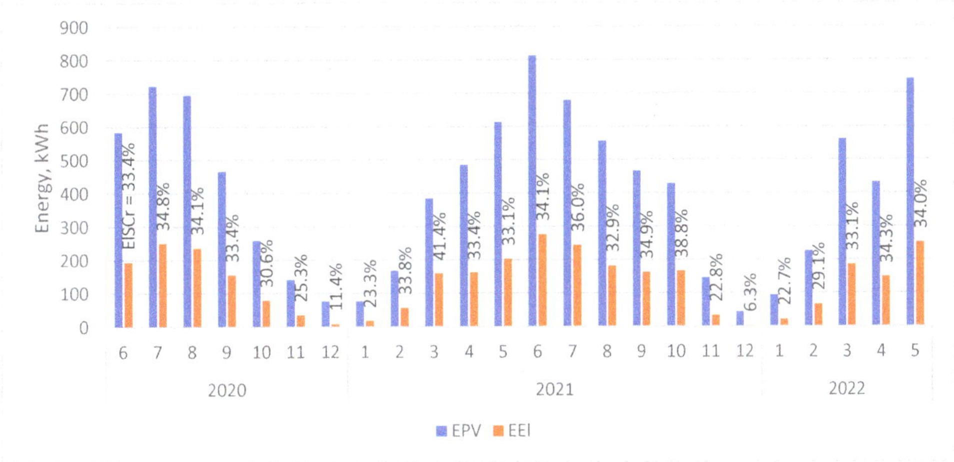

Figure 4. PV energy production (*EPV*), electrolyzer energy consumption on a monthly scale (*EEI*), and PV self-consumption ratio including electrolyzer consumption (*ElSCr*) for the period June 2020–May 2022.

The maximum monthly value of PV energy production was recorded in June 2021: 813 kWh, and the minimum in December 2021: 44 kWh. For electrolyzer energy consumption, the values were 277 kWh and 3 kWh, respectively. The maximum value of PV self-consumption ratio, including electrolyzer consumption (*ElSCr*), was observed in July 2020: 34.8%, and the minimum in December 2021: 6.3%. In the autumn and winter period, a lower share of energy directed to the electrolyzer was noticeable due to the higher level of direct self-consumption (before storage installation) and lower energy surpluses. The charging status of the storage with no capacity limitations is presented in the figure below (Figure 5).

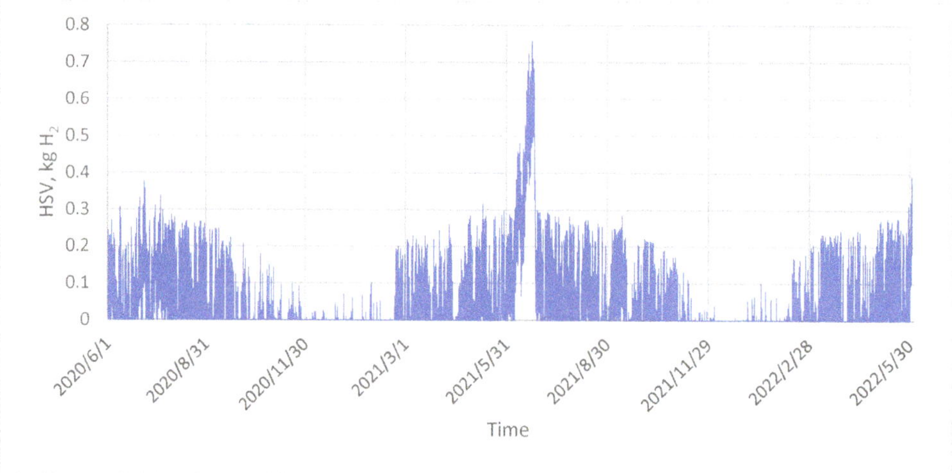

Figure 5. Hydrogen storage level (*HSV*—amount of hydrogen stored in the tank, kg) over two years, assuming unlimited storage capacity.

Despite the assumption of unlimited hydrogen tank capacity, the amount of hydrogen stored in the analyzed installation never exceeded 0.8 kg. Therefore, it can be concluded that for almost the entire two-year analysis period, the hydrogen energy storage system operated in daily cycles—the entire amount of hydrogen produced during midday hours was consumed in the evening and at night. The only exception occurred in June 2021, when a larger amount of hydrogen accumulated over several consecutive days—details are presented in Figure A2 in the Appendix A. For 99% of the time, the amount of hydrogen stored in the tank did not exceed 0.3 kg. Thus, for an installation of this scale, a tank of such capacity would be sufficient, as illustrated in the figure below. The observed daily charging and discharging behavior of the hydrogen tank suggests, on the one hand, that the photovoltaic and electrolyzer capacities should be significantly increased. On the other hand, such scaling would lead to less efficient use of the infrastructure (e.g., in terms of electrolyzer operating hours) and would negatively affect energy flows to and from the grid (e.g., as expressed by the *VDC* indicator)—Figure 6.

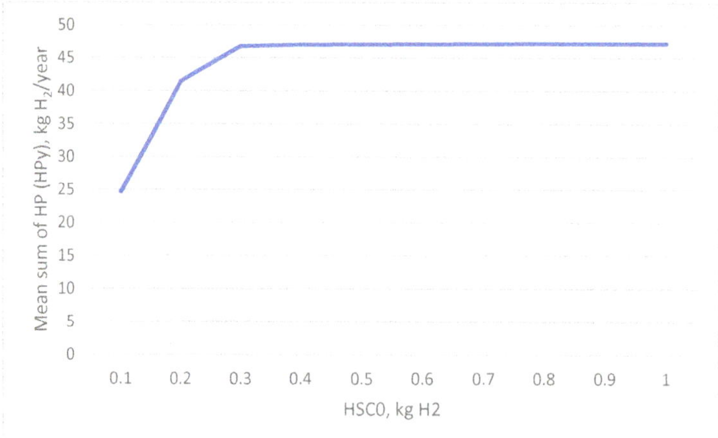

Figure 6. Total mass of hydrogen produced as a function of the useful capacity of hydrogen storage.

In the presented case study (research object), over a two-year period, increasing the hydrogen storage capacity beyond 0.7 kg resulted in no change in the amount of hydrogen produced annually (*HPy*). Additionally, the influence of storage on the maximum differences in energy flows was demonstrated (Figure 7).

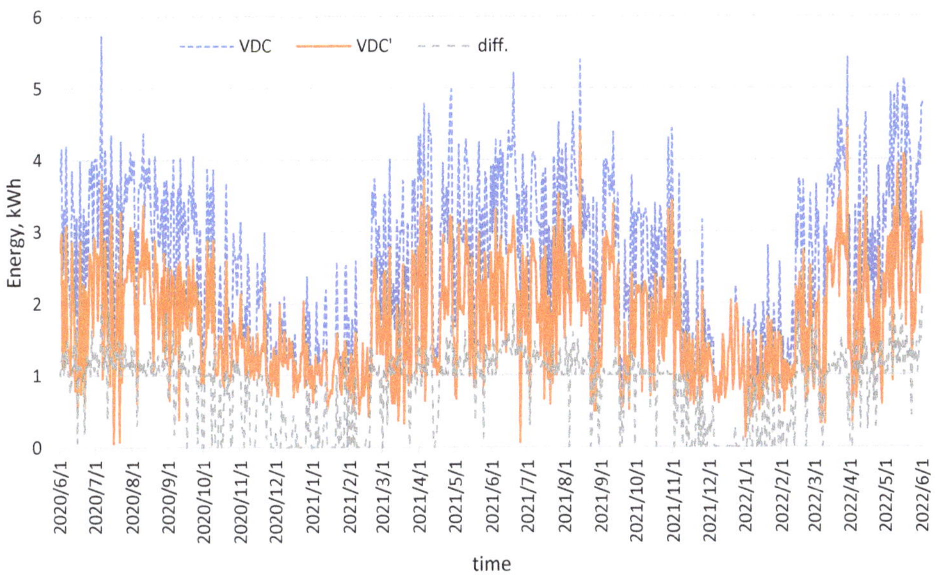

Figure 7. *VDC* value curve over two years, corrected by hydrogen storage (*VDC′*), and the difference before and after the hydrogen storage system.

Using energy storage in hydrogen, the maximum *VDC* value was reduced from 5.7 kWh to 3.8 kWh (6 July 2020), while over the two-year period, the maximum unevenness occurred on 14 August 2021, 4.4 kWh. The maximum daily reduction was 1.96 kWh.

Over the two years, the total energy produced in the home PV installation was 9.86 MWh, energy consumed was 9.49 MWh, and the amount of energy sent to the grid was 6.90 MWh and received was 6.53 MWh. Adding hydrogen storage reduced the amount of energy sent to the grid to 3.60 MWh (over two years) and reduced the amount of energy drawn from the grid to 5.50 MWh. As a result of these changes, the self-consumption rate (*SCr*) increased from 30% to 63.5% while the self-sufficiency rate (*SSr*) increased from 31.2% to 42.1%.

4. Discussion

The main results of the study consist of determining the functional relationships regarding the tank capacity required to store hydrogen at household scale, in order to utilize surplus energy (partly at the household scale), while taking into account the conditions of energy consumption and production. For an installed power of 1 kW, the useful hydrogen storage capacity needed is 0.7 kg. This is important from an economic point of view because it saves costs, resources, and infrastructure for storing hydrogen [35,36]. Safety issues are also important [37], mainly regarding hydrogen storage (under pressure in homes) and transport [38,39]. In terms of saving resources, in addition to the tanks, the power level is also relevant—in this case 1 kW. This amount of power enabled 2183 h of operation in the first year and 2025 h in the second year.

The dependence on the power conversion number of hours would be smaller and by reducing the power, power utilization would be greater. The selected system with a power of 1 kW and a capacity of 0.7 kg is able to cover 11.8% and 10.1% of electricity demand in the first and second year, respectively, under actual production and consumption conditions.

However, 1 kW of output power could theoretically cover 7900 h of demand annually. In turn, full coverage of the power generated from PV was not a priority in this computational analysis due to the differences in installed power and its potential use of 1 kW in the electrolyzer and 5.04 kWp in PV. The most controversial issue is the cost of storing energy in the amount of 40–50 kg of hydrogen per year. In the analyzed case, an energy storage system with a capacity below 3 kWh has total investment costs of EUR 1000, while the cost of hydrogen infrastructure exceeds EUR 20,000. This amount would suffice for 100 kWh of storage capacity in lithium-ion batteries. The benefit of using both technologies will be a drop in prices, which will reduce investment outlays, but in the case of hydrogen, it will be difficult to eliminate operational service costs related to hydrogen technology, as well as failure rates, water purification, and compression requirements [40].

As shown in the literature analysis, this topic has been developed for over 25 years [13], and this potential will be greater with decreasing Levelized Cost of Electricity (LCOE) for new PV installations, as well as improved efficiencies, lower investment outlays, and reduced energy storage costs. Developing hydrogen technologies for vehicles is an opportunity, but it also carries certain risks (such as resources competition). It should be noted that until 2022, energy storage in micro-installations was practically unattractive for economic and technical reasons, primarily due to the net-metering support program [41]. Since then, things have improved in terms of battery storage due to support schemes. While some benefits of hydrogen technologies can be expected in the future, the economic feasibility and market conditions may outweigh these potential advantages. However, long-term benefits include a lower number of charge/discharge cycles, which negatively impacts economic viability.

An analysis of the sensitivity of the results presented in Figures 5–7 was also conducted—see Figure 8.

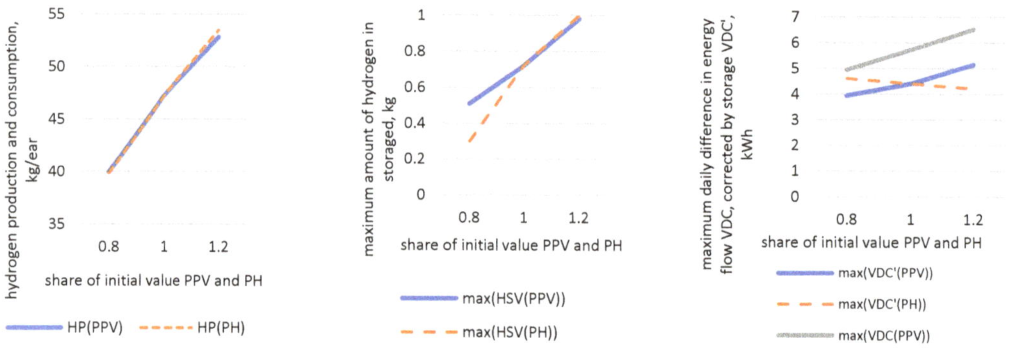

(a) (b) (c)

Figure 8. Sensitivity analysis of the following results: (**a**) average annual hydrogen production and consumption, (**b**) the maximum amount of hydrogen in storage, (**c**) the maximum daily difference in energy flow without storage (*VDC*) and with hydrogen storage (*VDC'*) as a function of the share of the initial values *PPV* = 5.04 kWp and *PH* = 1 kW.

An increase or decrease in the power of the PV system, electrolyzer, and fuel cell similarly affects the amount of hydrogen produced and consumed in the storage system (Figure 8a). The difference in impact occurs when reducing the power of the electrolyzer and fuel cell compared to reducing the power of the PV system (Figure 8b). A 20% decrease in *PH* results in a greater reduction in the maximum amount of stored hydrogen (from 0.72 to 0.3 kg) than a decrease in the power of the PV system (from 0.72 to 0.51 kg). The reason for this is, among other things, that the power of the PV system (despite its 20% reduction) is more than sufficient to produce hydrogen with the maximum power of the electrolyzer during periods of high solar radiation intensity. On the other hand, any decrease in the power of the electrolyzer or fuel cell significantly limits the possibilities of hydrogen production. Additionally, the amount of stored hydrogen increased above typical values (up to approx. 0.3 kg H_2—Figure 8) is relatively very sensitive to its occurrence (such a situation occurred only once in two years—Figure 8).

Sensitivity analysis for the calculated values of the largest unevenness of daily differences in energy flow to and from the installation on a yearly basis is based on the size of the PV installation, which determines the reference values of *VDC* ranging from 4.95 to 6.53 kWh (Figure 8c). These values refer to the day on which the largest unevenness in energy flows to and from the installation occurred over a 2-year period. In contrast, the maximum value obtained using hydrogen storage *VDC′* ranged from 4.42 to 5.17 kWh, depending on the *PPV* value. The result obtained also indicates that increasing the power of the electrolyzer and fuel cell (from 80% to 120% of the reference *PH* value of 1 kW) leads to a reduction in the maximum value of *VDC′*, from 4.62 to 4.22 kWh. In this sensitivity range of energy flow unevenness, the maximum values for the analysis period (max *VDC* and max *VDC′*) were used, as these are particularly important for the stability of the power system. The obtained reduction in the maximum *VDC* value for *PH* = 1.2 kW is comparable to the reduction achieved using two electric energy storages with a capacity of 12 kWh [8].

5. Conclusions

The potential for energy storage using hydrogen technology (from electrolyzer to fuel cell) was analyzed, focusing on the needs of a typical household in Poland. The study used real data from an existing PV installation, including recorded values of hourly energy production and actual energy consumption for a house located in the Małopolskie region.

The proposed energy storage system was designed with a capacity of 1 kW for both the input power of the electrolyzer and the output power of the fuel cell, assuming an overall energy storage efficiency of 30%. The analysis revealed that the electrolyzer operated for 2183 h (1742 h at nominal power) in the first year and 2025 h (1562 h at nominal power) in the second year, producing 49.8 kg and 44.6 kg of hydrogen, respectively. For 99% of the days, a hydrogen tank capacity of 0.33 kg was sufficient, while covering 100% of the days over two years would require a tank capacity of 0.7 kg. The implementation of this infrastructure increased the PV self-consumption rate from 30% to over 63%. However, due to the low efficiency of hydrogen-based energy storage, the energy self-sufficiency ratio only increased from 31% to 42%.

Hydrogen storage contributed to reducing the daily energy flow imbalance in the analyzed micro-installation, from over 5.7 kWh to approximately 4.4 kWh. Despite the relatively low power ratings of the devices (electrolyzer and fuel cells), their capacity factor remained below 20% annually. Increasing device utilization is achievable by expanding the PV installation's capacity; however, this would, in turn, increase the energy flow imbalance between the household and the power grid.

Additionally, the study highlights that the integration of hydrogen storage technology presents unique challenges and opportunities for decentralized energy systems. While the technology effectively increases self-consumption and reduces energy flow disparities, its scalability is hindered by high initial investment costs and low conversion efficiencies.

Nevertheless, hydrogen's long-term storage capability makes it a promising solution for addressing seasonal mismatches between energy supply and demand.

Furthermore, the implementation of hydrogen storage systems can contribute to the stabilization of the national grid by mitigating the "duck curve" phenomenon. By storing surplus energy during peak PV production periods and redistributing it during periods of high demand, hydrogen technology can support grid resilience and enable higher penetration of renewable energy sources.

Directions for further research:

- Analysis of alternative power ratings and larger household scales—extending the study to include different system sizes and a greater number of households.
- Considering the degradation of devices and their variable efficiency depending on the operating conditions, including load.
- Implementation of operational constraints—for example, preventing hydrogen storage discharge during periods of low electricity prices or at night (e.g., between 1:00 and 4:00 AM), as well as considering hydrogen storage charging at night.
- Estimation of water consumption—assessing water usage associated with electrolysis and other system processes.
- Assessment of thermal energy utilization—evaluating the potential for recovering and using heat generated during hydrogen storage processes, particularly from the fuel cell.

Author Contributions: Conceptualization, P.O. and D.M.; methodology, P.O.; formal analysis, P.O. and D.M.; investigation, P.O. and D.M.; software, P.O. and D.M.; visualization, P.O.; writing—original draft preparation, P.O. and D.M.; writing—review and editing, P.O. and D.M.; supervision, P.O.; project administration, P.O.; funding acquisition, P.O. and D.M. All authors have read and agreed to the published version of the manuscript.

Funding: This research was partially supported by the Ministry of Science and Higher Education, Poland, grant AGH number 16.16.210.476.

Data Availability Statement: The raw data supporting the conclusions of this article will be made available by the authors on request.

Conflicts of Interest: The authors declare no conflicts of interest.

Glossary

Nomenclature

AC	alternative current
CAPEX	capital expenditure
DC	direct current
DwPV	electricity demand in national grid scale, GWh
DwoPV	electricity demand in national grid scale minus electricity generated by photovoltaic, GWh
EC	hourly energy consumption in household installation, kWh
Eef	electrolyzer energy consumption, kWh/kg H_2
EEl	energy allocated to the electrolysis process, kWh
EF	value of hourly energy flow, kWh
EF′	energy flow values after hydrogen system installed, kWh
EFC1	energy obtained from the fuel cell as a first approximation
ElSCr	PV self-consumption ratio, including electorysel consumption
EPV	hourly energy production from PV, kWh
ER	hourly energy received from the grid to installation, kWh
ER′	hourly energy received from the grid to installation with hydrogen installation, kWh
ES	hourly energy sent to the grid from installation, kWh

ES′	hourly energy sent to the grid from installation with hydrogen installation, kWh
ESC	hourly energy self-consumed, kWh
ESC′	self-consumed energy with hydrogen installation, kWh
FC	fuel cell
FCC	fuel cell unit consumption, kg H_2/kWh
HC	hourly hydrogen consumption by fuel cell, kg H_2
HC1	estimated hourly mass value of hourly hydrogen consumption from the tank, kg H_2
HP	hourly value of mass of hydrogen production in the electrolyzer, kg
HPy	annual value of mass of hydrogen production in the electrolyzer, kg
HSC0	hydrogen storage useful capacity, kg
HSV	value of hydrogen storage in hydrogen tank, kg
LPSP	loss of power supply probability
NSGA-II	non-dominated sorting genetic algorithm II
OPEX	operating expense
PEMFC	proton exchange membrane fuel cell
PH	capacity of electrolyzer input and capacity of fuel cell output, kW
PPV	capacity of PV installation, kWp
PV	photovoltaic
RES	renewable energy source
SCr	self-consumption ratio
SCr′	self-consumption ratio with hydrogen installation
SSr	self-sufficiency ratio
SSr′	self-sufficiency ratio with hydrogen installation
VDC	maximum daily difference in energy flow without storage, kWh
VDC′	maximum daily difference in energy flow with hydrogen installation, kWh
year1	period from 06/2020 to 05/2021
year2	period from 06/2021 to 05/2022
Greek	
τ	hour

Appendix A

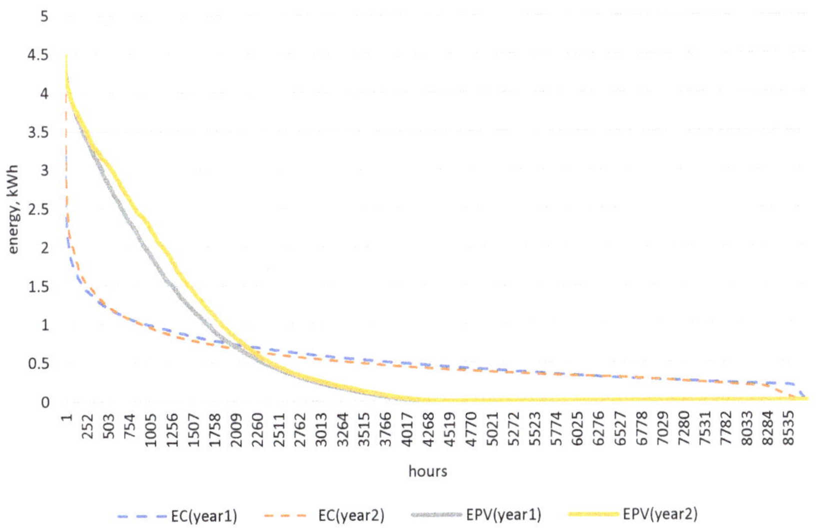

Figure A1. Energy production from PV (*EPV*), energy consumption (*EC*) for the analyzed household installation—real data from two-year period (2020–2022).

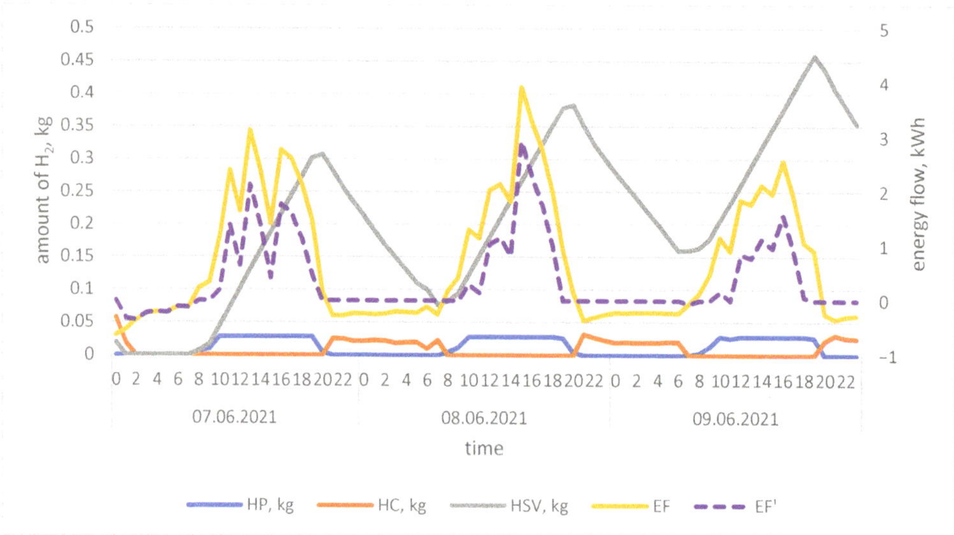

Figure A2. Volume of mass hydrogen production and consumption over a period of several days during which the total demand is lower than the total production.

Appendix A.1. Additional Equations

The formula for surplus PV energy to be utilized is based on the amount measured by a bidirectional energy meter:

$$EF(\tau) = ES(\tau) - ER(\tau) \tag{A1}$$

where:

EF—hourly energy flow, kWh
ES—energy sent to the grid (source: energy meter), kWh
ER—energy received from the grid (source: energy meter), kWh
τ—time hour

Energy self-consumed (*ESC*) is calculated using the following equation [18,42]:

$$ESC(\tau) = EPV(\tau) - ES(\tau) \tag{A2}$$

where:

EPV—hourly photovoltaic energy production, kWh
ES—energy sent to the grid, kWh

Energy allocated to the electrolysis process (provided that the maximum storage capacity is not exceeded—so that the tank can accept an additional amount of hydrogen):

$$EEl(\tau) = \begin{cases} 1 \; if \; EF(\tau) \geq 1 \\ EF(\tau) \; if \; 0.1 \leq EF(\tau) < 1 \\ 0 \; if \; EF(\tau) < 0.1 \end{cases} \tag{A3}$$

The total annual amount of hydrogen produced was determined according to the following formula:

$$HPy(year) = \sum\nolimits_{\tau} HP(\tau) \tag{A4}$$

Calculation of the energy obtained from the fuel cell as a first approximation of EFC1, based on the energy flow (*EF*):

$$EFC1(\tau) = \begin{cases} 1\ if\ EF(\tau) \leq -1 \\ -EF(\tau)\ \ if -0.1 \geq EF(\tau) > -1 \\ 0\ if\ EF(\tau) > -0.1 \end{cases} \tag{A5}$$

On this basis, the first estimated hourly mass value of hourly hydrogen consumption from the tank can be calculated:

$$HC1(\tau) = EFC1(\tau)/FCC \tag{A6}$$

where:

FCC—fuel cell unit consumption, kg H_2/kWh

The correction of hourly hydrogen consumption according to the mass amount of hydrogen accumulated in the tank is as follows:

$$HC(\tau) = \begin{cases} HC1(\tau)\ \ if\ HSV(\tau-1) > HC1 \\ HSV(\tau)\ \ if\ HSV(\tau-1) \leq HC1 \\ 0\ if\ HSV(\tau-1) = 0 \end{cases} \tag{A7}$$

The stored mass amount of hydrogen in the tank is calculated according to the formula:

$$HSV(\tau, HSC0) = HSV(\tau-1) + HP(\tau) - \mathrm{HC}(\tau), \mathrm{HSV} \in [0, HSC0] \tag{A8}$$

Conditions—correction due to hydrogen storage capacity:

$$HSV(\tau, HSC0) = HSC0\ if\ HSV(\tau, HSC0) > HSC0 \tag{A9}$$

$$HSV(\tau, HSC0) = 0\ if\ HSV(\tau, HSC0) < 0 \tag{A10}$$

Appendix A.2. Self Consumption Ratio and Self-Sufficiency Ratio

The energy sent to the grid with the hydrogen storage system was calculated by the following equation:

$$ES\prime(\tau) = ES(\tau) - EEl(\tau, HSC0) \tag{A11}$$

where:

ES—hourly energy sent to the grid from installation, kWh
EEl—hourly energy allocated to the electrolysis process, kWh

The energy received from the grid with the hydrogen storage system was calculated using the following equation:

$$ER\prime(\tau) = ER(\tau) - \frac{HC(\tau, HSC0)}{FCC} \tag{A12}$$

where:

ER—hourly energy received from the grid to installation, kWh
HC—hourly hydrogen consumption by fuel cell, kg H_2
HSC0—hydrogen storage useful capacity, kg H_2
FCC—fuel cell unit consumption, kg H_2/kWh

The PV self-consumption ratio including electrolyzer consumption (*ElSCr*) was calculated by following equation:

$$ElSCr = \frac{\sum_{\tau} EEl(\tau,\ HSC0)}{\sum_{\tau} EPV(\tau)} \tag{A13}$$

where:

EEl—hourly energy allocated to the electrolysis process, kWh
HSC0—hydrogen storage useful capacity, kg

The self-consumption ratio of energy over two years was calculated by the following:

$$ESCr = \frac{\sum_{\tau}(EPV(\tau) - ES(\tau)}{\sum_{\tau} EPV(\tau)} \tag{A14}$$

where:

ES—hourly energy sent to the grid from installation, kWh
EPV—hourly PV energy production, kWh

The self-consumption ratio of energy with the hydrogen storage system was calculated using the following:

$$ESCr\prime(HSC0) = \frac{\sum_{\tau}(EPV(\tau) - ES\prime(\tau,\ HSC0)}{\sum_{\tau} EPV(\tau)} \tag{A15}$$

where:

EPV—hourly PV energy production, kWh
HSC0—hydrogen storage useful capacity, kg
ES′—hourly energy sent to the grid from installation with hydrogen installation, kWh

The self-sufficiency ratio over two years was calculated by the following:

$$ESSr = \frac{\sum_{\tau}(EC(\tau) - ER(\tau)}{\sum_{\tau} EC(\tau)} \tag{A16}$$

where:

EC—hourly energy consumption in household installation, kWh
ER—hourly energy received from the grid to installation, kWh

The self-sufficiency ratio over two years with the hydrogen storage system was calculated by the following:

$$ESSr\prime = \frac{\sum_{\tau}(EC(\tau) - ER\prime(\tau, HSC0)}{\sum_{\tau} EC(\tau)} \tag{A17}$$

where:

HSC0—hydrogen storage useful capacity, kg
EC—hourly energy consumption in household installation, kWh
ER′—hourly energy received from the grid to installation with hydrogen installation, kWh

References

1. Szczerbowski, R.; Ceran, B. Polityka Energetyczna Polski w Aspekcie Wyzwań XXI Wieku. *Polityka Energ.-Energy Policy J.* **2017**, *20*, 17–28.
2. Zdonek, I.; Tokarski, S.; Mularczyk, A.; Turek, M. Evaluation of the Program Subsidizing Prosumer Photovoltaic Sources in Poland. *Energies* **2022**, *15*, 846. [CrossRef]
3. Wróblewski, P.; Niekurzak, M. Assessment of the Possibility of Using Various Types of Renewable Energy Sources Installations in Single-Family Buildings as Part of Saving Final Energy Consumption in Polish Conditions. *Energies* **2022**, *15*, 1329. [CrossRef]

4. PTPiREE. Micro-Installations in Poland. Available online: http://www.ptpiree.pl/energetyka-w-polsce/energetyka-w-liczbach/mikroinstalacje-w-polsce (accessed on 5 November 2023).
5. Kryzia, D.; Olczak, P.; Wrona, J.; Kopacz, M.; Kryzia, K.; Galica, D. Dampening Variations in Wind Power Generation Through Geographical Diversification. *IOP Conf. Ser. Earth Environ. Sci.* **2019**, *214*, 012038. [CrossRef]
6. Wróbel, J.; Sołtysik, M.; Rogus, R. Selected Elements of the Neighborly Exchange of Energy—Profitability Evaluation of the Functional Model. *Polityka Energ.* **2019**, *22*, 53–64. [CrossRef]
7. Kaczmarzewski, S.; Matuszewska, D.; Sołtysik, M. Analysis of Selected Service Industries in Terms of the Use of Photovoltaics before and during the COVID-19 Pandemic. *Energies* **2022**, *15*, 188. [CrossRef]
8. Olczak, P. *Magazynowanie Energii Elektrycznej w Prosumenckich Mikroinstalacjach Fotowoltaicznych*; Mokrzycki, E., Ed.; IGSMiE PAN: Kraków, Poland, 2022.
9. Olczak, P.; Matuszewska, D. Energy Storage Potential Needed at the National Grid Scale (Poland) in Order to Stabilize Daily Electricity Production from Fossil Fuels and Nuclear Power. *Energies* **2023**, *16*, 6054. [CrossRef]
10. Szablicki, M.; Rzepka, P.; Soltysik, M.; Czapaj, R. The Idea of Non-Restricted Use of LV Networks by Electricity Consumers, Producers, and Prosumers. *E3S Web Conf.* **2019**, *84*, 2014. [CrossRef]
11. Olczak, P.; Koval, V.; Yanovska, V.; Lomachynska, I. The Use of Electricity Storage on the Path to Prosumer Energy Self-Sufficiency: Eastern Europe Case Study. In *Green Energy and Technology*; Springer: Cham, Switzerland, 2023; pp. 19–31. [CrossRef]
12. Kupecki, J.; Motylinski, K.; Jagielski, S.; Wierzbicki, M.; Brouwer, J.; Naumovich, Y.; Skrzypkiewicz, M. Energy Analysis of a 10 kW-Class Power-to-Gas System Based on a Solid Oxide Electrolyzer (SOE). *Energy Convers. Manag.* **2019**, *199*, 111934. [CrossRef]
13. Hollmuller, P.; Joubert, J.-M.; Lachal, B.; Yvon, K. Evaluation of a 5 KWp Photovoltaic Hydrogen Production and Storage Installation for a Residential Home in Switzerland. *Int. J. Hydrogen Energy* **2000**, *25*, 97–109. [CrossRef]
14. Cader, J.; Koneczna, R.; Olczak, P. The Impact of Economic, Energy, and Environmental Factors on the Development of the Hydrogen Economy. *Energies* **2021**, *14*, 4811. [CrossRef]
15. Lewis, N.S. Research Opportunities to Advance Solar Energy Utilization. *Science* **2016**, *351*, aad1920. [CrossRef] [PubMed]
16. Koohi-Fayegh, S.; Rosen, M.A. A Review of Energy Storage Types, Applications and Recent Developments. *J. Energy Storage* **2020**, *27*, 101047. [CrossRef]
17. Kalamaras, E.; Belekoukia, M.; Lin, Z.; Xu, B.; Wang, H.; Xuan, J. Techno-Economic Assessment of a Hybrid Off-Grid DC System for Combined Heat and Power Generation in Remote Islands. *Energy Procedia* **2019**, *158*, 6315–6320. [CrossRef]
18. Lacko, R.; Drobnič, B.; Mori, M.; Sekavčnik, M.; Vidmar, M. Stand-Alone Renewable Combined Heat and Power System with Hydrogen Technologies for Household Application. *Energy* **2014**, *77*, 164–170. [CrossRef]
19. Tiam Kapen, P.; Medjo Nouadje, B.A.; Chegnimonhan, V.; Tchuen, G.; Tchinda, R. Techno-Economic Feasibility of a PV/Battery/Fuel Cell/Electrolyzer/Biogas Hybrid System for Energy and Hydrogen Production in the Far North Region of Cameroon by Using HOMER Pro. *Energy Strateg. Rev.* **2022**, *44*, 100988. [CrossRef]
20. Attemene, N.S.; Agbli, K.S.; Fofana, S.; Hissel, D. Optimal Sizing of a Wind, Fuel Cell, Electrolyzer, Battery and Supercapacitor System for off-Grid Applications. *Int. J. Hydrogen Energy* **2020**, *45*, 5512–5525. [CrossRef]
21. Budak, Y.; Devrim, Y. Comparative Study of PV/PEM Fuel Cell Hybrid Energy System Based on Methanol and Water Electrolysis. *Energy Convers. Manag.* **2019**, *179*, 46–57. [CrossRef]
22. Hashimoto, K.; Kawahara, K. Optimal Operation of Fuel Cell and Electrolyzer in Household Hybrid System by Stochastic Programming. In Proceedings of the 2017 IEEE Innovative Smart Grid Technologies—Asia (ISGT-Asia), Auckland, New Zealand, 4–7 December 2017; pp. 1–6. [CrossRef]
23. Hassan, Q. Optimisation of Solar-Hydrogen Power System for Household Applications. *Int. J. Hydrogen Energy* **2020**, *45*, 33111–33127. [CrossRef]
24. Török, L.; Nielsen, C.K.; Munk-Nielsen, S.; Rømer, C.; Flindt, P. High Efficiency Electrolyser Power Supply for Household Hydrogen Production and Storage Systems. In Proceedings of the 2015 17th European Conference on Power Electronics and Applications (EPE'15 ECCE-Europe), Geneva, Switzerland, 8–10 September 2015; pp. 1–9. [CrossRef]
25. Ceran, B.; Mielcarek, A.; Hassan, Q.; Teneta, J.; Jaszczur, M. Aging Effects on Modelling and Operation of a Photovoltaic System with Hydrogen Storage. *Appl. Energy* **2021**, *297*, 117161. [CrossRef]
26. Abdolmaleki, L.; Berardi, U. Hybrid Solar Energy Systems with Hydrogen and Electrical Energy Storage for a Single House and a Midrise Apartment in North America. *Int. J. Hydrogen Energy* **2024**, *52*, 1381–1394. [CrossRef]
27. Abdolmaleki, L.; Berardi, U. Dynamic Simulation of a Hydrogen-Fueled System for Zero-Energy Buildings Using TRNSYS Software. *E3S Web Conf.* **2023**, *396*, 04005. [CrossRef]
28. Jahanbin, A.; Abdolmaleki, L.; Berardi, U. Techno-Economic Feasibility of Integrating Hybrid-Battery Hydrogen Energy Storage in Academic Buildings. *Energy Convers. Manag.* **2024**, *309*, 118445. [CrossRef]
29. Shao, Z.; Cao, X.; Zhai, Q.; Guan, X. Risk-Constrained Planning of Rural-Area Hydrogen-Based Microgrid Considering Multiscale and Multi-Energy Storage Systems. *Appl. Energy* **2023**, *334*, 120682. [CrossRef]

30. Zhuk, A.Z.; Shkolnikov, E.I.; Borodina, T.I.; Valiano, G.E.; Dolzhenko, A.V.; Kiseleva, E.A.; Kochanova, S.A.; Filippov, E.D.; Semenova, V.A. Aluminium—Water Hydrogen Generator for Domestic and Mobile Application. *Appl. Energy* **2023**, *334*, 120693. [CrossRef]
31. Nastasi, B.; Mazzoni, S.; Groppi, D.; Romagnoli, A.; Astiaso Garcia, D. Optimized Integration of Hydrogen Technologies in Island Energy Systems. *Renew. Energy* **2021**, *174*, 850–864. [CrossRef]
32. Olczak, P.; Jaśko, P.; Kryzia, D.; Matuszewska, D.; Fyk, M.I.; Dyczko, A. Analyses of Duck Curve Phenomena Potential in Polish PV Prosumer Households' Installations. *Energy Rep.* **2021**, *7*, 4609–4622. [CrossRef]
33. Kazanecka, E.; Olczak, P. A Specific Yield Comparison of 2 Photovoltaic Installations—Polish Case Study. *Polityka Energ.-Energy Policy J.* **2023**, *26*, 129–148. [CrossRef]
34. Groenemans, H.; Saur, G.; Mittelsteadt, C.; Lattimer, J.; Xu, H. Techno-Economic Analysis of Offshore Wind PEM Water Electrolysis for H2 Production. *Curr. Opin. Chem. Eng.* **2022**, *37*, 100828. [CrossRef]
35. Sezer, D.; Ada, E.; Kazançoğlu, Y. Investigating the Key Drivers in the Transition to Sustainable Hydrogen Transportation Fuel. *Econ. Ecol. Socium* **2024**, *8*, 16–26. [CrossRef]
36. NRC; NAE. Transportation, Distribution, and Storage of Hydrogen. In *The Hydrogen Economy: Opportunities, Costs, Barriers, and R&D Needs*; National Research Council and National Academy of Engineering, Ed.; The National Academies Press: Washington, DC, USA, 2004. [CrossRef]
37. Hrinchenko, H.; Antonenko, N.; Khomenko, V.; Artiukh, S. Assessment of Power Equipment Operational Safety in the Sustainable Management of Residual Lifespan. *Econ. Ecol. Socium* **2024**, *8*, 78–91. [CrossRef]
38. Stecuła, K.; Olczak, P.; Kamiński, P.; Matuszewska, D.; Duong Duc, H. Towards Sustainable Transport: Techno-Economic Analysis of Investing in Hydrogen Buses in Public Transport in the Selected City of Poland. *Energies* **2022**, *15*, 9456. [CrossRef]
39. Dawood, F.; Anda, M.; Shafiullah, G.M. Hydrogen Production for Energy: An Overview. *Int. J. Hydrogen Energy* **2020**, *45*, 3847–3869. [CrossRef]
40. Genovese, M.; Piraino, F.; Kovač, A.; Marciuš, D.; Pagnotta, L.; Fragiacomo, P. Integration of Hydrogen Compressors and Turbines into Current and Future Hydrogen Infrastructure. *J. Power Sources* **2025**, *629*, 235965. [CrossRef]
41. Trela, M.; Dubel, A. Net-Metering vs. Net-Billing from the Investors Perspective—Impacts of Changes in RES Financing in Poland on the Profitability of a Joint Photovoltaic Panels and Heat Pump System. *Energies* **2022**, *15*, 227. [CrossRef]
42. Roberts, M.B.; Sharma, A.; MacGill, I. Efficient, Effective and Fair Allocation of Costs and Benefits in Residential Energy Communities Deploying Shared Photovoltaics. *Appl. Energy* **2022**, *305*, 117935. [CrossRef]

energies

MDPI

Article

Investigation of Flame Structure and PAHs' Evolution in a Swirl-Stabilized Spray Flame at Elevated Pressure

Wenyu Wang [1], Runfan Zhu [1], Siyu Liu [1], Yong He [1,2,*], Wubin Weng [1,2], Shixing Wang [3], William L. Roberts [3] and Zhihua Wang [1,2]

1 State Key Laboratory of Clean Energy Utilization, Zhejiang University, Hangzhou 310027, China
2 Qingshanhu Energy Research Center, Zhejiang University, Hangzhou 311300, China
3 Clean Energy Research Platform, King Abdullah University of Science and Technology, Thuwal 23955-6900, Saudi Arabia
* Correspondence: heyong@zju.edu.cn

Abstract

Swirl spray combustion has attracted significant attention due to its common usage in gas turbines. However, the high pressure in many practical applications remains a major obstacle to the deep understanding of flame stability and pollutant formation. To address this concern, this study investigated a swirl spray flame fueled with n-decane at elevated pressure. Planar laser-induced fluorescence (PLIF) of OH and polycyclic aromatic hydrocarbons (PAHs) were used simultaneously, enabling the distinction of the locations of OH, PAHs, and mixtures of them, providing detailed information on flame structure and evolution of PAHs. The effects of swirl number and ambient pressure on reaction zone characteristics and PAHs' formation were studied, with the swirl number ranging from 0.30 to 1.18 and the pressure ranging from 1 to 3 bar. The data suggest that the swirl number changes the flame structure from V-shaped to crown-shaped, as observed at both atmospheric and elevated pressures. Additionally, varying swirl numbers lead to the initiation of flame divergence at distinct pressure levels. Moreover, PAHs of different molecular sizes exhibit significant overlap, with larger PAHs able to further extend downstream. The relative concentration of PAH increased with pressure, and the promoting effect of pressure on producing larger PAHs was significant.

Keywords: spray combustion; swirl; elevated pressures; OH; PAH; PLIF

1. Introduction

Modern combustors, such as aero-engines and gas turbines, are required to achieve higher combustion efficiency, lower pollution emissions, and a wider operating range (turn down ratio) and operate under high temperature and pressure conditions to increase thermodynamic efficiency. Swirl is a common feature of many practical combustors, which can enhance flame stabilization and minimize pollutant formation by inducing a recirculation region to improve the evaporation rate and fuel-oxidizer mixing.

Swirl flames have been extensively studied, often employing optical diagnostics that enable in situ and non-intrusive measurements. The structure of swirling flames is known to be influenced by various factors including inlet fuel/air temperature [1,2], fuel–air equivalence ratio [3,4], burner geometry [5,6], and combustor pressure [7,8]. Boxx et al. [9] performed 3 kHz particle image velocimetry (PIV) and OH planar laser-induced fluorescence (PLIF) measurements of flame dynamics in a lean, partially premixed swirl-stabilized CH_4 flame at elevated pressure up to 5 bar. A similar study has also been conducted by Chterev and Boxx [10] for H_2-enriched C_2H_4 lean swirl-stabilized flames at 5 bar to study heat release distribution, combustion instability, and flow dynamics. Subsequently, W. Agostinelli et al. [11] used large eddy simulations (LES) to further analyze the combustion dynamics of the flame described in [10]. Their results showed that both H_2 addition and elevated pressure modified the flame shape, as flame height was reduced and became more compact. In addition, pressure altered the characteristic time scales of the turbulent reactive flow due to the increase in Reynolds and Damköhler numbers, along with a decrease in laminar flame speed, thermal flame thickness, and chemical time scale.

While the aforementioned studies mainly focus on gas-phase combustion, the phenomena present in aero-engines where swirl-stabilized spray combustion is prevalent are not fully captured [12]. Cavaliere et al. [13] compared flame structures and relative blow-off behaviors of swirling premixed and non-premixed methane and heptane swirl spray flames using OH* chemiluminescence and OH-PLIF. The results indicated that the structure of spray flame differed from that of methane flames and a reduced probability of local extinction was noticed. The combustion characteristics of swirl spray flames with various liquid fuels such as rapeseed methyl esters, Jet A-1, palm biodiesel/methyl esters (PME), acetone-butanol-ethanol (ABE), and glycerol were examined by Chong and Kumar et al. [14–17]. However, their work was conducted under atmospheric pressure conditions, which significantly differ from the actual operating environments of aero-engines that typically run under pressurized conditions. Extra effort is required to fulfill experimental measurement under high-pressure conditions. Recently, Slabaugh et al. [18] conducted experiments on a piloted, partially premixed, liquid-fueled swirl burner at 1 MPa. They performed 100 kHz PIV measurements to analyze the influence of combustion instability on the interaction between the flow and combustion. A novel procedure for separating OH/fuel-PLIF signal in Jet-A swirl spray flame up to 5 bar was proposed by Chterev et al. [19], which was helpful for quantifying statistics about the relative locations of fuel, OH, and a mixture of fuel and OH. To enhance combustion efficiency and reduce pollutants, Malbois and Legros et al. [20–22] proceeded with a series of experiments utilizing various laser diagnostics and LES within the HERON (high-pressure facility for aero-engines combustion) combustion facility fed with liquid kerosene. The distribution of droplets, flame structure, and OH/kerosene/NO PLIF were studied under pressure conditions ranging from 0.4 to 1.8 MPa. The results showed that most NO-PLIF signals distributed in the inner recirculation zone (IRZ), and the mole fraction of NO decreased with elevated pressure. Nevertheless, studies on swirl-stabilized spray flames remain limited owing to system complexities. In addition, considering the pivotal role of swirl intensity in flame dynamics, comprehensive understanding of flame behavior under varied swirl conditions is indispensable.

Furthermore, most studies of spray swirl-stabilized flames only focus on flame structures under high-pressure conditions. Given climate change and stringent emission regulations, a greater demand for understanding the soot formation mechanisms was posed, especially for systems containing hydrocarbons [23,24], e-fuels [25–27], and their blends [28–31]. It has been widely accepted that polycyclic aromatic hydrocarbons (PAHs) formed from the incomplete combustion of hydrocarbons are the precursors of soot particles. Therefore, studies on PAHs' formation and the conversion from PAHs to soot are valuable for comprehending the mechanism of soot formation. Bouvier et al. [32] revealed the distinct regions related to the inception, growth, and oxidation of soot in swirl stratified premixed C_2H_4/air flames using PIV, OH-PLIF, PAH-PLIF, and laser-induced incandescence (LII). Wang and Jain et al. [33,34] observed that within the piloted liquid-spray flames, smaller PAHs were enveloped by the flame fronts and transformed into larger PAHs and nascent soot as flame height increased. The LIF intensity of PAHs with different molecular sizes was compared to evaluate the effect of pressure, equivalent ratio, and iso-octane addition on the evolution of PAHs in a laminar jet flame at a pressure of 1–5 bar [35]. However, studies on the formation and evolution of PAHs in a high-pressure swirl spray flame remain insufficient, especially regarding the evolution pathway of PAHs with different sizes. In addition, the interference of PAH fluorescence on OH-PLIF signal can not be ignored under high pressure conditions.

To address the above issue, n-decane (n-$C_{10}H_{22}$), a well-known surrogate component of jet fuels [36–38], was employed in a laboratory-scale high-pressure spray swirl-stabilized burner in this study. OH-PLIF and PAH-PLIF were employed simultaneously to visualize reaction zone and PAHs' development. The effect of swirl number and pressure on flame structure and PAHs' formation was investigated to provide insights for an in-depth understanding of soot formation in multiphase turbulent flames at elevated pressure.

2. Experimental Apparatus and Diagnostic Methods

2.1. Experimental Apparatus

A laboratory-scale spray swirl-stabilized burner integrated into a high-pressure vessel was designed for the experiments, and the schematic of the combustor is depicted in Figure 1. The cylindrical combustion chamber, designed with an inner diameter of 200 mm and a total height of 400 mm, is capable of operating at pressures up to 2 MPa. The chamber is equipped with four quartz windows for flame observation and optical measurements. Within the chamber, a cylindrical quartz glass with a diameter of 170 mm and a height of 200 mm is placed, and nitrogen is introduced into the small holes around the glass to prevent contamination from soot and water condensation. The pressure of the chamber is monitored by a pressure sensor (Keller, 33X) with an accuracy of ±0.01% and controlled by a combination of a manual pressure valve and a digital PID controller.

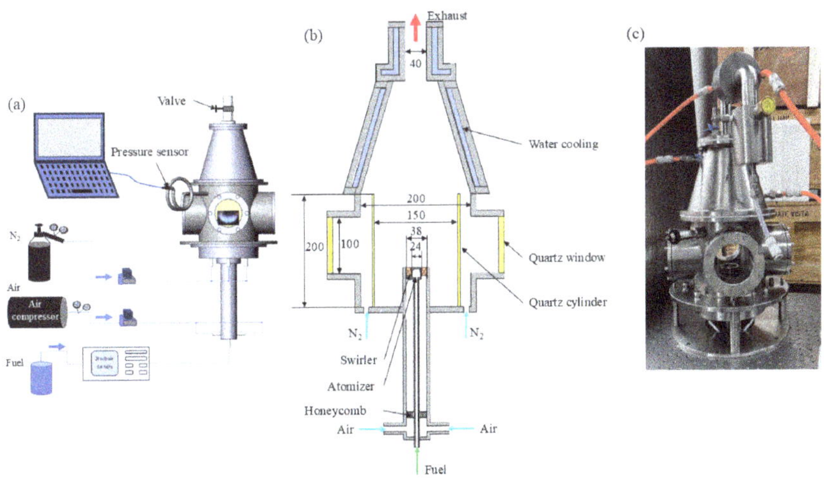

Figure 1. Schematic of the high-pressure spray swirl-stabilized combustor (**a**), the main parameters of the combustor (**b**), and the real picture of the combustor (**c**).

The fuel is supplied by a constant flow pump (Beijing Xiangyue Huanyu,) with an accuracy of ±0.5%, regulating the fuel mass flow rate into the pressure atomizer (Danfoss, 80°, hollow). The flows of air and nitrogen (N_2) are controlled by mass flow controllers (Alicat), each with an accuracy of ±0.1%. An axial swirler with eight vans is mounted at the exit of the burner, and the structure of the burner is carefully designed to allow for the replacement of the swirlers with different swirl numbers (S_N). These swirlers share identical hub diameter (19 mm), tip diameter (38 mm), and vane thickness (2 mm) but feature different vane angles corresponding to different S_N as listed in Table 1.

Table 1. The investigated Swirl Flame conditions.

Flame Case	Fuel Mass Flow Rate (g/s)	Air Mass Flow Rate (g/s)	Vane Angle	Swirl Number (S_N)	Pressure (bar)
S1			20°	0.30	1
S2			30°	0.48	1–3 (increment of 0.5 bar)
S3	0.2	6.17	43°	0.77	1–3 (increment of 0.5 bar)
S4			50°	0.99	1
S5			55°	1.18	1–3 (increment of 0.5 bar)

2.2. Flame Conditions

N-decane was used as the fuel and mixed with dry and clean air supplied by an air compressor. The swirl intensity can be classified by swirl number into 3 parts: weak swirl ($S_N \leq 0.4$), medium swirl ($0.4 < S_N \leq 0.6$), and high swirl ($S_N > 0.6$) [39,40]. Five swirlers with vane angles of 20°, 30°, 43°, 50°, and 55° were used to study the influence of swirl intensity on flame structure and PAH formation, and the geometry of the swirlers is shown in Figure 2.

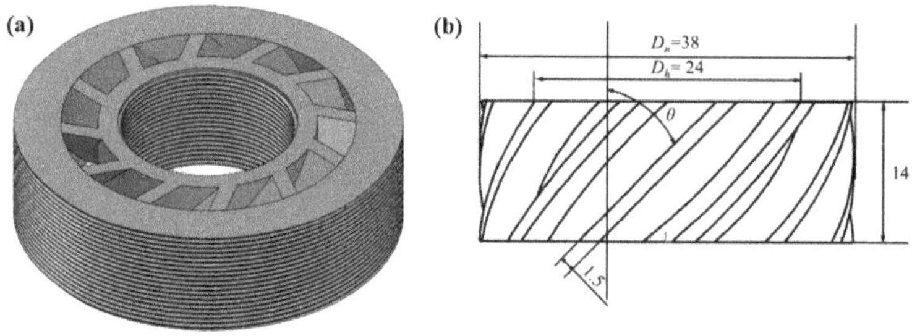

Figure 2. (**a**) A 3D diagram of the swirler; (**b**) the geometric dimension of the swirler.

The corresponding S_N is estimated based on the following equation [41,42]:

$$S_N = \frac{2}{3}\left(\frac{1-\left(\frac{D_h}{D_n}\right)^3}{1-\left(\frac{D_h}{D_n}\right)^2}\right) tan\theta \quad (1)$$

where D_h and D_n are the hub and tip diameters (mm) of the swirler geometry, and θ is the swirler vane angle. The operating condition is summarized in Table 1, and the S_N of the experiments which covered the swirl intensity range from weak to strong was similar to that in the Refs. [5,13,43]. Moreover, S2, S3, and S5 were selected to further investigate the effect of pressure on the swirl spray flame. The experiments were conducted at pressures ranging from 1 to 3 bar, with pressure increments of 0.5 bar.

2.3. Laser Diagnosis

The optical measurements were performed 7 mm above the burner exit to avoid interference from laser scattering caused by liquid droplets. OH/PAH PLIF is used to measure flame structures and PAHs' formation, and a schematic of the PLIF setup is shown in Figure 3. A frequency-doubled Nd:YAG laser (Continuum, Powerlite DLS 8010) with a laser wavelength of 532 nm, a frequency of 10 Hz, and energy of 600 mJ was used to pump a dye laser (Continuum, Vista FX-1). The output of the dye laser, with Rhodamine

590 dissolved in ethanol, was frequency-doubled to produce pulses at 283.35 nm, corresponding to the $Q_1(8)$ transition in the A2∑←X2∏ (1-0) band of the OH molecule. The same laser was utilized to excite PAH species. The uniformly distributed laser was shaped into a laser sheet (height = 5 mm, width = 500 μm) by a combination of a cylindrical lens (focus length = −30 mm) and a spherical lens (focus length = 500 mm). The pulse energy was 6 mJ, thus the laser fluence was estimated to be 0.02 J/cm^2, with the fluorescence operated within the linear regime. Such low laser fluence could reduce the interference of soot [44,45].

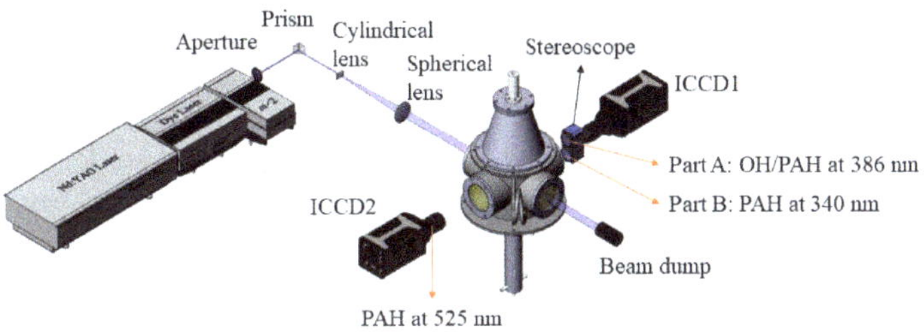

Figure 3. Schematic of the experimental setup for OH-PLIF and PAH-PLIF measurements.

Because PAHs may also fluoresce in the ranges of 280–650 nm [46], interference from PAHs' fluorescence in the OH collection system is inevitable, with a more pronounced impact at elevated pressures. Therefore, the OH and smaller PAH were simultaneously acquired using an emICCD (Princeton Instruments, PIMAX4) equipped with a UV lens (105 mm, $f/4.5$ Nikon) and a stereoscope (Lavision, VZ-image) placed perpendicular to the laser sheet to collect the PLIF signals. The fluorescence of OH was captured in the range of 305–315 nm using a 310 nm narrow-band filter with FWHM of 10 nm (Alluxa) and a 305 nm long-pass filter (Andover, 305FG01-50). A 340 nm narrow-band filter with FWHM of 26 nm (Edmund) was used to collect the signals originating from two and three aromatic rings (A2 and A3) [47].

Since the fluorescence wavelengths of the PAHs are correlated with the molecular sizes of PAHs, the detection wavelength bands were determined based on [47,48]. Part A mounted a 386 nm narrow-band filter with FWHM of 27 nm (Edmund) to capture pyrene (A4) fluorescence. Part B of the stereoscope mounted a 340 nm narrow-band filter with FWHM of 26 nm (Edmund) to capture fluorescence of A2 and A3 both imaged by ICCD1 (emICCD, Princeton Instruments, PIMAX4). The ICCD2 (Princeton Instruments, PIMAX4) was equipped with a 525 nm narrow-band filter with FWHM of 25 nm (Edmund) to obtain the PLIF signals of acenaphthylene (A2R5) fluorescence. Details of cameras and filters are shown in Table 2.

Table 2. Filters for simultaneous imaging.

Combination	Object	Camera	Range (nm)
OH+PAH (340 nm)	OH	ICCD1: Part A	305–315
	PAH at 340 nm	ICCD1: Part B	314–366
PAHs	PAH at 340 nm	ICCD1: Part B	314–366
	PAH at 386 nm	ICCD1: Part A	359–413
	PAH at 525 nm	ICCD2	500–550

The in-plane spatial resolutions for emICCD and ICCD were 0.22 × 0.22 mm^2 and 0.10 × 0.10 mm^2, respectively. The emICCD and ICCD were triggered by a digital Delay/Pulse Generator (Stanford Research Systems, DG535), and the gate widths of both emICCD and ICCD were set as 50 ns.

2.4. Image Processing

For each swirl spray flame, 200 simultaneous OH/PAH PLIF images were recorded, followed by 200 PAHs-PLIF images. Each set of simultaneously obtained images was corrected for the vertical variation in laser sheet intensity and aligned by a calibration image via pixel calibration. Subsequently, background subtraction was employed to remove the extraneous noise, and then a Gaussian filter was used to further improve the signal-to-noise ratio (SNR). The thresholds of images were determined using Otsu's method [49] and then the images were binarized to find the contour of OH/PAH. The calculated thickness of the OH layer was based on these contours. The image size presented in this study is 200 × 507 pixels (approximately 100 × 50 mm).

3. Results and Discussion

3.1. Flame Structures

The time-averaged (200 shots) and instantaneous OH-PLIF with different S_N are illustrated in Figure 4, combined with the simultaneously obtained PAH-PLIF (A2 and A3) contours, providing insights into the structures and the main reaction zone of swirl spray flames. The presented PAH-LIF profiles here serve to illustrate the interference of PAH on OH-PLIF spatially, and the intensity within the white contour was attributed to the combination of OH-PLIF and fluorescence of A2 and A3. The formation of PAH will be discussed in Section 3.3.

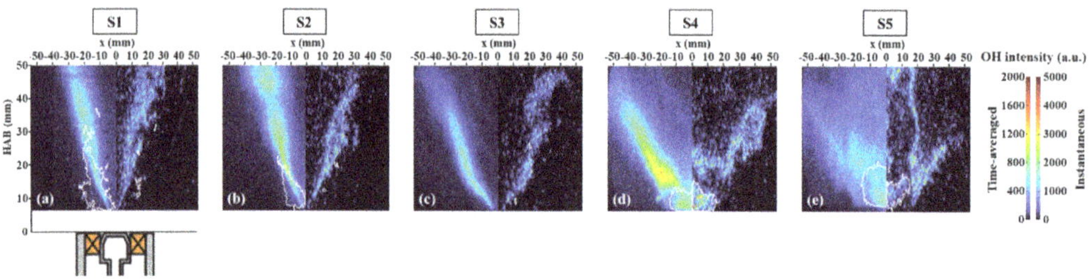

Figure 4. Time-averaged (**left**) and instantaneous (**right**) images of OH-PLIF signal combined with PAH contour under different S_N at 1 bar, (**a**–**e**) refers to the S_N varied from 0.30 to 1.18. The contour of simultaneous fluorescence of A2 and A3 is depicted by white lines, and HAB denotes height above the burner.

In general, the flame shape changed from being thin and elongated to wider and shorter. Specifically, the flame structures were V-shaped with OH primarily distributed in the periphery of the spray jet, and an almost complete absence of OH signals inside the IRZ was observed except for S4 and S5. For S1–S3, the absence of OH signals in the IRZ indicates that the fuel droplets were essentially consumed in the flame branches, and the roll-up at the outside of the shear layer (SL) is easily discernible from instantaneous OH-PLIF images.

The flame opened with the increase in S_N, consequently altering the flame structures. OH fluorescence was observed to appear within the IRZ as the value of S_N exceeded 0.99 (i.e., S4). For S4 and S5, the shapes of the flame featured one or two upward bulges of the OH layer inside the "V" shape, ultimately transforming the flame shape into a crown-shaped configuration. This phenomenon might be due to the more intense evaporation and mixing process, the widening of the spray angle, and the higher concentration of large droplets near the core of the swirling with the increase in S_N as indicated by [40,43].

The medium and strong S_N were selected to study the influence of pressure on flame structure, as illustrated in Figure 5. It is evident that the OH region shortened for all three S_N as the pressure increased from 1 to 3 bar, attributed to faster three-body recombination reactions depleting the OH concentration at higher pressure [7]. As the pressure increased, the OH distribution exhibited more frequent breaks and closures compared to atmospheric pressure. Although the time evolution is not depicted here, consecutive images revealed the potential for the closing of the breaks (i.e., reignition) and the development of new local extinctions. Furthermore, the increase in pressure altered OH structures and thinned the OH layer in three cases. As pressure increased, the area with higher OH signals gradually moved upstream, indicating that the primary reaction zone was closer to the nozzle, suggesting that increased pressure promoted a higher reaction rate. In the case of V-shaped flames (S2 and S3), an observation is that each OH layer separated into two branches, as indicated by the red circle in Figure 5c. Simultaneously obtained PAH-PLIF distributions demonstrated the increased formation of PAH located inside the OH layers. A similar feature can be seen in S5, where the outer branch also diverged under higher pressure (red circle in Figure 5m), and this bifurcation feature will be discussed in detail in Section 3.2.

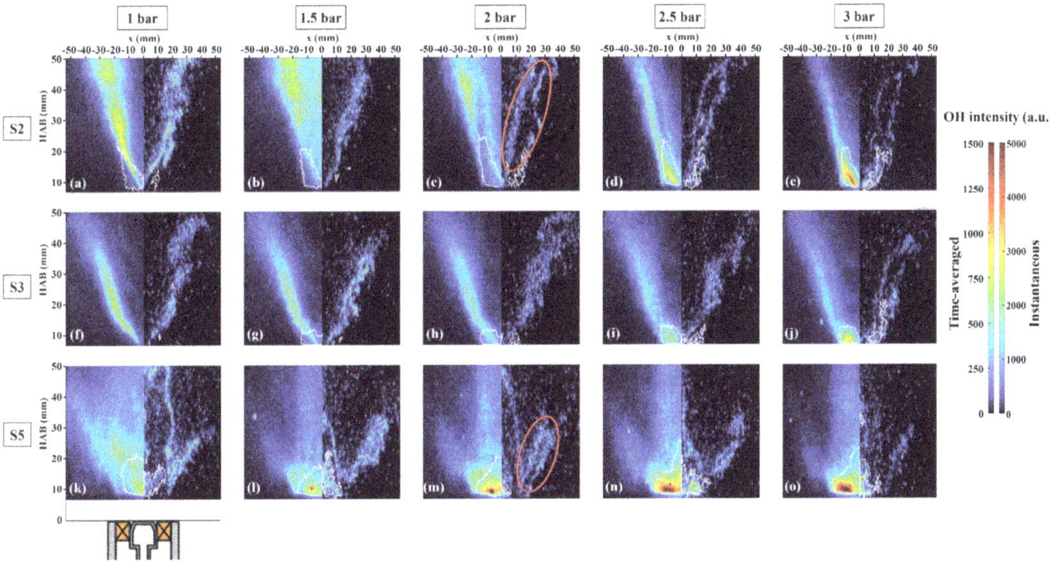

Figure 5. Time-averaged (**left**) and instantaneous (**right**) images of OH-PLIF under different pressures and S_N. Labels (**a–e**) correspond to case S2 with the pressures ranging from 1 to 3 bar, (**f–j**) and (**k–o**) show cases S3 and S5, respectively, under the same pressure range. The contour of PAH-PLIF (A2 and A3) is depicted with white lines.

3.2. *Effect of Swirl Number and Pressure on OH Distribution*

Figure 6 depicts the radical distribution of time-averaged OH-PLIF intensity at different pressures and HABs, normalized based on the maximum signal intensity at each pressure. The bimodal and symmetrical distributions of OH intensities along radical distance are easily observed. The slight asymmetry distribution of OH intensity at higher pressure was likely due to the laser sheet energy attenuation. The spikes of OH signal encircled by the green dashed line in Figure 6 were a result of the interference from droplets' Raman-scattering, as the spotted intense signals in Figure 5.

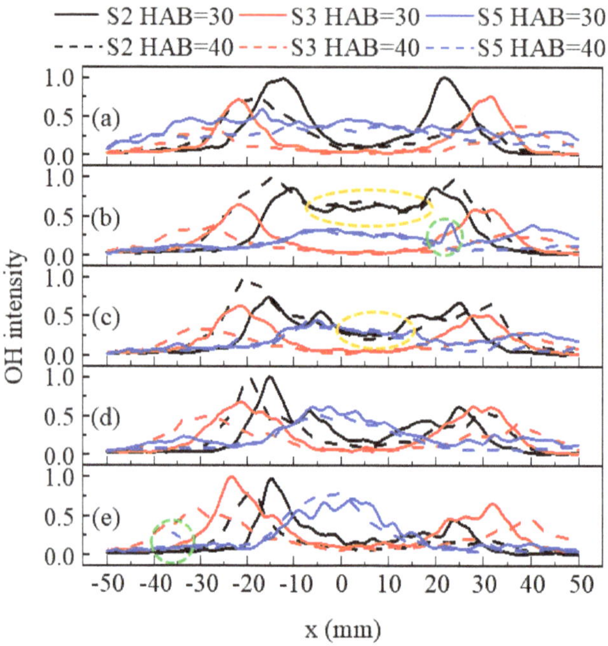

Figure 6. The normalized time-averaged OH intensity at different HABs under pressures of (**a**) 1 bar, (**b**) 1.5 bar, (**c**) 2 bar, (**d**) 2.5 bar and (**e**) 3 bar.

In the case of S2 and S3 (V-shaped flames), the OH was mainly distributed in the periphery of the spray jet near the nozzle, and four peaks emerged with increasing pressure corresponding to the bifurcation structure. However, in the case of S2, the OH signal appeared in the IRZ at 1.5 bar and 2 bar, as encircled by the yellow dashed line in Figure 6, corresponding to the time-averaged part of Figure 5b and differing from the instantaneous one. Upon a detailed observation of images for S2 at 1.5 bar, the distinct distribution of time-averaged OH signals was attributed to the flame instability. Furthermore, the comparable OH intensity for S2 at HAB = 30 mm and HAB = 40 mm revealed that with a smaller S_N, the reaction zone extended downstream owing to the poorer mixing quality of droplets and air. The lower mixing quality impacted the evaporation rate of the droplets [50,51], resulting in a higher OH signal intensity observed at a greater HAB in most cases with smaller S_N as shown in Figures 4 and 5. Additionally, the flame angles remained unchanged with increasing pressure, since the radical positions of the maximum OH intensity at different pressures were almost the same.

At 1 bar, the variation in the radical distribution of OH of S5 was smaller than S2 and S3, and the maximum OH intensity of S5 was the lowest. With the increase in S_N, the increased centrifugal force on droplets promoted disturbance on the liquid sheet or ligament, causing the droplets to distribute more uniformly radially at an earlier stage, increasing the evaporation rate. Moreover, with the increase in pressure, the OH distribution of S5 became more compact, primarily within the IRZ. This is explained as being due to the movement of the droplets towards the IRZ with elevated pressure [52].

A double structure of the OH layer under higher pressure can be observed in Figure 5. The proportion of the bifurcated structure is shown in Figure 7, calculated by selecting the bifurcation feature of 100 instantaneous images to quantify the effect of pressure on the flame structure. The locations of newly generated OH branches are compared to those under lower pressures using the image process described in Section 2.4. It has been found that the newly generated branches were located in the inner region of the OH layer.

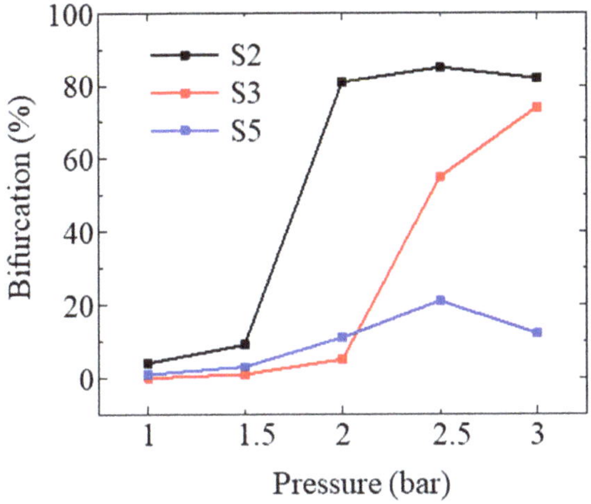

Figure 7. The proportion of bifurcated structure in 100 instantaneous images under different pressures.

For medium S_N, the OH layer exhibited a bifurcated structure at 1.5 bar, and this structure became more pronounced with increasing pressure, whereas the branch structures occurred at higher pressure with the increase in S_N. The reduced proportion of bifurcation in S2 and S5 at 3 bar was probably due to the attenuation of the laser sheet and local extinction caused by flame instability. This combustion phenomenon resembled that of the spray structure under high pressure as described in [52]. One possible explanation for this phenomenon is that the negative correlation between pressure and the atomization cone angle led to poorer mixing quality, and the droplet distribution transitioned from hollow cone to solid cone as pressure increased [53]. However, the elevated pressure also resulted in smaller droplets [54], which were more prone to moving with the airflow. Consequently, unburned droplets, caused by poor mixing quality, were carried by the airflow to the IRZ and burned in this region. Additionally, higher S_N with a more uniform radial distribution of droplets could alleviate the negative impact of increased pressure on mixing quality. Therefore, the bifurcation feature occurred at higher pressures as S_N increased.

In premixed flames, the lifetime of the OH signal is too long to indicate the thickness of the reaction zone, whereas, in diffusion flames, OH radicals could quickly disappear on either side of the reaction zone [55], providing the ability to analyze the variation of reaction thickness with pressure. Probability density functions (PDFs) of the OH layer were computed for 100 instantaneous images and plotted in Figure 8 for the three S_N under different pressures. To obtain the thickness of the OH layer, the edge of the OH region was identified using the same method as described in Section 2.4, then the longest line across the OH region at a certain HAB was defined as the OH layer thickness. It is evident that the OH layer is much thinner at higher HAB since droplets were gradually consumed as the flame developed downstream. The OH layer thickness decreased as pressure increased, which demonstrated a similar trend to the spray flame described in [56].

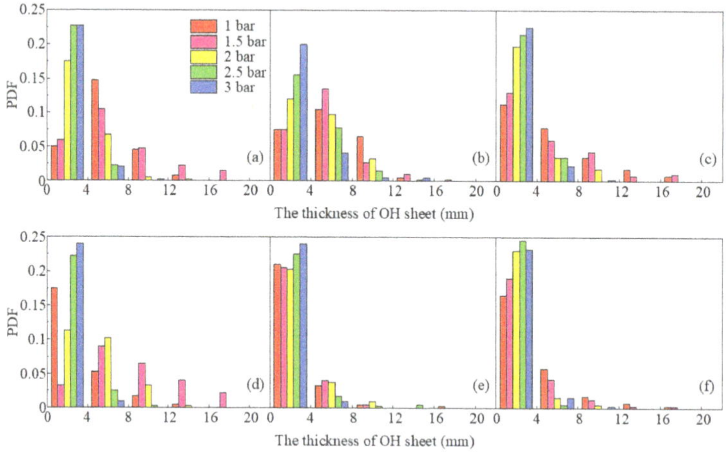

Figure 8. PDF of the thicknesses of OH layer at different pressures, (**a**,**d**) is S2, (**b**,**e**) is S3, (**c**,**f**) is S5. The upper row is at HAB = 30 mm, and lower row is HAB = 40 mm.

3.3. *Laser-Induced PAH Fluorescence*

Figure 9 shows the time-averaged OH-PLIF and PAH-PLIF images for S1–S5, where the PAH-PLIF were acquired simultaneously using different central wavelength filters, and OH-PLIF was obtained from the same flame. As described previously, there is a positive correlation between the emitted wavelength and the size of PAH molecules, thus the PAH-LIF signal represents various PAH molecules with differing molecular sizes. The PAH-PLIF signals were observed upstream of the flame, and the distribution areas of these PAH signals exhibited significant overlap. By analyzing the contours of PAHs at different wavelengths, it is found that PAHs at larger wavelengths extended further downstream. Similar to OH, the PAHs displayed a V-shaped distribution in S1–S3 but were mainly situated on the outer side of the OH branches. This observation illustrates that PAHs formed along the fresh fuel/air stream [32]. However, with an increase in S_N, the upstream PAHs extended radically towards the axis, which might be attributed to the increasing fuel concentration in the IRZ. In the cases of S1–S3, the PLIF of A2R5 showed no detectable fluorescence. As the increasing S_N altered the structure of OH-PLIF, the distribution of PAHs also changed, with a sudden intensification in PAH-PLIF signals.

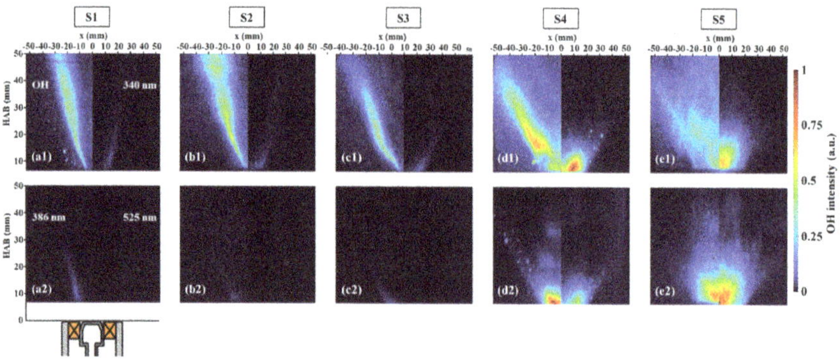

Figure 9. Time-averaged images of OH-PLIF and PAH-PLIF with different S_N at 1 bar, and (**a–e**) correspond to the cases S1–S5. (**a1–e1**) show OH-PLIF (**left**) and PAH-PLIF at 340 nm (**right**). (**a2–e2**) show PAH-PLIF at 386 nm (**left**) and PAH-PLIF at 525 nm (**right**).

The effect of pressure on PAH formation is depicted in Figure 10, and each molecular PLIF intensity is normalized by the maximum intensity among these cases. At elevated pressure, the area of PAH distribution still had considerable overlap, and the distribution area of each PAH increased with pressure. The increase in pressure did not change the V-shape feature of PAH in S2 and S3, but it promoted the PAH formation along the outer side of SL downstream and inside the IRZ near the nozzle. This enhanced effect of pressure was more significant on the formation of A2R5 compared to the other two PAHs. While for S5, the rising pressure changed the PAH generation region, the PAH branch along the SL gradually disappeared and occurred at the axis. This might be attributed to the poor mixing quality and more droplets moving to the IRZ.

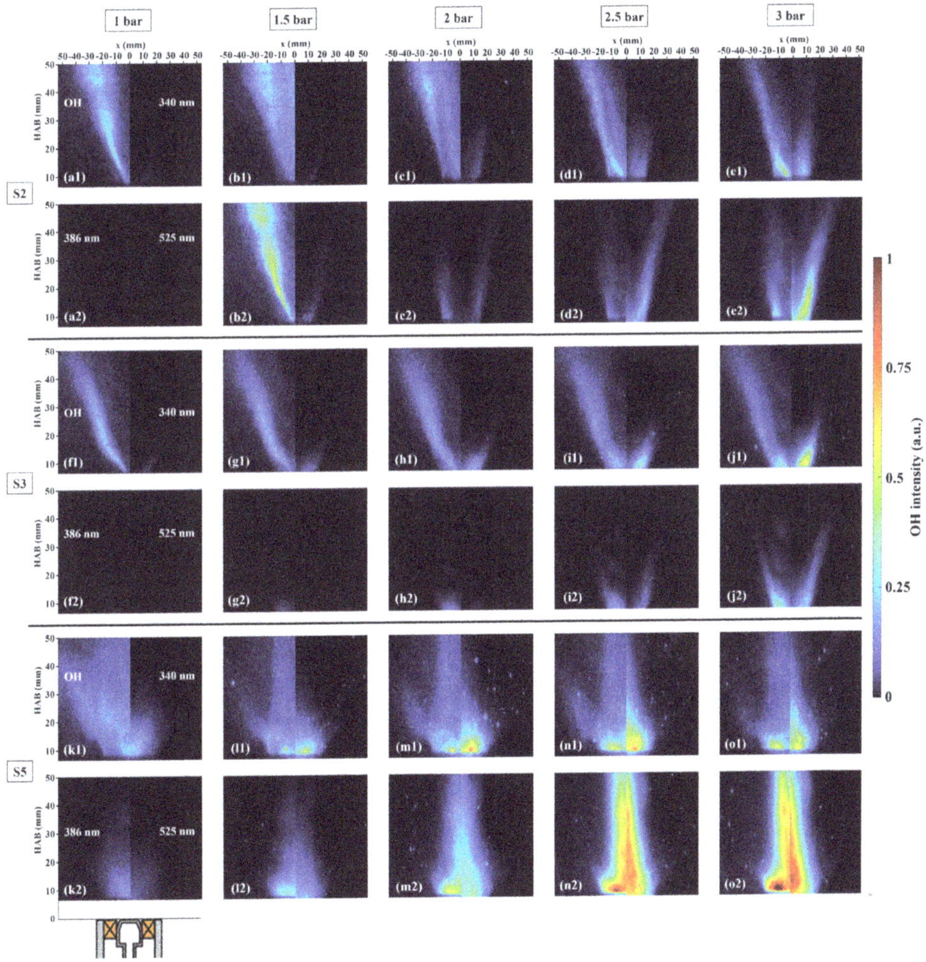

Figure 10. Time-averaged images of OH-PLIF and PAH-PLIF at 1.0–3.0 bar. Labels (**a–e**) correspond to case S2 with the pressures ranging from 1 to 3 bar: (**a1–e1**) show OH-PLIF (**left**) and PAH-PLIF at 340 nm (**right**), while (**a2–e2**) display PAH-PLIF at 386 nm (**left**) and PAH-PLIF at 525 nm (**right**). Labels (**f1–j1,f2–j2**) and (**k1–o1,k2–o2**) represent cases S3 and S5, respectively, under the same pressure range and configuration.

The LIF intensity is influenced by various factors in addition to component concentration, such as laser energy, laser sheet size, temperature, and pressure. In this experiment, the LIF intensity was dependent on temperature, pressure, and component concentration. According to the simulation results from [57], changes in ambient pressure had no significant impact on temperature distribution. Besides, the negative correlation between pressure and LIF intensity implies that if the LIF intensity increases with pressure, the concentration of the component must increase strongly with pressure. Additionally, there is no evidence proving the quenching for a large PAH's LIF is weaker than that of a smaller PAH's LIF [35]. Hence, it is feasible to compare the LIF intensity of PAHs at different wavelengths.

As depicted in Figure 10, it is evident that the pressure enhanced the formation of three molecular sizes of PAHs. To further quantify the effect of pressure on PAHs with different molecular sizes, the ratios of maximum intensity of PAH-PLIF at a certain wavelength divided by that at another wavelength are calculated, as shown in Figure 11. The promoting effect of pressure on PAH-PLIF intensity at both 386/340 (A4/A2 and A3) and 525/340 (A2R5/A2 and A3) in S2 was more significant as pressure increased from 1.5 bar to 2.5 bar, but the growth rates of these two larger PAHs slowed down when the pressure reached 3 bar. The growth rates of PAHs at 340 nm and 386 nm were similar in S3, resulting in two inflection points on the curve, but overall, the growth rate of PAH at 386 nm was still greater than that of at 340 nm. For PAH at 525 nm, in both cases S3 and S5, there was no apparent increase in the ratio of 525/386, but when the pressure increased from 2 bar to 3 bar, the growth rate of A2R5 significantly accelerated. Moreover, the ratios of 386/340 and 525/340 increased by a maximum of 1.3 and 35.68 in S2, respectively. In S3, they were 0.37 and 4.78, and in S5, they were 0.92 and 2.46. This indicated that pressure had different degrees of promotion on PAHs with different molecular sizes, and the contribution of pressure to the PAH-PLIF intensity was more notable at longer wavelengths [35].

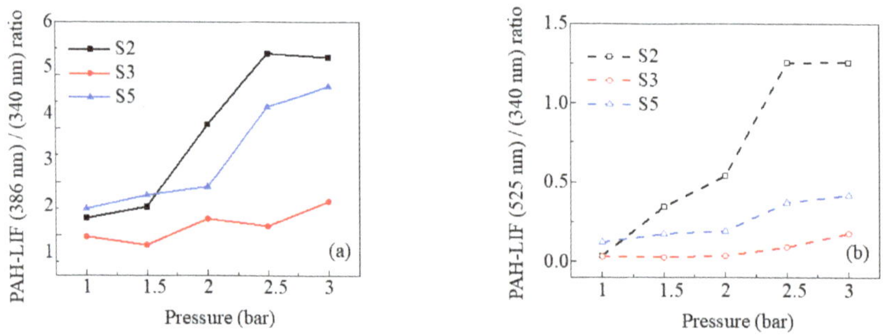

Figure 11. Ratios of maximum PAH-LIF signals for case S2, S3, and S5 as a function of pressure. (**a**) 386 nm/340 nm, (**b**) 525 nm/340 nm.

4. Conclusions

The flame structures and evolution of PAHs with different molecular sizes were investigated in the swirl spray flames at elevated pressures with different swirl numbers, employing simultaneous OH-PLIF and PAH-PLIF measurements to eliminate PAH-induced interference in OH detection. The major findings in this study are as follows:

The influence of swirl number, ranging from weak to strong swirl, was systematically investigated. The flames exhibited two different structures, V-shaped and crown-shaped. In V-shaped flames, the distribution of PAHs was located on the outer side of the OH branches, while in crown-shaped flames, PAHs were more radically distributed and moved towards the axis. In both V-shaped and crown-shaped flames, the PAHs at different wavelengths overlapped significantly, but the PAH of different molecular sizes extended further downstream.

The pressure effect on the flame structure with different swirl numbers was different. Cases of S2, S3, and S5 were selected to study the pressure effects on the flame structure and PAHs' formation characteristics. In V-shaped and crown-shaped flames, an increase in pressure led to a slight increase in the flame angle and a decrease in the thickness of the reaction layer. The OH layer exhibited a bifurcation structure with increasing pressure, and the newly generated OH branch was located in the inner region of the original OH layer.

Pressure enhanced PAH formation with stronger effects for PAHs with large molecular sizes. With increasing pressure, PAHs gradually extended downstream, and the upstream PAHs stretched towards the axis. In all three cases, PAH concentrations at three wavelengths increased with pressure, with larger PAHs exhibiting stronger pressure-enhanced phenomena.

Future work will focus on conducting experiments at higher pressures to enable a more comprehensive investigation of swirl spray flames, including spray characteristics, temperature distribution, and soot formation at elevated pressures. These studies aim to provide deeper insights into the fundamental mechanism of swirl-stabilized flame under high pressure conditions.

Author Contributions: Investigation, W.W. (Wenyu Wang), R.Z. and S.L.; Data curation, W.W. (Wenyu Wang); Writing—original draft, W.W. (Wenyu Wang); Writing—review & editing, W.W. (Wenyu Wang), R.Z., S.L., Y.H., W.W. (Wubin Weng), S.W., W.L.R. and Z.W.; Supervision, Y.H., W.W. (Wubin Weng) and Z.W.; Project administration, Z.W.; Funding acquisition, Z.W. All authors have read and agreed to the published version of the manuscript.

Funding: This work was supported by the National Key R&D Program of China (2023YFE0114400), the Fundamental Research Funds for the Central Universities (2022ZFJH04), and the ZJU-KAUST Joint Laboratory for Clean Combustion of E-fuels (JLACCE).

Data Availability Statement: The original contributions presented in the study are included in the article, further inquiries can be directed to the corresponding author.

Conflicts of Interest: The authors declare no conflict of interest.

Nomenclature

ABE	Acetone-butanol-ethanol
HAB	Height above the burner
IRZ	Inner recirculation zone
LES	Large eddy simulations
LII	Laser-induced incandescence
PAHs	Polycyclic aromatic hydrocarbons
PDFs	Probability density functions
PIV	Particle image velocimetry
PLIF	Planar laser-induced fluorescence
PME	Palm biodiesel/methyl esters
PVC	Processing vortex core
SL	Shear layer
SNR	Signal to noise ratio
D_h	Hub diameter of the swirler (mm)
D_n	Tip diameter of the swirler (mm)
S_N	Swirl number
θ	Swirler vane angle

References

1. Philo, J.J.; Shahin, T.T.; McDonald, C.T.; Gejji, R.M.; Lucht, R.P.; Slabaugh, C.D. Effect of fuel temperature on the structure of a high-pressure liquid-fueled swirl flame. *Fuel* **2023**, *354*, 129142. [CrossRef]
2. Xin, S.; Wang, W.; He, Y.; Zhu, Y.; Wang, Z. Effect of low fuel temperature on combustion deterioration of kerosene swirling spray flames using OH-PLIF. *Fuel* **2024**, *358*, 130098. [CrossRef]
3. Oberleithner, K.; Stöhr, M.; Im, S.H.; Arndt, C.M.; Steinberg, A.M. Formation and flame-induced suppression of the precessing vortex core in a swirl combustor: Experiments and linear stability analysis. *Combust. Flame* **2015**, *162*, 3100–3114. [CrossRef]
4. Grader, M.; Eberle, C.; Gerlinger, P.; Aigner, M. LES of a Pressurized, Sooting Aero-Engine Model Combustor at Different Equivalence Ratios with a Sectional Approach for PAHs and Soot. In *Turbo Expo: Power for Land, Sea, and Air*; American Society of Mechanical Engineers: New York, NY, USA, 2018.
5. Guiberti, T.F.; Zimmer, L.; Durox, D.; Schuller, T. Experimental Analysis of V- to M-Shape Transition of Premixed CH4/H2/Air Swirling Flames. In *Turbo Expo: Power for Land, Sea, and Air*; American Society of Mechanical Engineers: New York, NY, USA, 2013.
6. Chterev, I.; Foley, C.W.; Foti, D.; Kostka, S.; Caswell, A.W.; Jiang, N.; Lynch, A.; Noble, D.R.; Menon, S.; Seitzman, J.M.; et al. Flame and Flow Topologies in an Annular Swirling Flow. *Combust. Sci. Technol.* **2014**, *186*, 1041–1074. [CrossRef]
7. Strakey, P.; Woodruff, S.; Williams, T.; Schefer, R. OH-PLIF measurements of high-pressure, hydrogen augmented premixed flames in the simval combustor. In Proceedings of the 45th AIAA Aerospace Sciences Meeting and Exhibit, Reno, NV, USA, 8–11 January 2007; American Institute of Aeronautics and Astronautics: Washington, DC, USA, 2007; p. 980.
8. Stopper, U.; Meier, W.; Sadanandan, R.; Stöhr, M.; Aigner, M.; Bulat, G. Experimental study of industrial gas turbine flames including quantification of pressure influence on flow field, fuel/air premixing and flame shape. *Combust. Flame* **2013**, *160*, 2103–2118. [CrossRef]
9. Boxx, I.; Slabaugh, C.; Kutne, P.; Lucht, R.P.; Meier, W. 3kHz PIV/OH-PLIF measurements in a gas turbine combustor at elevated pressure. *Proc. Combust. Inst.* **2015**, *35*, 3793–3802. [CrossRef]
10. Chterev, I.; Boxx, I. Effect of hydrogen enrichment on the dynamics of a lean technically premixed elevated pressure flame. *Combust. Flame* **2021**, *225*, 149–159. [CrossRef]
11. Agostinelli, P.W.; Laera, D.; Chterev, I.; Boxx, I.; Gicquel, L.; Poinsot, T. Large eddy simulations of mean pressure and H2 addition effects on the stabilization and dynamics of a partially-premixed swirled-stabilized methane flame. *Combust. Flame* **2023**, *249*, 112592. [CrossRef]
12. Rault, T.M.; Vishwanath, R.B.; Gülder, Ö.L. Influence of m-xylene addition to Jet A-1 on spray structure, flow field and soot production in turbulent swirl-stabilized spray flames in a model combustor. *Combust. Flame* **2020**, *219*, 258–267. [CrossRef]
13. Cavaliere, D.E.; Kariuki, J.; Mastorakos, E. A Comparison of the Blow-Off Behaviour of Swirl-Stabilized Premixed, Non-Premixed and Spray Flames. *Flow Turbul. Combust.* **2013**, *91*, 347–372. [CrossRef]
14. Chong, C.T.; Hochgreb, S. Spray flame structure of rapeseed biodiesel and Jet-A1 fuel. *Fuel* **2014**, *115*, 551–558. [CrossRef]
15. Chong, C.T.; Chiong, M.-C.; Ng, J.-H.; Tran, M.-V.; Valera-Medina, A.; Józsa, V.; Tian, B. Dual-Fuel Operation of Biodiesel and Natural Gas in a Model Gas Turbine Combustor. *Energy Fuels* **2020**, *34*, 3788–3796. [CrossRef]
16. Xie, T.; Chong, C.T.; Wang, S.; Seljak, T.; Ng, J.-H.; Tran, M.-V.; Karmakar, S.; Tian, B. Flow field, flame structure and emissions quantifications of oxygenated glycerol in a swirl flame combustor. *Fuel* **2022**, *321*, 124052. [CrossRef]
17. Kumar, M.; Karmakar, S.; Chong, C.T. Investigation on combustion characteristics of acetone-butanol-ethanol/Jet A-1 mixture in a Swirl-stabilized combustor for its potential application in gas turbine engines. *Fuel* **2023**, *340*, 127610. [CrossRef]
18. Philo, J.J.; Frederick, M.D.; Slabaugh, C.D. 100 kHz PIV in a liquid-fueled gas turbine swirl combustor at 1 MPa. *Proc. Combust. Inst.* **2021**, *38*, 1571–1578. [CrossRef]
19. Chterev, I.; Rock, N.; Ek, H.; Emerson, B.; Seitzman, J.; Jiang, N.; Roy, S.; Lee, T.; Gord, J.; Lieuwen, T. Simultaneous imaging of fuel, OH, and three component velocity fields in high pressure, liquid fueled, swirl stabilized flames at 5 kHz. *Combust. Flame* **2017**, *186*, 150–165. [CrossRef]
20. Malbois, P.; Salaün, E.; Rossow, B.; Cabot, G.; Bouheraoua, L.; Richard, S.; Renou, B.; Grisch, F. Quantitative measurements of fuel distribution and flame structure in a lean-premixed aero-engine injection system by kerosene/OH-PLIF measurements under high-pressure conditions. *Proc. Combust. Inst.* **2019**, *37*, 5215–5222. [CrossRef]
21. Salaün, E.; Frindt, F.; Cabot, G.; Renou, B.; Richard, S.; Cazalens, M.; Malbois, P.; Grisch, F. Experimental Investigation on NO Pollutant Formation in High-Pressure Swirl-Stabilized Kerosene/Air Flames Using NO-, OH- and Kerosene-PLIF and PIV Laser Diagnostics. In *Turbo Expo: Power for Land, Sea, and Air*; American Society of Mechanical Engineers: New York, NY, USA, 2020.
22. Legros, S.; Brunet, C.; Domingo-Alvarez, P.; Malbois, P.; Salaun, E.; Godard, G.; Caceres, M.; Barviau, B.; Cabot, G.; Renou, B.; et al. Combustion for aircraft propulsion: Progress in advanced laser-based diagnostics on high-pressure kerosene/air flames produced with low-NOx fuel injection systems. *Combust. Flame* **2021**, *224*, 273–294. [CrossRef]
23. Lyu, Z.; Yan, T.; Qian, Y.; Cen, L.; Jin, Z.; Zhou, D.; Lu, X. Experimental and numerical investigation on soot formation characteristics in n-decane diffusion flames at elevated pressures. *Proc. Combust. Inst.* **2024**, *40*, 105628. [CrossRef]

24. Wang, L.-Y.; Bauer, C.K.; Gülder, Ö.L. Soot and flow field in turbulent swirl-stabilized spray flames of Jet A-1 in a model combustor. *Proc. Combust. Inst.* **2019**, *37*, 5437–5444. [CrossRef]
25. Al Sadi, K.; Nadimi, E.; Wu, D. Catalytic Ammonia Combustion: Legacy Catalytic Burner Designs and Catalyst Requirements for In Situ Hydrogen Production. *Energies* **2025**, *18*, 3505. [CrossRef]
26. Shi, J.; Yang, P.; Ren, X.; Yang, X.; Yan, H.; Tan, Y.; Lei, Z. Study on the Mechanism of Soot Inhibition in Methanol-Ethylene Mixed Combustion under High-Pressure Conditions. *Fuel* **2025**, *391*, 134697. [CrossRef]
27. Chen, C.; Liu, D. Review of effects of zero-carbon fuel ammonia addition on soot formation in combustion. *Renew. Sustain. Energy Rev.* **2023**, *185*, 113640. [CrossRef]
28. Zeng, W.; Liu, J.; Liu, Y.; Chen, B.; Liu, A. The Effect of Hydrogen Addition on the Combustion Characteristics of RP-3 Kerosene/Air Premixed Flames. *Energies* **2017**, *10*, 1107. [CrossRef]
29. Hui, X.; Zhang, C.; Xia, M.; Sung, C.-J. Effects of hydrogen addition on combustion characteristics of n-decane/air mixtures. *Combust. Flame* **2014**, *161*, 2252–2262. [CrossRef]
30. Yan, Z.; Zhu, T.; Xue, X.; Liu, H.; Li, Q.; Huang, Z. Effects of NH3 and H2 addition on morphology, nanostructure and oxidation of soot in n-decane diffusion flames. *Fuel Process. Technol.* **2024**, *253*, 108003. [CrossRef]
31. Li, R.; Yang, D.; Liu, F.; Hu, Q.; Liu, Q.; Yue, H.; Meng, Y.; Mei, Y. Study on in-cylinder soot formation process of F-T diesel/methanol dual-fuel diesel engine. *Fuel* **2025**, *390*, 134745. [CrossRef]
32. Bouvier, M.; Cabot, G.; Yon, J.; Grisch, F. On the use of PIV, LII, PAH-PLIF and OH-PLIF for the study of soot formation and flame structure in a swirl stratified premixed ethylene/air flame. *Proc. Combust. Inst.* **2021**, *38*, 1851–1858. [CrossRef]
33. Jain, A.; Wang, Y.; Schweizer, C.; Kulatilaka, W.D. Investigation of Flow-Flame Interactions in Kerosene Piloted Liquid-Spray Flames Using Simultaneous OH and PAH PLIF. In Proceedings of the AIAA Scitech 2020 Forum, Orlando, FL, USA, 6–10 January 2020; AIAA SciTech Forum. American Institute of Aeronautics and Astronautics: Washington, DC, USA, 2020.
34. Wang, Y.; Jain, A.; Schweizer, C.; Kulatilaka, W.D. OH, PAH, and sooting imaging in piloted liquid-spray flames of diesel and diesel surrogate. *Combust. Flame* **2021**, *231*, 111479. [CrossRef]
35. Liang, S.; Li, Z.; Gao, J.; Ma, X.; Xu, H.; Shuai, S. PAHs and soot formation in laminar partially premixed co-flow flames fuelled by PRFs at elevated pressures. *Combust. Flame* **2019**, *206*, 363–378. [CrossRef]
36. Wan, K.; Huang, Y.; Gao, Z.; He, Y.; Jiang, C. Large-Eddy Simulation of a Swirling Aviation Kerosene Spray Flame Using 3-Component Surrogate Fuels. In Proceedings of the 32nd Congress of the International Council of the Aeronautical Sciences, Shanghai, China, 6–10 September 2021.
37. Yi, R.; Chen, X.; Chen, C.P. Surrogate for Emulating Physicochemical and Kinetics Characteristics of RP-3 Aviation Fuel. *Energy Fuels* **2019**, *33*, 2872–2879. [CrossRef]
38. Liu, J.; Hu, E.; Yin, G.; Huang, Z.; Zeng, W. An experimental and kinetic modeling study on the low-temperature oxidation, ignition delay time, and laminar flame speed of a surrogate fuel for RP-3 kerosene. *Combust. Flame* **2022**, *237*, 111821. [CrossRef]
39. Balakrishnan, P.; Srinivasan, K. Influence of swirl number on jet noise reduction using flat vane swirlers. *Aerosp. Sci. Technol.* **2018**, *73*, 256–268. [CrossRef]
40. Sharma, S.; Ghate, K.; Sundararajan, T.; Sahu, S. Effects of air swirler geometry on air and spray droplet interactions in a spray chamber. *Adv. Mech. Eng.* **2019**, *11*, 1687814019850978. [CrossRef]
41. Gupta, A.K.; Lilley, D.G.; Syred, N. *Swirl Flows*; Tunbridge Wells: Kent, UK, 1984.
42. Huang, Y.; Yang, V. Dynamics and stability of lean-premixed swirl-stabilized combustion. *Prog. Energy Combust. Sci.* **2009**, *35*, 293–364. [CrossRef]
43. Bren˜a de la Rosa, A.; Wang, G.; Bachalo, W.D. The Effect of Swirl on the Velocity and Turbulence Fields of a Liquid Spray. *J. Eng. Gas Turbines Power* **1992**, *114*, 72–81. [CrossRef]
44. Mulla, I.A.; Renou, B. Simultaneous imaging of soot volume fraction, PAH, and OH in a turbulent n-heptane spray flame. *Combust. Flame* **2019**, *209*, 452–466. [CrossRef]
45. Liao, Y.-H.; Hermanson, J.C. OH-PLIF Imaging of the Reaction Zone in Swirled, Strongly-Pulsed Jet Diffusion Flames with a Low Reynolds Number. *Combust. Sci. Technol.* **2018**, *190*, 615–631. [CrossRef]
46. Beretta, F.; D'Alessio, A.; D'Orsi, A.; Minutolo, P.U.V. and Visible Laser Excited Fluorescence from Rich Premixed and Diffusion Flames. *Combust. Sci. Technol.* **1992**, *85*, 455–470. [CrossRef]
47. Liu, P.; He, Z.; Hou, G.-L.; Guan, B.; Lin, H.; Huang, Z. The Diagnostics of Laser-Induced Fluorescence (LIF) Spectra of PAHs in Flame with TD-DFT: Special Focus on Five-Membered Ring. *J. Phys. Chem. A* **2015**, *119*, 13009–13017. [CrossRef] [PubMed]
48. Gao, Z.; Ren, F.; Cheng, X.; Wang, L.; Zhu, L.; Huang, Z. The effects of 1-methylnaphthalene addition to n-dodecane on the formation of soot and polycyclic aromatic hydrocarbons in laminar coflow diffusion flames. *Fuel* **2022**, *329*, 125378. [CrossRef]
49. Otsu, N. A Threshold Selection Method from Gray-Level Histograms. *IEEE Trans. Syst. Man Cybern.* **1979**, *9*, 62–66. [CrossRef]
50. Pakhomov, M.A.; Terekhov, V.I. Numerical modeling of turbulent flow structure and heat transfer in a droplet-laden swirling flow in a pipe with a sudden expansion. *Numer. Heat Transf. Part A Appl.* **2017**, *71*, 721–736. [CrossRef]

51. Pakhomov, M.; Terekhov, V. The effect of droplets thermophysical properties on turbulent heat transfer in a swirling separated mist flow. *Int. J. Therm. Sci.* **2020**, *149*, 106180. [CrossRef]
52. Shin, D.; Satija, A.; Lucht, R.P. Spray characteristics of standard and alternative aviation fuels at high ambient pressure conditions. *Exp. Therm. Fluid Sci.* **2022**, *130*, 110511. [CrossRef]
53. Jasuja, A.K.; Lefebvre, A.H. Influence of Ambient Air Pressure on Pressure-Swirl Atomizer Spray Characteristics. In *Turbo Expo: Power for Land, Sea, and Air*; American Society of Mechanical Engineers: New York, NY, USA, 2001.
54. Chigier, N.A. The Atomization and Burning of Liquid Fuel Sprays. In *Energy and Combustion Science*; Chigier, N.A., Ed.; Pergamon: New York, NY, USA, 1979; pp. 183–200.
55. Cessou, A.; Stepowski, D. Planar Laser Induced Fluorescence Measurement of [OH] in the Stabilization Stage of a Spray Jet Flame. *Combust. Sci. Technol.* **1996**, *118*, 361–381. [CrossRef]
56. Allen, M.G.; McManus, K.R.; Sonnenfroh, D.M.; Paul, P.H. Planar laser-induced-fluorescence imaging measurements of OH and hydrocarbon fuel fragments in high-pressure spray-flame combustion. *Appl. Opt.* **1995**, *34*, 6287–6300. [CrossRef] [PubMed]
57. Dekterev, A.A.; Lobasov, A.S.; Dekterev, A.A. Numerical simulation of unsteady combustion at elevated pressure and initial temperature of the mixture for the model combustion chamber. *J. Phys. Conf. Ser.* **2020**, *1677*, 012036. [CrossRef]

energies

Article

Hybrid Small Modular Reactor—Renewable Systems for Smart Cities: A Simulation-Based Assessment for Clean and Resilient Urban Energy Transitions

Nikolay Hinov [1,2]

[1] CoE "National Center of Mechatronics and Clean Technologies", 1000 Sofia, Bulgaria; hinov@tu-sofia.bg
[2] Department of Computer Systems, Faculty of Computer Systems and Technologies, Technical University of Sofia, 1000 Sofia, Bulgaria

Abstract

The global transition to clean energy necessitates integrated solutions that ensure both environmental sustainability and energy security. This paper proposes a scenario-based modeling framework for urban hybrid energy systems combining small modular reactors (SMRs), photovoltaic (PV) generation, and battery storage within a smart grid architecture. SMRs offer compact, low-carbon, and reliable baseload power suitable for urban environments, while PV and storage enhance system flexibility and renewable integration. Six energy mix scenarios are evaluated using a lifecycle-based cost model that incorporates both capital expenditures (CAPEX) and cumulative carbon costs over a 25-year horizon. The modeling results demonstrate that hybrid SMR–renewable systems—particularly those with high nuclear shares—can reduce lifecycle CO_2 emissions by over 90%, while maintaining long-term economic viability under carbon pricing assumptions. Scenario C, which combines 50% SMR, 40% PV, and 10% battery, emerges as a balanced configuration offering deep decarbonization with moderate investment levels. The proposed framework highlights key trade-offs between emissions and capital cost and seeking resilient and scalable pathways to support the global clean energy transition and net-zero commitments.

Keywords: carbon neutrality; distributed generation; hybrid systems; small modular reactors; smart grids; urban energy strategies

1. Introduction

The rapid urbanization observed in the 21st century has created an unprecedented demand for clean, secure, and adaptive energy systems in cities. Smart cities—defined by their integration of digital technologies into urban infrastructure—are emerging as a transformative response to challenges such as climate change, energy security, and quality of life. These cities rely on real-time data, interconnected systems, and intelligent control to optimize energy use across sectors including residential buildings, industry, transportation, and public services [1].

In this context, the global shift toward net-zero emissions by mid-century has become a central objective of international climate and energy policy. According to the International Energy Agency (IEA), achieving net zero by 2050 requires a complete transformation of the global energy system, including a rapid scaling of low-carbon power generation technologies [2]. The Intergovernmental Panel on Climate Change (IPCC) also emphasizes the vital role of both renewable energy and nuclear power in deep mitigation scenarios compatible with the 1.5 °C goal [3]. At the European level, the REPowerEU plan outlines an ambitious framework for reducing dependence on fossil fuels by accelerating renewable deployment and supporting innovative technologies such as small modular nuclear reactors (SMRs) [4]. These initiatives collectively underscore the need for integrated, flexible, and urban-compatible energy solutions that can combine the stability of nuclear baseload with the variability of renewables, thereby enabling resilient and climate-aligned smart city infrastructures.

One of the central pillars of a smart city is the smart energy system: a decentralized, flexible, and data-driven grid capable of integrating a high share of renewable energy sources (RES), managing load fluctuations, and empowering consumers through real-time feedback. However, the practical realization of smart cities introduces multiple challenges: high initial investment in digital infrastructure, interoperability between legacy and new systems, energy storage constraints, data privacy and cybersecurity, and ensuring reliable baseload generation to support dynamic energy flows [5].

Smart cities must support increased electrification of transport, heating, and industry, resulting in significantly higher electricity consumption and more complex demand profiles. Peak load conditions, seasonal variations, and intermittent RES production require resilient and flexible energy strategies [6].

In the context of smart cities, the implementation of smart electrical grids plays a key role in ensuring sustainable, efficient, and adaptive energy resource management. With increasing urbanization and growing electricity demand, traditional energy infrastructures are proving insufficient to meet the needs of modern urban systems. Smart grids, through the integration of digital technologies, sensors, automation, and two-way communication between providers and consumers, enable better responsiveness to changes in consumption, load balancing, and the integration of renewable energy sources. This not only enhances energy efficiency and security but also contributes to reducing carbon emissions—an essential aspect in building sustainable and environmentally oriented smart cities [7,8].

To address these issues, small modular reactors (SMRs) are gaining global attention as a viable addition to the smart urban energy mix. SMRs offer clean, continuous baseload power with compact footprints, passive safety systems, and modular deployment. Unlike traditional nuclear reactors, they are designed for urban proximity, remote siting, or industrial integration, offering opportunities for combined heat and power (CHP), hydrogen production, or district heating—key features in smart city ecosystems [9,10].

By integrating SMRs with advanced smart grid infrastructures, cities can benefit from both stability and flexibility, ensuring that low-carbon energy systems remain resilient to load volatility and external disruptions. Furthermore, such integration supports dynamic energy pricing, peak shaving, and demand-side management—core functionalities of intelligent grid systems [11,12].

The aim of this study is to evaluate the long-term economic and environmental performance of hybrid SMR–renewable urban energy systems under realistic operational conditions and to identify cost-effective decarbonization pathways using lifecycle-based scenario modeling.

The paper is structured as follows: Section 1 provides an introduction to the context of the global clean energy transition, the challenges of smart urban electrification, and the potential role of hybrid SMR–renewable systems in supporting resilient and low-carbon infrastructures. Section 2 provides a comprehensive review of the latest advancements in nuclear energy, with a particular focus on small modular reactors (SMRs). Section 3 presents the proposed hybrid energy architecture that integrates SMRs with solar power, battery storage, and intelligent digital control mechanisms. Section 4 evaluates the system-level performance, including emissions reduction, cost estimation, and grid flexibility. Section 5 concludes the paper and discusses the practical implications of the proposed system. Additional sections focus on scenario-based emissions modeling, capital investment calculations, and their relevance for urban energy decision-makers and planners.

2. Overview of Current SMR Research and Development

SMRs are an emerging category of nuclear technologies that offer a number of advantages, including smaller size, modularity, integrated safety, and flexible installation options. They are seen as a key part of future energy systems, especially in the context of smart cities, where decentralized and sustainable energy production is critical. In recent years, numerous SMR concepts have been proposed worldwide, some of which are still in the conceptual phase, while others have already passed the engineering design stage and are now undergoing pre-legitimization activities. In this sense, several projects have already started the construction of pilot or demonstration reactors. On this basis, the following classification and systematization is described:

2.1. Conceptual Phase

In the conceptual phase, many SMR projects are still in the early stages of development, focusing mainly on theoretical models and feasibility studies. The main objective of these projects is to explore new concepts for nuclear reactors that can be integrated into decentralized energy systems and provide safe and efficient solutions for future clean energy requirements.

Examples of conceptual SMR technologies [13,14]:

- Terrestrial Energy IMSR (Integrated Small Salt Reactor): A concept that uses molten salt as a fuel and coolant. This reactor is designed to be much more efficient than traditional nuclear technologies, focusing on high operating temperatures and low pressures, making it suitable for a variety of applications, including hydrogen production.
- Seaborg CMSR (Modular Salt Reactor): A technology that also uses molten salt and aims to offer a safe and cost-effective way to produce nuclear energy. The CMSR is characterized by high efficiency and low pressure, making it suitable for use in coastal and remote areas [15,16].

2.2. Engineering Design and Licensing Preparation

SMR designs that have passed the engineering design phase are beginning to prepare for licensing and potential construction. These technologies are often more advanced and are undergoing the more stringent safety and durability requirements required to begin construction [17–19].

Examples of SMRs in the engineering design phase:

- NuScale VOYGR (USA): NuScale is among the first companies to receive license approval from the US Nuclear Regulatory Commission (NRC) for its SMR design. The design proposes a 77 MW reactor per module that can be integrated into larger power plants. It is the first SMR to receive such approval and is already scheduled for construction.
- GE Hitachi BWRX-300 (USA/Canada): Based on boiling water reactor (BWR) technology, this 300 MW SMR is in the engineering design phase. GE Hitachi plans to complete this project as part of a global effort to deploy safe, compact nuclear technologies.

2.3. Construction Phase of SMR Deployment

Several SMR projects have already moved into the construction phase, with some in the early stages of pilot or demonstration plants. Although this phase is still early, it shows the promising potential of SMR for commercial and industrial deployment.

Examples of SMRs in construction [19,20]:

- Rolls-Royce SMR (UK): In recent years, Rolls-Royce has launched its small modular Reactor project in the UK. The project is in the early stages of construction and is expected to provide 470 MW of capacity. The technology aims to reduce carbon emissions and provide stable energy production in line with the UK's climate goals.
- Brest-OD-300 (Russia): A leader in the development of ultra-high temperature SMRs that use liquid lead as a coolant. Russia has started construction of its first SMR under the Brest-OD-300 project, with the reactor expected to be fully operational by the end of 2025.

2.4. Classification of SMR Technologies

SMR technologies can be classified based on their coolant, pressure, and fuel type, each offering unique characteristics [21–23].

Pressurized Water Reactor (PWR): This is the most common SMR technology, such as the NuScale VOYGR. It has a proven safety and stability of operation.

Boiling Water Reactor (BWR): This simpler and cheaper technology, such as the GE Hitachi BWRX-300, generates steam directly in the reactor and powers the turbine.

High-temperature Gas-cooled Reactor (HTGR): This system used helium cooling and offers high thermal efficiency, suitable for hydrogen production and process heating.

Molten Salt Reactor (MSR): Molten salt reactors offer unique safety characteristics and efficiency at high temperatures; examples include the Seaborg CMSR and Terrestrial IMSR.

Lead-cooled Fast Reactor (LFR): This system uses liquid lead for cooling, allowing it to operate at high temperatures and reducing radiation waste; BREST-OD-300 is an example.

2.5. Global Efforts and Innovations

As SMR technologies develop, many countries are beginning to adopt nuclear energy innovation strategies, with a significant number launching national programs to develop SMR solutions. SMR projects not only provide a solution to future energy needs, but also represent an important aspect of global carbon reduction policies [23,24].

USA: SMR initiatives in the USA are based on projects such as NuScale, which has already completed licensing processes and is planning to build commercial reactors.

Canada: The Canadian company Terrestrial Energy is working on IMSR as part of efforts to develop new nuclear infrastructure.

UK: The UK is leading the way in SMR development with the Rolls-Royce SMR project, with the aim of achieving energy autonomy and reducing carbon emissions.

Denmark: Seaborg Technologies also has leading developments in SMR technologies, such as CMSR, which will be tested in the future.

Small modular reactors (SMRs) represent a transformative class of advanced nuclear power technologies. With an electrical capacity typically up to 300 MW per module, they are designed to be factory-fabricated, easily transported, and rapidly deployed, offering scalable and safer nuclear solutions for modern energy systems. Unlike conventional gigawatt-scale nuclear plants, SMRs are optimized for flexibility, modularity, and integration into distributed energy infrastructures [13,14].

Table 1 systematizes the main technologies, advantages and disadvantages of small modular reactors.

The development and implementation of SMRs are accompanied by a complex balance of technological, economic, and social considerations. On one hand, SMRs offer several compelling advantages such as enhanced safety, modularity, and suitability for integration with renewable energy sources. On the other hand, they also face notable barriers including high capital costs, regulatory uncertainties, and public acceptance issues. Economic advantages of SMR [23–25]:

Table 1. Comparative summary of SMR technologies [17–20].

Reactor Type	Coolant	Key Benefits	Challenges	Example(s)
PWR	Pressurized Water	Mature technology; strong safety track record	High pressure systems; complex containment	NuScale VOYGR, Rolls-Royce SMR
BWR	Boiling Water	Simple design; lower capital cost	Radioactive steam; limited isolation	GE Hitachi BWRX-300
HTGR	Helium Gas	High efficiency; suitable for process heat	Large heat exchangers; low heat transfer gas	X-Energy Xe-100
MSR	Molten Salt	Low pressure; high heat transfer	Material corrosion; immature licensing	Seaborg CMSR, IMSR
LFR	Liquid Lead/Lead-Bismuth	High temperature operation; waste burning	Heavy coolant; slow heat transfer	BREST-OD-300

- Reduced initial investment: One of the main economic aspects of SMRs is that they require significantly lower initial investment compared to traditional large nuclear reactors. This is a result of their modular design, which allows for the production and construction of factory-made parts, reducing costs and construction time. This means that the design and construction of SMRs can be significantly faster and cheaper compared to traditional nuclear power plants.
- Scalability and flexibility: SMRs offer the possibility of adding additional modules if there is a need to increase capacity in the future. This flexibility allows operators to start with smaller investments and increase production depending on the growing demand for energy, without having to build large and expensive plants.
- Improved safety and reduced operating costs: These features are due to the use of passive safety systems. SMRs have passive safety systems that reduce the need for active interventions in the event of an accident. This reduces the risk of accidents and the associated economic losses. These systems also reduce the safety costs traditionally associated with large nuclear reactors. Furthermore, the small scale of SMRs allows for more efficient management and optimization of operating processes, which can lead to reduced operating costs in the long term. This is particularly important when it comes to operating costs, which can traditionally be high in the nuclear industry.
- Competitiveness against renewables and gas technologies: SMRs have a stable and continuous production process energy. Although renewable sources such as solar and wind power are becoming increasingly competitive, they cannot provide a stable baseload due to their intermittency. SMR, as a nuclear source, can provide a stable, continuous and reliable baseload, which is critical for energy security and grid stability. Despite the high initial investment, nuclear power and SMR in particular offer significantly lower carbon emissions compared to coal and natural gas plants. This is essential for achieving the climate goals of many countries and for minimizing the negative economic consequences of climate change.
- Improved public acceptance and integration opportunities in urban areas. Traditional nuclear reactors are often located in remote areas, which can raise questions about safety and public perception. SMRs, thanks to their smaller size and the possibility

of modular expansion, can be integrated into more densely populated urban areas, which reduces the need to build new facilities in remote areas.

The conclusion is that the economic advantages of SMRs consist not only in the reduced initial investment, but also in their flexibility, safety, and lower operating costs compared to traditional nuclear technologies. These advantages, combined with the ability to provide a stable baseload, make SMRs economically competitive in the context of modern energy needs and decarbonization efforts.

Table 2 summarizes the most relevant benefits and challenges associated with the deployment of SMRs across different technological platforms.

Table 2. The most relevant benefits and barriers to SMR deployment [12–25].

Advantages	Challenges
Passive safety and low meltdown risk	High upfront capital costs per kW
Modular, fast construction	Lack of established licensing paths
Low operational CO_2 emissions	Nuclear waste and long-term liability
Urban/remote deployment suitability	Public acceptance concerns
Supports load-following with renewables	Supply chain readiness and industrial scale-up

Over the past decade, the global interest in SMRs has led to a growing portfolio of designs under various stages of development, licensing, and deployment. These projects reflect the diversity of SMR technologies and national strategies aimed at achieving clean, reliable, and decentralized energy generation. Table 3 provides an overview of selected SMR initiatives worldwide, highlighting their developer, technological type, power output, current development status, and country of origin. The table emphasizes the international nature of SMR innovation and the variety of approaches pursued to meet future energy needs.

Table 3. Overview of current SMR initiatives worldwide [26–30].

Developer	Reactor	Power	Status	Country
NuScale (Corvallis, OR, USA)	VOYGR	77 MW	NRC Licensed	USA
GE Hitachi (Wilmington, NC, USA; Peterborough, ON, Canada)	BWRX-300	300 MW	Design phase	USA/Canada
Rolls-Royce (Derby, UK)	UK SMR	470 MW	Planning stage	UK
Terrestrial Energy (Oakville, ON, Canada)	IMSR	192 MW	Conceptual	Canada
Seaborg (Copenhagen, Denmark)	CMSR	~100 MW	Prototype	Denmark

Each SMR technology presents unique characteristics. While PWR and BWR types are closest to commercial deployment due to regulatory familiarity, advanced designs like MSR and HTGR offer higher efficiency and new capabilities for future smart city energy systems.

A nuclear generator works by converting the thermal energy produced by nuclear reactions into electrical energy. Nuclear fission processes occur in the core of the reactor, which generate a significant amount of heat. This heat is transferred to a coolant—usually water under high pressure—which circulates in a closed primary circuit, preventing it from boiling [18–20].

The heated coolant transfers the thermal energy to a steam generator, in which a secondary circuit of non-radioactive water is heated and converted into steam. The resulting steam drives a turbine, which transforms the thermal energy into mechanical energy. The turbine, in turn, powers an electric generator, which produces electrical energy.

After passing through the turbine, the steam is taken to a condenser, where it is cooled and condensed back into liquid water. This water is returned to the cycle, re-entering the steam generator. The entire process is managed by a centralized control room, which provides monitoring and regulation of all key systems, while ensuring the safety of operators by physically separating them from the reactor area.

The system is protected by multiple levels of shielding and containment structures that prevent radiation leakage and provide resilience against internal damage and external threats. This integrated and compact architecture is typical of small modular reactors (SMRs), which are characterized by a high degree of safety, factory assembly, and the ability to be flexibly deployed in various energy environments [29,30].

SMRs offer a strategic path toward carbon-neutral, resilient energy systems—especially in the context of smart cities and decentralized grids. While technological, regulatory, and societal challenges remain, the growing portfolio of pilot projects and supportive policy landscapes signal strong momentum for their adoption.

3. Justification and Structure of a Combined SMR–Renewable Urban Power System

3.1. System Configuration

The implementation of a hybrid energy system that integrates a small modular reactor (SMR), photovoltaic (PV) solar generation, and energy storage technologies is strategically justified by the evolving energy requirements of modern smart cities. Although SMRs offer a stable and continuous baseload power supply, they alone are insufficient to address the dynamic demands of contemporary urban energy systems, which require not only reliability but also operational flexibility [31,32].

The key arguments in support of this combined configuration are as follows:

- Operational Flexibility and Peak Load Management:

SMRs are optimized for continuous, steady-state operation and are not designed to ramp output rapidly in response to fluctuations in demand. During periods of peak load, this lack of agility can lead to system stress or the need for supplementary power sources. In contrast, solar power—when available—and energy storage systems can be rapidly dispatched to meet sudden increases in consumption. This enhances the overall flexibility and responsiveness of the energy system.

- Mitigation of Renewable Intermittency:

While solar energy contributes significantly to decarbonization goals, its intermittent nature limits its ability to provide consistent power, especially during nighttime or unfavorable weather conditions. However, during periods of high solar irradiance, PV systems can reduce the burden on the grid and decrease reliance on nuclear output or external sources. Furthermore, coupling solar generation with energy storage allows excess energy to be captured and deployed during low-generation periods, thus stabilizing supply.

- Economic and Environmental Benefits:

The declining cost of solar photovoltaic technologies has made them increasingly economically viable. When integrated with SMRs, solar generation can contribute to a reduction in total system capital expenditures, while simultaneously improving the environmental profile of the system. Energy storage plays a pivotal role by enabling time-shifting of solar output, ensuring that renewable energy contributes not only during peak production periods but also during times of increased demand or reduced generation.

In summary, the synergistic combination of a SMR, solar power, and energy storage addresses the core challenges of modern energy systems—ensuring stability, flexibility, and sustainability. This integrated approach is particularly well-suited for deployment in smart urban environments, where energy systems must dynamically adapt to variable loads and contribute to long-term decarbonization objectives.

Although SMR provides a stable and safe baseload, the cost of building such a reactor and its operation is significantly higher compared to renewable energy sources. Solar power and energy storage can provide the necessary flexibility of the system and reduce the overall cost of building and operating the energy infrastructure. By combining SMR with solar panels and batteries, it is possible to reduce operating costs by using solar energy during sunny weather and also to improve the resilience of the system by providing additional sources of energy during supply interruptions or peak loads. In addition, it is possible to ensure better compliance with climate goals by reducing carbon emissions from the energy mix [33,34].

The actual operational time of solar energy depends on the geographical location and atmospheric conditions; however, for temperate regions such as Central Europe, solar energy is usually available for about 4–6 h a day during sunny days, which represents about 30–40% of the daytime period. For the rest of the time, when the sun is not shining or the conditions are not favorable, solar energy will not be able to provide enough power, making the combination with energy storage and baseload from SMRs necessary.

The decision to include solar energy and energy storage in the system configuration is motivated by the need to ensure not only a stable baseload, but also flexibility during peak loads and intermittency of renewable sources. Combining SMRs with renewable sources such as solar panels and the use of energy storage technologies allows for a better economic and environmental result, while ensuring flexibility and resilience of the energy system.

The proposed smart urban energy system consists of the following components:

- A SMR acting as a baseload, low-carbon generator;
- Solar photovoltaic (PV) array, representing variable renewable energy;
- Lithium-ion battery storage unit for peak shaving and load balancing;
- Smart grid control system with demand-response logic and time-of-use pricing;
- Dynamic load model for residential, industrial, and electric vehicle (EV) charging profiles.

The system is designed to serve a daily urban load of approximately 100 MWh, representative of a mid-size smart city district. The structure of the proposed system is shown in Figure 1. In this study, all power values for reactors, PV generators, and batteries refer to electrical power (MWe), unless explicitly stated otherwise.

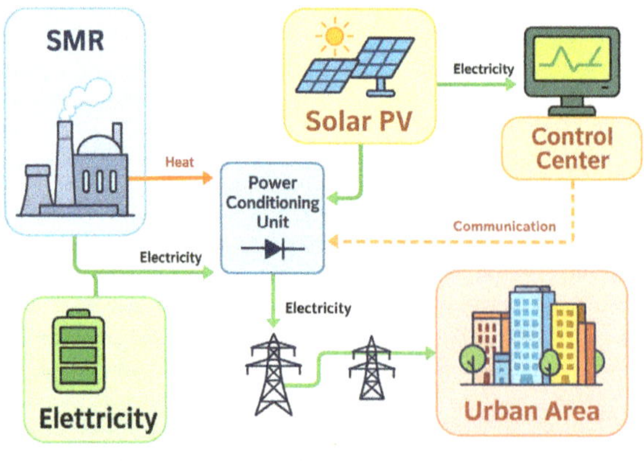

Figure 1. The structure of the proposed hybrid system.

Figure 1 illustrates the architecture of a smart hybrid energy system tailored for urban environments. It integrates a variety of distributed energy resources with intelligent control and bidirectional communication to enable flexible, low-emission, and resilient electricity supply in the context of smart cities.

On the generation side, the system combines a SMR, responsible for providing continuous baseload power, with solar photovoltaic (PV) units that contribute renewable electricity under favorable irradiance conditions. A battery energy storage system, positioned between the generators and the loads, stores excess energy from the PV array and discharges it during peak demand or supply deficits, thereby enhancing system flexibility and reliability.

At the core of the system lies the smart grid controller, which manages the energy flows in real time based on demand forecasts, price signals, and operating constraints. The controller also interacts with a demand-response logic module, allowing the system to dynamically modulate consumption in industrial, residential, and electric vehicle (EV) charging domains according to grid conditions.

On the consumption side, the system delivers energy to several types of urban end-users. EV charging infrastructure is managed adaptively to avoid grid stress during peak hours. Industrial loads benefit from the stable supply provided by the SMR, while residential loads are controlled using smart meters and home energy management systems. All energy flows and control commands are coordinated through a two-way digital communication infrastructure, ensuring responsiveness, transparency, and optimization. This configuration reflects a realistic and forward-looking model for urban energy systems, in which nuclear, renewable, and storage technologies are harmonized under the principles of digitalization and sustainability.

3.2. Emissions and Cost Estimation

To evaluate the sustainability of the proposed hybrid energy system, two parallel assessments were conducted: (1) carbon emissions estimations and (2) capital expenditure (CAPEX) calculations. Both analyses were applied to multiple operational scenarios defined by varying shares of nuclear, renewable, and storage sources.

3.2.1. Emissions Model

Carbon emissions were computed using standardized emission factors (EF) expressed in grams of CO_2 per kilowatt-hour (gCO_2/kWh) generated. For each scenario, the total daily electricity generated from each source was multiplied by its respective EF, and the sum was normalized by the total energy demand (100 MWh/day) to obtain the average emissions in gCO_2/kWh. This normalization is performed using the formula:

$$Average\ Emissions = \frac{\sum_{i=1}^{n} (EF_i \times E_i)}{D}, \quad (1)$$

where

EF_i is the energy produced by source i (in kWh),

E_i is the emission factor of source i (in gCO_2/kWh),

D is the total daily energy demand (100,000 kWh).

This approach ensures that the emissions result is a comparable metric across scenarios, reflecting how much CO_2 is emitted per kilowatt-hour of electricity delivered to the city. These values, based on the literature [31,32] and international reports, represent average life-cycle or operational emissions for each energy source and are systematized in Table 4.

Table 4. Emission factors used for each energy source in scenario-based modeling [31–35].

Source	Emission Factor (gCO_2/kWh)	Source Type
SMR (nuclear)	12	Life cycle (IEA 2023) [32]
Solar PV	45	Life cycle
Wind	15	Life cycle
Battery Storage	30	Charged from RES/SMR
Natural Gas	450	Operational
Coal	950	Operational

For nuclear power (SMR), we now apply a life cycle emission factor of 12 gCO_2/kWh, based on data from the International Energy Agency (IEA, 2023). This value includes emissions from uranium mining, enrichment, reactor construction, operations, and decommissioning. Battery storage is assigned a life-cycle emission factor of 30 gCO_2/kWh, reflecting upstream impacts from material extraction, manufacturing, and end-of-life processing. Although charged from low-carbon sources in the model, storage introduces indirect emissions through its embedded energy footprint.

3.2.2. CAPEX Model

Investment costs were estimated based on typical installation costs per capacity unit, derived from contemporary vendor and institutional data. The following values were applied, summarized in Table 5.

Table 5. Capital cost assumptions for each technology used in scenario calculations [36–41].

Technology	CAPEX Estimate	Reference
SMR	5000 EUR/kW	IAEA, NuScale, RR-UK
Solar PV	800 EUR/kW	IRENA (2023) [39]
Wind	1200 EUR/kW	IRENA (2022) [40]
Battery Storage	400 EUR/kWh	Bloomberg NEF (2022) [41]
Natural Gas Plant	1000 EUR/kW	US DOE (2021) [38]
Coal Plant	1500 EUR/kW	US DOE (2021) [37]

Capital expenditure (CAPEX) values are expressed per unit of installed capacity (EUR/kW or EUR/kWh). The cost model enables comparative evaluation of required investments across mixed energy portfolios designed for smart urban applications.

For each scenario, the technology mix was applied proportionally to 100 MW of installed capacity. The CAPEX for each technology was computed using the formula:

$$CAPEX_{scenario} = \sum_{i=1}^{n} \left(\frac{P_i}{100} \times C_i \right) \times 100000 \text{, (in EUR)} \quad (2)$$

where Pi is the share of technology i in percent, Ci is the cost in EUR/kW (or EUR/kWh for battery), and 100,000 is the base capacity in kW.

This enabled a comparative assessment of trade-offs between emissions performance and required capital investments for smart urban deployment.

The current CAPEX-based cost model does not explicitly include long-term cost components such as nuclear insurance, decommissioning, or radioactive waste management for SMRs. These costs are known to vary significantly across jurisdictions, often being internalized through national regulatory frameworks or dedicated state-managed funds. While their exclusion simplifies cross-scenario comparison, it represents a modeling limitation. Future iterations of the model will aim to integrate these externalities to provide a more comprehensive lifecycle economic assessment.

3.3. Scenario Design Rationale

The definition of the six energy supply scenarios (A–F) was driven by the need to establish a robust analytical framework that explores a wide spectrum of real-world energy transition pathways, particularly applicable to smart urban environments. Each scenario was constructed to reflect varying degrees of technological maturity, emissions performance, investment requirements, and operational feasibility. The goal is to enable comparative analysis not only of the technical configurations, but of their policy, economic, and environmental implications.

A. Principles of Scenario Construction.

Three guiding principles shaped the scenario selection process:

- Technological Representativeness: The scenarios cover a full set of currently viable and emerging technologies, including fossil fuels (natural gas and coal), renewable sources (solar PV and wind), advanced nuclear (SMR), and battery energy storage. This ensures that the model reflects plausible near- and long-term energy mixes, grounded in both deployment trends and strategic planning documents from the EU, IEA, and national frameworks;
- Emissions Performance Spectrum: The configurations were explicitly chosen to span the full spectrum of carbon intensities—from highly carbon-intensive baselines (Scenarios A and F), through partially decarbonized hybrid systems (Scenario B) to highly decarbonized or near-zero emission configurations (Scenarios C, D, and E). This stratification supports a nuanced understanding of how technology portfolios affect net emissions at the city scale;
- Policy and Investment Realism: Each scenario is not hypothetical but reflects either a current status quo (fossil-dominant scenario), a transitional phase (hybrid with renewables), or a long-term strategic target (SMR-centered scenario with resilience enhancements). The chosen mixes are intended to match plausible deployment stages and investment profiles found in current EU energy transition policies, including REPowerEU, Net Zero Industry Act, and National Energy and Climate Plans (NECPs).

B. Justification for Percentage Allocations

The internal distribution of technologies within each scenario was selected based on three key parameters: (1) operational complementarity; (2) cost-emissions trade-offs; and (3) empirical alignment with modeled urban loads.

Scenario A (100% Natural Gas): This scenario serves as a reference baseline for moderate emissions and low CAPEX and represents the status quo in many mid-income urban environments.

Scenario B (70% Gas, 30% Solar): This scenario reflects early-stage transition strategies where renewables are integrated without large structural overhaul. Here, 30% PV is a realistic penetration level for cities with moderate irradiance and limited storage.

Scenario C (50% SMR, 40% Solar, 10% Battery): This scenario represents a balanced and feasible hybrid system. Here, 50% SMR ensures continuous baseload supply, 40% PV utilizes daytime solar potential, and10% battery provides limited but cost-effective flexibility. The 10% storage was chosen to match the battery capacity needed for peak shaving under typical irradiance patterns in Central Europe.

Scenario D (70% SMR, 20% Wind, 10% Battery): This scenario prioritizes system resilience. Here, 70% SMR allows maximum decarbonization while incorporating 20% wind to address nocturnal supply gaps and 10% battery to enhance operational flexibility.

Scenario E (80% SMR, 10% Solar, 10% Battery): This scenario represents an ultra-low emission strategy designed for high-reliability districts such as medical or industrial zones. This configuration assumes significant capital availability and long-term policy commitment.

Scenario F (100% Coal): This scenario is used as a worst-case reference, highlighting the maximum emissions and outdated planning trajectory still applicable in legacy systems globally.

C. Analytical Purpose

This spectrum of scenarios allows for the following:

- Quantifying trade-offs between emissions and investment;
- Assessing how incremental integration of advanced technologies (e.g., SMRs) affects carbon intensity and capital needs;
- Supporting decision-makers in identifying cost-effective pathways under carbon taxation or emissions trading schemes;
- Stress-testing the impact of variable renewable energy penetration under solar irradiance conditions typical of Central Europe.

D. Validation and Comparability

The scenario structure is consistent with practices in multi-criteria energy systems modeling, including those used in EU JRC studies, the IEA Energy Technology Perspectives, and IPCC pathway analysis. The shares are aligned with documented technical limits and policy targets (e.g., the expected SMR share in EU 2035 forecasts ranges between 20% and 70% depending on scenario).

By constructing scenarios that are both analytically diverse and policy-relevant, the study provides a stable foundation for comparative simulation and optimization of future-ready urban energy systems.

3.4. *Performance, Emissions, and Investment Assessment of Hybrid Urban Energy Systems*

The performance of the proposed smart hybrid energy system was evaluated using detailed simulations and comparative scenario analysis. The objective was to assess how varying combinations of SMRs, renewable energy sources, and battery storage influence energy reliability, carbon emissions, and investment costs in a smart urban context. This section presents the outcomes of the simulation models, analyzes the emission and economic performance of six strategic energy mixes, and interprets their implications for future urban energy planning.

3.4.1. Performance of the Hybrid Energy System

The simulation results demonstrate that the integration of a SMR into a smart grid architecture provides significant improvements in energy stability, flexibility, and carbon performance. The SMR delivers continuous baseload power, while solar PV contributes intermittent generation during daylight hours. The battery storage unit efficiently balances short-term fluctuations, reducing reliance on fossil backup. Figure 2 illustrates the energy contribution of each source over a 24-h cycle, confirming that the system maintains stable power output under varying demand.

The smart grid controller successfully manages energy dispatch based on real-time load and pricing signals. Demand peaks during the early morning and evening hours are efficiently covered through a combination of SMR generation and battery discharge. Solar input reduces mid-day grid stress and charges the battery during surplus periods.

3.4.2. Scenario-Based Emission and Investment Analysis

To quantify the trade-offs between environmental impact and required investment, six energy supply scenarios were evaluated. Each scenario assumes a total urban energy demand of 100 MWh/day, supplied through different combinations of SMR, solar PV, wind, battery storage, and conventional generation. The results are summarized in Table 6.

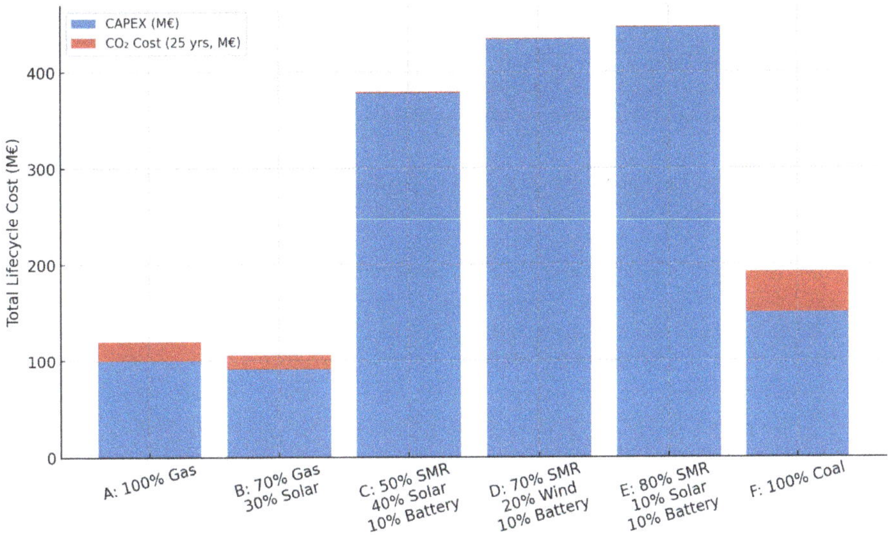

Figure 2. Total lifecycle cost by scenario (25 Years).

Table 6. Summary of emissions and capital expenditure estimates for each energy supply scenario.

Scenario	Emissions	Estimated CAPEX	Comment
Scenario A—100% Natural Gas	450 gCO_2/kWh	EUR 100 million	Highest emissions, lowest cost
Scenario B—70% Natural Gas, 30% Solar	346 gCO_2/kWh	EUR 91 million	Moderate emissions reduction
Scenario C—50% SMR, 40% Solar, 10% Battery	27 gCO_2/kWh	EUR 378 million	Balanced sustainability and cost
Scenario D—70% SMR, 20% Wind, 10% Battery	21 gCO_2/kWh	EUR 434 million	Low emissions, high resilience
Scenario E—80% SMR, 10% Solar, 10% Battery	18 gCO_2/kWh	EUR 446 million	Lowest carbon footprint, highest cost
Scenario F—100% Coal	950 gCO_2/kWh	EUR 150 million	Maximum emissions, outdated strategy

The selection of scenarios was motivated by the need to explore a broad spectrum of technological pathways for meeting the energy demand of a smart urban environment. The six scenarios were designed to capture representative energy mixes that range from fossil-dominated (Scenarios A and F) to hybrid (Scenarios B, C, D) and nuclear-intensive configurations (Scenario E). These scenarios were not only chosen for their technical feasibility but also for their relevance to real-world policy and investment considerations.

Scenario A (100% natural gas) and Scenario F (100% coal) serve as baseline references, representing widely deployed but carbon-intensive legacy systems. Scenario B introduces partial decarbonization by supplementing fossil generation with solar PV. Scenarios C and D illustrate balanced hybrid models with varying contributions from SMR, renewables, and battery storage, offering both emissions reductions and investment manageability. Scenario E was included as a future-oriented configuration, prioritizing deep decarbonization through a dominant SMR share, suitable for high-resilience and climate-neutral urban planning.

This scenario set enables a comprehensive assessment of technological trade-offs, informing decision-makers on the cost-effectiveness and sustainability potential of different smart energy strategies.

In the context of urban energy planning, it is essential to evaluate not only the upfront investment in infrastructure but also the long-term economic and environmental implications of different energy configurations. With increasing regulatory pressure through mechanisms such as carbon pricing, scenarios that appear financially viable in the short term may become less attractive when cumulative emissions costs are considered.

To this end, we introduce a lifecycle-based cost model that combines initial capital expenditure (CAPEX) with the cumulative cost of carbon dioxide emissions over a standard operational lifetime. This approach allows for a more accurate comparison between traditional fossil fuel-based systems and emerging low-carbon technologies such as small modular reactors (SMRs) and renewable energy sources.

The total cumulative carbon cost over the 25-year period is calculated using the following equation.

Cumulative carbon cost computed as follows:

$$CO_2COST_{Total} = \frac{g_{CO2}}{kW/h} \times E_{day} \times Day \times T \times C_{CO2}\,,\ (\text{in Euro}) \tag{3}$$

where

g_{CO2} is the average carbon intensity of the scenario (grams CO_2 per kWh);

E_{day} is daily electricity consumption (in kWh) = 100,000;

Day = 365 is the number of days per year;

T = 25 is the operational lifetime in years;

C_{CO2} = 80 is the carbon price (EUR/ton CO_2),

The stacked bar chart below (Figure 2) illustrates the distribution of total cost over 25 years for each scenario, combining the initial capital investment (CAPEX) and cumulative carbon cost. The chart clearly shows how high-emission scenarios accumulate significant environmental penalties, while SMR-based hybrid systems maintain low total costs over time despite higher upfront investments.

The extended cost analysis demonstrates that short-term capital cost considerations alone are insufficient for informed energy planning. When cumulative carbon costs are accounted for over a standard 25-year operational horizon, high-emission scenarios (A and F) become economically unattractive, while hybrid systems with strong SMRs and renewable integration (C, D, and E) show superior long-term performance. Notably, Scenario C emerges as a balanced option with significantly reduced emissions and acceptable total cost. These findings emphasize the strategic importance of lifecycle-based assessment frameworks for sustainable urban energy infrastructure design.

3.4.3. Interpretation

The results indicate that hybrid energy architectures combining SMRs with renewables and storage offer significant advantages in reducing emissions and enhancing grid reliability. Although scenarios with higher SMR shares require greater upfront investment, they enable long-term decarbonization, support 24/7 energy availability, and simplify grid management.

Scenario C emerges as a particularly promising configuration, balancing ecological impact and investment feasibility, making it ideal for cities in transition. On the other hand, Scenario E represents a future-proof approach for fully climate-neutral smart cities with robust infrastructure budgets.

4. Simulating Different Power Distribution Profiles

To evaluate the operational dynamics of the proposed hybrid energy system for smart urban environments, a system-level modeling approach was developed. This section describes the mathematical representation, control strategies, and simulation setup that underpin the performance assessment.

4.1. Mathematical Model Structure

To realize the general power balance model, the hybrid energy system's individual elements, including SMRs, solar panels, and battery storage, are described by the following component-specific equations, capturing their dynamic behavior over time.

- For SMR power generation, the following equation is used:

$$P_{SMR}(t) = C_{SMR} \times E_{SMR} \times T(t), \quad (4)$$

where $P_{SMR}(t)$ is the power generated by the SMR at time t, C_{SMR} is the SMR efficiency coefficient, E_{SMR} is its rated capacity, and $T(t)$ represents the time-dependent demand/load factor.

- For photovoltaic solar panels, the following equation is used:

$$P_{PV}(t) = \eta_{pv} \times G(t) \times A_{solar} \times [1 - \beta \times (T_{cell}(t) - T_{ref})], \quad (5)$$

where $P_{PV}(t)$ is the power produced by the solar array; η_{PV} is nominal efficiency of the PV module under standard test conditions (STC), dimensionless (e.g., 0.2); $G(t)$ global solar irradiance incident on the panel surface at time t [W/m^2]; A_{solar} is the total panel surface area [m^2]; $T_{cell}(t)$ is cell temperature at time t [°C], and T_{ref} is the reference temperature [°C], typically 25 °C.

- For battery charge/discharge dynamics, the following equation is used:

$$E_{batt}(t) = \eta_{charge} \times [P_{PV}(t) + P_{SMR}(t) - P_{load}(t)], \quad (6)$$

where $E_{batt}(t)$ represents the battery energy level at time t, η_{charge} is the charge/discharge efficiency, and $P_{load}(t)$ is the total energy demand.

The hybrid system consists of three primary energy sources: a SMR, a PV solar array, and a lithium-ion battery energy storage system (BESS). These are integrated using a smart grid controller that performs real-time optimization of power flows based on load demand, resource availability, and economic signals.

The energy balance equation for the system is defined as follows:

$$P_{gen}(t) = P_{SMR}(t) + P_{PV}(t) + P_{batt}(t) = P_{load}(t) + P_{grid}(t), \quad (7)$$

where

- output power of the SMR—$P_{SMR}(t)$ is assumed constant or semi-variable for advanced SMRs with load-following capability;
- power generated by the PV array—$P_{PV}(t)$ is modeled as a function of solar irradiance $Ir(t)$ and panel characteristics;
- $P_{batt}(t)$ is net power from the battery system (positive when discharging, negative when charging);
- $P_{load}(t)$ is aggregated load profile of residential, industrial, and EV consumption;
- $P_{grid}(t)$ is import/export power exchanged with the external grid (positive for import);

In the model, power exchange with the external grid, represented by $P_{grid}(t)$, is governed by system-level balance constraints. Grid import (positive $P_{grid}(t)$) is triggered only when local resources—SMR output, solar PV, and battery discharge—are insufficient to meet instantaneous load. Grid export (negative $P_{grid}(t)$) is allowed only when there is surplus generation and the grid has the technical capacity to accept power.

The system is modeled with bounded grid exchange capacity (±20 MW), reflecting realistic infrastructure constraints for remote or semi-islanded smart city contexts. Furthermore, the optimization incorporates economic penalties for importing electricity (e.g., high marginal grid cost) and low or capped compensation for export, which aligns with real-world market limitations and promotes local energy autonomy.

Battery dynamics are governed by the state-of-charge (SoC) as follows:

$$SoC(t+1) = SoC(t) + [\eta_{ch} \times P_{ch}(t) - P_{dis}(t)/\eta_{dis}] \times \Delta t/C_{batt} \quad (8)$$

where η_{ch} and η_{dis} are the charge and discharge efficiencies; P_{ch} and P_{dis} are the charging and discharging powers, respectively; and C_{batt} is the total battery capacity.

A hierarchical control strategy is adopted:

- Primary control ensures baseload is met by the SMR.
- Secondary control dispatches PV generation based on real-time irradiance.
- Tertiary control schedules battery operation to minimize grid interaction, prioritize self-consumption, and reduce peak loads.

The smart controller utilizes a rule-based energy management system (EMS) supplemented with price signals, weather forecasts, and demand predictions to coordinate resources.

Hourly load profiles were generated using synthetic data models calibrated to match the energy usage of a mid-sized urban district (100 MWh/day). These include household, commercial, and EV charging patterns. Solar generation profiles were derived from irradiance datasets typical for Central European conditions.

Three solar scenarios were considered:

- Low irradiance (overcast)
- Moderate irradiance (partly cloudy)
- High irradiance (clear skies)

Each scenario was simulated over a 24-h period with 1-h time steps. Simulations were conducted using a Python-based model [42] the performance of the proposed hybrid urban energy system. The SMR was modeled as a constant-output generator supplying 50 MW of baseload power. The solar photovoltaic (PV) array followed a typical daily irradiance curve representative of Central Europe. Battery energy storage was managed using a rule-based energy management system (EMS), which prioritized renewable surplus for charging and dispatched stored energy during high-demand periods.

Hourly load profiles were generated using synthetic demand models for residential, industrial, and electric vehicle (EV) consumers. These profiles reflect real-world variability and peak load clustering throughout the day. The simulation timeframe covered 24 h with 1-h resolution.

4.2. Power Balance Profiles Under Scenario C

Several tables summarize the model parameters used to simulate the results for Scenario C at three solar radiation levels: bad, medium, and good. The model represents a hybrid energy system serving a medium-sized smart city. Table 7 provides general model parameters used in the simulation.

Table 7. General model parameters.

Parameter	Value	Description
Total Energy Demand	100 MWh/day	Representative of a mid-sized urban area
Time Interval	1 h	Simulation covers 24 h
SMR Power Output	50 MW (constant)	Baseload energy supply
Battery Capacity	10 MWh	Maximum energy storage capacity
Battery Power Limit	±10 MW	Charge/discharge limit
Battery Efficiency	95%	Charge/discharge efficiency
PV Area and Efficiency	100,000 m^2 @ 20%	Effective area and panel efficiency
Battery Initial State	10 MWh	Fully charged at 00:00
Demand Response	Disabled	No dynamic load reduction

Table 8 presents the hourly solar irradiance values applied in the simulation of Scenario C under three different weather conditions: poor, average, and good sunlight. These profiles were used to model the variable photovoltaic (PV) generation throughout the day. The values reflect typical seasonal patterns observed in Central European regions and serve as input to assess system behavior under fluctuating renewable availability.

Table 8. Solar irradiation used to implement the studied scenario C.

Hour Range	Poor Sunlight (kW/m^2)	Average Sunlight (kW/m^2)	Good Sunlight (kW/m^2)
5–7	0.05	0.10	0.15
8–10	0.10	0.30	0.45
11–14	0.20	0.50	0.70
15–17	0.10	0.30	0.45
18–19	0.05	0.10	0.15
Other	0.00	0.00	0.00

The battery charges during midday hours when solar generation is available and discharges during peak demand periods in the morning and evening. In the poor sunlight scenario, battery usage is limited due to insufficient solar input, increasing dependence on the grid. In contrast, the good sunlight scenario results in full battery charging and significant reduction of grid import. The SMR operates continuously in all scenarios, ensuring a stable base power supply.

The following figures illustrate the hourly power balance of the proposed smart hybrid energy system under Scenario C, which combines 50% SMR, 40% solar PV, and 10% battery storage. Each figure corresponds to a different solar irradiance condition: bad, average, and good sunlight. The system is designed to meet a 24-h urban load profile. When SMR and PV generation are insufficient, the battery compensates up to its capacity (10 MW/h), and any remaining demand is supplied by grid imports. Figure 3 presents simulation results on a bad sunny day. In this case, the photovoltaic contribution is minimal, requiring significant battery discharge and significant grid input.

Figure 4 shows simulation results for an average sunny day. In this case, PV generation reduces battery and grid reliance during midday.

Figure 5 shows the simulation results on a very sunny day. In this case, photovoltaic generation significantly supports the load, minimizing battery usage and eliminating the need for grid imports, for most of the day.

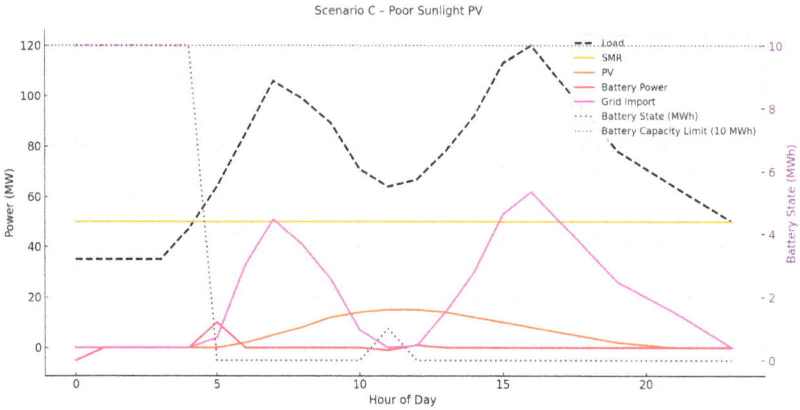

Figure 3. Power balance on a day with poor sunlight conditions.

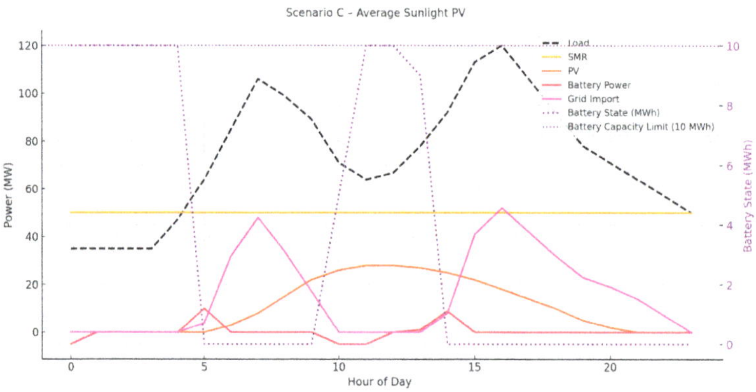

Figure 4. Power balance on a day with moderate sunlight. PV generation reduces battery and grid reliance during midday.

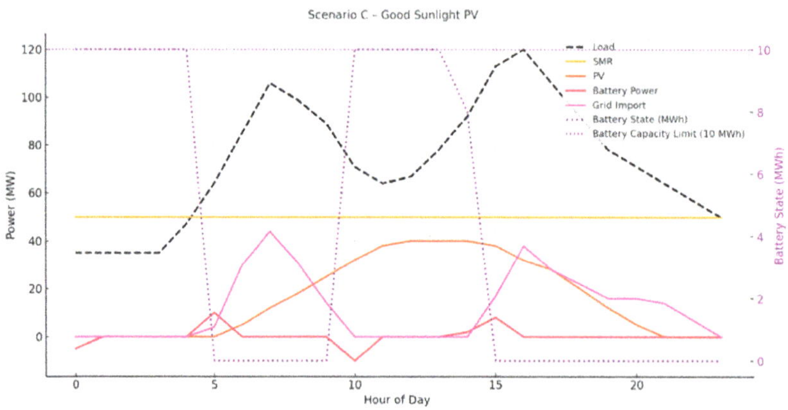

Figure 5. Power balance on a day with strong sunlight. PV generation significantly supports the load, minimizing battery use and eliminating the need for grid imports.

4.3. Conclusions on Grid Dependency and Mitigation Strategies

The simulation results confirm that even well-configured hybrid systems composed of baseload nuclear, batteries and solar photovoltaic generators, when applying optimization based on economic and emission indicators, have a high degree of grid dependence, caused by the variability of a part of the generated energy. The degree of this dependence is strongly influenced by the availability of solar energy:

- Under poor sunlight conditions, the system relies heavily on grid imports due to insufficient photovoltaic production and depleted battery reserves. Despite the constant power of the SMR, demand exceeds the available local supply during the morning and evening peaks.
- During average sunlight, the photovoltaic systems contribute significantly around noon, partially charging the battery and slightly reducing the dependence on the grid. However, the battery is not able to fully cover both peaks, leading to moderate grid use.
- In a scenario with good sunlight, the PV system reaches its optimal contribution, allowing the battery to be fully charged and discharged during peak demand periods. Grid input is minimal and limited to very early and late hours.

These results highlight the need for strategic solutions to improve grid independence, especially in a variable weather environment. Based on this, the following mitigation strategies are recommended [41,43]:

- Increasing battery storage capacity

Expanding the installed capacity beyond the current 10 MWh ($\approx$10% of daily load) allows for better use of excess solar energy and extends coverage during cloudy or winter days.

- Implementing predictive energy management

Integrating weather forecasts and data-driven demand forecasting can optimize battery usage and SMR scheduling, minimizing reactive grid drawdowns.

- Diversify the renewable energy portfolio

Adding wind power complements photovoltaic systems by generating power during nighttime hours or when solar radiation is low, thereby stabilizing the overall renewable energy production.

- Implementing flexible demand response

Enabling flexible loads, especially in the industrial and commercial sectors, can help shift demand away from critical off-peak hours, relieving pressure on local generation and storage.

- Enabling load-following SMR modes

Where technologically feasible, configuring SMRs to adjust their production partially based on residual demand can smooth grid interactions and reduce dependence on external imports.

- Introducing seasonal thermal or hydrogen energy storage

These technologies can store excess energy from photovoltaics or small marine systems (SMRs) during periods of low demand and feed it back during prolonged deficits, improving seasonal energy autonomy.

- Using vehicle-to-grid (V2G) integration

Using electric vehicle batteries for two-way energy exchange allows distributed and mobile storage assets to contribute during peak loads or solar power shortages, effectively acting as additional flexibility without additional stationary capacity.

Through a combination of these strategies, a smart hybrid energy system can further reduce its dependence on centralized electricity supply, increase its operational flexibility and significantly contribute to decarbonization goals while maintaining energy security [44–46].

4.4. Minimum Configuration for Full Grid Autonomy

To evaluate the minimum required configuration for achieving complete grid independence, simulations were performed by varying SMR output, battery capacity, and battery charge/discharge power. The goal was to find the lowest combination of these parameters that results in zero grid import under each PV condition. The results are summarized below in Table 9.

Table 9. The lowest combination of these parameters that leads to zero import from the grid under different photovoltaic conditions.

PV Scenario	SMR Power (MW)	Battery Capacity (MWh)	Battery Power (MW)	Grid Import
Good Day	64	20	20	0 MWh
Average Day	64	35	25	0 MWh
Poor Day	86	60	40	0 MWh

These data illustrate that grid independence is technically feasible across all weather conditions, provided that appropriate scaling of SMR output and battery infrastructure is implemented:

- On good sunlight days, a moderately sized SMR (64 MW) combined with 20 MWh of battery capacity is sufficient to fully meet demand, thanks to strong PV contribution during the day and effective energy shifting by the battery.
- Under average sunlight conditions, PV generation is less pronounced, necessitating an increase in battery capacity (35 MWh) and discharge rate (25 MW) to maintain full autonomy, while SMR output remains unchanged.
- Under poor sunlight conditions, where PV contribution is minimal, the system must compensate by significantly increasing SMR output (86 MW) and battery storage (60 MWh at 40 MW) to fully cover demand fluctuations and avoid grid import.

These findings confirm that solar intermittency can be mitigated entirely through flexible and adaptive system design. Instead of depending on the external grid, the system leverages higher local generation (via SMR) and expanded storage to ensure uninterrupted supply. Notably, the grid is never used, even in the worst-case irradiance scenario.

This supports the article's core proposition that hybrid nuclear–renewable systems, when properly dimensioned, can serve as resilient, autonomous, and decarbonized energy solutions for smart urban environments.

5. Discussion

The results of the simulation and scenario analysis highlight several critical insights regarding the design of hybrid energy systems for smart urban environments. The integration of SMRs, renewable sources, and intelligent control mechanisms offers not only environmental benefits but also operational flexibility and long-term energy security. However, the adoption of such architectures raises complex trade-offs between capital investment, technological maturity, and regulatory acceptance.

5.1. Trade-Offs Between Emissions and Cost

The scenario comparison reveals a clear inverse relationship between carbon emissions and investment costs. Scenarios dominated by fossil fuels (A and F) are economically attractive in the short term but incompatible with decarbonization goals. In contrast, scenarios with high SMR penetration (D and E) achieve dramatic reductions in CO_2 emissions (10 g/kWh) at the cost of significantly higher upfront capital expenditure. Scenario C presents a middle ground, offering a 97% reduction in emissions compared to the natural gas baseline while maintaining financial feasibility.

This trade-off underscores the importance of policy instruments such as carbon pricing, green public financing, and investment tax credits, which can shift the cost-benefit balance in favor of cleaner solutions.

5.2. Role of SMRs in Urban Energy Planning

The results confirm that SMRs can play a vital role in future urban energy systems by supplying stable baseload power, reducing dependency on volatile renewables, and enabling grid stability. Their compact size and modular construction make them suitable for urban deployment, district heating, and even hydrogen production. Moreover, SMRs can enhance the resilience of critical infrastructure during extreme weather events or geopolitical disruptions, a growing concern in energy policy [44,45].

However, the deployment of SMRs must be accompanied by transparent regulatory frameworks, community engagement, and lifecycle waste management strategies to gain public trust and ensure sustainable operation.

5.3. Grid Flexibility and Demand Response

The implementation of smart grid controls and demand-response logic within the simulation framework demonstrated measurable benefits in balancing load, shifting peak demand, and improving energy use efficiency. These digital tools are essential to maximizing the value of intermittent renewables and minimizing battery discharge cycles. In urban contexts with high EV adoption and diverse consumer profiles, demand-side flexibility will become a strategic asset for system operators.

The findings support the integration of real-time pricing models, smart metering, and predictive analytics into urban energy platforms, which can further reduce costs and enhance user engagement [46,47].

5.4. Vehicle-to-Grid and Long-Term Storage Strategies

The integration of vehicle-to-grid (V2G) technology in the proposed architecture represents a transformative approach to distributed energy balancing. In scenarios with high EV penetration, coordinated control of bidirectional charging allows urban fleets to support the grid during critical load periods, particularly in the evenings when solar production drops. Unlike fixed battery storage, V2G assets are mobile, distributed, and scalable through user participation. However, successful implementation depends on the availability of standardized charging infrastructure, communication protocols, and incentives for vehicle owners.

In parallel, long-duration storage technologies such as thermal storage (linked to district heating systems) and hydrogen production (via electrolysis) can act as energy buffers on seasonal time scales. These solutions increase resilience and reduce long-term dependency on the main grid, especially under extended periods of low renewable input.

5.5. Policy and Market Integration

As demonstrated by the system-level simulations presented in Section 4, the successful deployment of hybrid SMR–renewable energy systems at scale requires not only technological maturity and cost competitiveness, but also supportive policy frameworks and well-designed market mechanisms. As cities seek to achieve carbon neutrality while maintaining energy security and affordability, public policy will play a central role in enabling the integration of advanced nuclear technologies with intermittent renewables.

One of the most effective instruments to incentivize low-emission systems is the implementation of carbon pricing schemes, either through emissions trading systems (ETS) or carbon taxes. When lifecycle carbon costs are internalized—such as in the modeling presented in this study—SMR-based configurations become significantly more attractive compared to fossil-fueled baselines. Scenario C, for instance, demonstrates favorable cost-emissions trade-offs under a carbon price of EUR 80/ton CO_2, suggesting that regulatory alignment with decarbonization goals can directly affect investment decisions.

Green finance mechanisms, such as tax credits, investment guarantees, and public–private partnerships, are also critical to offset the high upfront capital costs typically associated with SMR deployment. In the United States, for example, the Inflation Reduction Act (IRA) provides nuclear-specific clean energy tax credits, while the European Union is actively exploring regulatory inclusion of SMRs within the sustainable finance taxonomy. Such measures can de-risk capital-intensive infrastructure and accelerate the scale-up of hybrid nuclear–renewable systems in urban areas.

Furthermore, regulatory innovation is needed to address the licensing challenges of SMRs, particularly in urban or distributed environments. Current nuclear regulations in most countries are oriented toward large-scale reactors and centralized siting. Fast-tracked, risk-informed licensing processes—already being piloted in Canada, the UK, and the US—could be adapted for modular urban deployment, provided that safety, environmental, and community concerns are transparently addressed.

On the market side, capacity remuneration mechanisms and flexibility services markets can incentivize SMRs to operate synergistically with renewables and storage, rather than competing against them. Through dynamic pricing, ancillary services compensation, and demand-response programs, operators can optimize multi-source portfolios and enhance overall system efficiency.

Lastly, strong community engagement and public acceptance will be essential for urban SMR projects to succeed. Public perception of nuclear energy remains cautious, and transparent communication, participatory planning, and long-term waste management strategies must be integrated into all phases of deployment.

To ensure the practical implementation of the proposed policy instruments, we outline several actionable pathways. First, regulatory innovation such as the introduction of licensing sandboxes can enable pilot deployment of SMRs under flexible oversight, expediting time-to-market while maintaining safety standards. Second, targeted fiscal incentives, including investment tax credits or accelerated depreciation schemes for hybrid systems, can reduce upfront capital costs and de-risk private investment. Third, capacity-building programs aimed at municipal energy planners and regulators can accelerate institutional readiness for advanced energy system integration. These measures should be embedded within national climate and energy strategies to ensure alignment with long-term decarbonization goals and urban resilience planning.

In sum, a coherent policy and market ecosystem is essential to unlock the full potential of SMR–renewable systems in the global clean energy transition. Integrating these technologies into national climate strategies, municipal energy plans, and green investment frameworks can catalyze resilient, low-carbon urban energy infrastructures at scale.

5.6. Limitations and Future Research

While the model provides valuable insights, it is based on simplified representations of generation, demand, and storage dynamics. Real-world deployment would require more granular time resolution, inclusion of seasonal variations, market dynamics, and reliability constraints. Additionally, social, legal, and environmental dimensions such as land use, supply chain resilience, and nuclear waste disposal need to be considered in future studies.

The results obtained in this study are consistent with findings from other scientific groups working on hybrid energy systems that incorporate Small Modular Reactors (SMRs).

In [18] conducted real-time simulation studies of SMR-integrated microgrids and demonstrated similar improvements in system stability and emissions reduction when SMRs are combined with renewable energy and storage technologies. In [19,20] provided a comprehensive review of nuclear–renewable synergies and highlighted the role of SMRs as key enablers of flexible, low-carbon energy systems—an outcome that aligns with the performance of Scenario C in this work. Furthermore, ref. [48] offered a detailed overview of modeling and control strategies for SMR-based applications, supporting the relevance of simplified simulation frameworks such as the one used here. Although differences exist in terms of modeling depth, control strategies, and geographic assumptions, the general conclusion is consistent: SMRs significantly enhance the adaptability, reliability, and decarbonization potential of future smart urban energy infrastructures.

While the proposed hybrid system demonstrates strong potential for decarbonization and grid resilience, the introduction of SMRs into urban or distributed energy systems also raises important long-term considerations. One critical issue is the management of spent nuclear fuel and radioactive waste. Although SMRs are designed with advanced safety and fuel efficiency features, they still produce high-level nuclear waste that requires secure and sustainable treatment and storage solutions. Future research should explore the life-cycle implications of SMR deployment, including fuel recycling, long-term repository strategies, and the development of compact on-site waste management systems. Relevant studies such as those by [49–51] provide valuable frameworks for assessing the environmental and regulatory challenges associated with SMR-generated waste. Incorporating such aspects into integrated planning models will be essential to ensure the long-term sustainability and public acceptance of SMR-based energy solutions.

While the proposed hybrid SMR–renewable system offers promising performance in terms of reliability and emissions, several limitations of the current model must be acknowledged.

First, the simulation framework uses deterministic inputs with fixed solar irradiance profiles and predefined urban load patterns. In reality, both solar availability and electricity demand are inherently stochastic and time-variant. The absence of probabilistic modeling limits the ability to assess the system's performance under uncertainty or to quantify risks related to underperformance or overinvestment.

Second, the time resolution of one hour is adequate for observing daily trends but may not capture fast dynamics, especially during transient demand spikes or sudden drops in renewable input. The model also assumes idealized operation of the SMR as a constant output source and does not simulate thermal inertia, reactivity feedback, or ramping limitations, which are important for system-level dispatch coordination.

Third, the absence of statistical or sensitivity analysis prevents deeper insights into how variation in input parameters (e.g., solar area, battery size, SMR capacity) influences system performance. This restricts the robustness of the conclusions drawn from Scenario C or the minimum autonomy configuration. Additionally, the cost model does not include uncertainty bounds, market fluctuations, or long-term degradation factors for storage and PV systems.

Finally, the environmental and regulatory implications of SMR deployment—such as nuclear waste management, decommissioning costs, and social acceptance—are only briefly mentioned and not quantitatively integrated into the evaluation [52].

Table 10 summarizes key policy and market instruments that can support the deployment of SMR–renewable hybrid systems, along with their intended impacts and real-world implementation examples. These mechanisms, when aligned with national decarbonization goals, provide a robust foundation for enabling the clean energy transition at the urban level.

Table 10. Policy instruments supporting hybrid SMR–renewable systems. (Note: Arrows indicate the expected direction of impact—"↑" denotes increase or positive effect, and "↓" denotes decrease or reduction).

Policy Instrument	Target Effect	Example Initiative
Carbon Pricing	↓ Lifecycle emissions, ↑ Competitiveness of low-carbon tech	EU ETS, Canada Carbon Tax
Green Finance and Tax Incentives	↓ Initial CAPEX, ↑ Investment Attractiveness	IRA (USA), EU Innovation Fund
Regulatory Acceleration for SMRs	↓ Time-to-market, ↑ Deployment Feasibility	CNSC (Canada), UK ONR Generic Design Process
Capacity Remuneration Schemes	↑ Flexibility, ↑ Reliability during peak demand	UK Capacity Market, PJM (USA)
Sustainable Taxonomy Inclusion	↑ Investor confidence, ↓ Cost of capital	EU Sustainable Finance Taxonomy (under review)
Public Engagement Strategies	↑ Acceptance, ↓ Project opposition	SMR Community Engagement Framework (OECD NEA)

Future work will address these limitations by incorporating stochastic simulations (e.g., Monte Carlo or Latin Hypercube sampling), scenario-based sensitivity analysis, and time-series variability in both generation and demand. Moreover, a dynamic thermal model of the SMR, combined with real-time control co-simulation, will be developed to assess operational flexibility under real-world constraints. The introduction of multi-objective optimization techniques will allow a more comprehensive exploration of trade-offs among emissions, cost, reliability, and autonomy, thereby advancing the model toward practical urban planning applications.

6. Conclusions

This study delivers a comprehensive simulation-based assessment of a hybrid urban energy system architecture that integrates small modular reactors (SMRs), photovoltaic (PV) solar generation, battery storage, and smart grid control. The primary objective was to evaluate the long-term economic and environmental performance of such systems using lifecycle-based modeling, with the aim of identifying viable decarbonization strategies that are aligned with global clean energy transition goals.

Simulation results across diverse solar irradiance conditions lead to several key insights with direct relevance for urban energy planning:

- Full energy autonomy is technically feasible—even under low solar availability—by appropriately scaling SMR output and battery storage capacity;
- Scenario C (50% SMR, 40% PV, 10% battery) emerges as a balanced configuration, offering a 97% reduction in lifecycle carbon emissions compared to fossil-dominated systems, while maintaining cost competitiveness;
- Dynamic coordination of SMR and battery systems enables complete elimination of grid imports, supporting full decentralization and urban energy resilience;
- Intelligent control strategies—including predictive energy management and demand-response—enhance system stability, optimize asset utilization, and reduce peak loads;
- The integration of vehicle-to-grid (V2G) functionality and long-duration storage (thermal or hydrogen-based) adds further flexibility, enabling seasonal balancing and reducing reliance on centralized infrastructure.

These findings confirm that hybrid SMR–renewable systems, when combined with adaptive digital control and supported by favorable policy frameworks, can serve as technically viable, economically robust, and strategically scalable solutions for resilient low-carbon urban energy infrastructures. Their ability to provide stable baseload power complements renewable intermittency and strengthens the path toward net-zero and autonomy targets in smart cities.

Future research should advance this work by developing an integrated multi-objective optimization model that identifies optimal configurations of hybrid systems under real-world constraints. Such a framework should consider the following:

- Realistic techno-economic parameters for SMRs, PV, battery storage, and potentially wind or hydrogen components;
- Dynamic, sector-specific electricity demand, including residential, industrial, and EV loads;
- Local resource availability (irradiance/wind), market prices, and regulatory policy environments.

By applying evolutionary or Pareto-based optimization techniques, the framework can generate a set of trade-off solutions that guide decision-makers in designing future-proof urban energy systems. This approach provides valuable support for municipalities, energy planners, and policymakers engaged in implementing cost-effective, scalable, and climate-aligned infrastructures within the broader context of the global clean energy transition.

Funding: This work was supported by European Regional Development Fund under "Research Innovation and Digitization for Smart Transformation" program 2021–2027 under the Project BG16RFPR002-1.014-0006 "National Centre of Excellence Mechatronics and Clean Technologies" and the APC was funded by Project BG16RFPR002-1.014-0006.

Data Availability Statement: The data are contained in this article OR the data are available upon request from a corresponding author.

Acknowledgments: The author would like to express their sincere gratitude to the anonymous reviewers for their insightful comments and constructive suggestions, which significantly contributed to improving the quality and clarity of this manuscript.

Conflicts of Interest: The author declare no conflicts of interest.

References

1. Zaman, M.; Puryear, N.; Abdelwahed, S.; Zohrabi, N. A Review of IoT-Based Smart City Development and Management. *Smart Cities* **2024**, *7*, 1462–1501. [CrossRef]
2. International Energy Agency (IEA). *Net Zero by 2050: A Roadmap for the Global Energy Sector*; IEA: Paris, France, 2021. Available online: https://www.iea.org/reports/net-zero-by-2050 (accessed on 6 July 2025).
3. Intergovernmental Panel on Climate Change (IPCC). *Climate Change 2023: Synthesis Report. Contribution of Working Groups I, II and III to the Sixth Assessment Report of the Intergovernmental Panel on Climate Change*; Lee, H., Romero, J., Eds.; IPCC: Geneva, Switzerland, 2023. Available online: https://www.ipcc.ch/report/ar6/syr/ (accessed on 6 July 2025).
4. European Commission. *REPowerEU Plan—COM (2022) 230 Final*; European Commission: Brussels, Belgium, 2022. Available online: https://eur-lex.europa.eu/legal-content/EN/TXT/?uri=CELEX:52022DC0230 (accessed on 6 July 2025).
5. Gracias, J.S.; Parnell, G.S.; Specking, E.; Pohl, E.A.; Buchanan, R. Smart Cities—A Structured Literature Review. *Smart Cities* **2023**, *6*, 80. [CrossRef]
6. Kolotouchkina, O.; Ripoll González, L.; Belabas, W. Smart Cities, Digital Inequalities, and the Challenge of Inclusion. *Smart Cities* **2024**, *7*, 3355–3370. [CrossRef]
7. Einabadi, N.; Kazerani, M. Nanogrids in Modern Power Systems: A Comprehensive Review. *Smart Cities* **2025**, *8*, 11. [CrossRef]
8. Santoro, D.; Delmonte, N.; Simonazzi, M.; Toscani, A.; Rocchi, N.; Sozzi, G.; Cova, P.; Menozzi, R. Local Power Distribution—A Review of Nanogrid Architectures, Control Strategies, and Converters. *Sustainability* **2023**, *15*, 2759. [CrossRef]
9. Housni, H.; Majit, N.; Amrous, N.; Daoudi, N.; Malzi, M.J. Energy transition and digitalization: Small modular reactors and smart grids. In Proceedings of the 2023 7th IEEE Congress on Information Science and Technology (CiSt), Agadir—Essaouira, Morocco, 16–22 December 2023; pp. 405–410. [CrossRef]

10. Gabbar, H.A.; Abdussami, M.R. Feasibility Analysis of Grid-Connected Nuclear-Renewable Micro Hybrid Energy System. In Proceedings of the 2019 IEEE 7th International Conference on Smart Energy Grid Engineering (SEGE), Oshawa, ON, Canada, 12–14 August 2019; pp. 294–298. [CrossRef]
11. International Atomic Energy Agency. *Electric Grid Reliability and Interface with Nuclear Power Plants*; IAEA Nuclear Energy Series No. NG-T-3.8; IAEA: Vienna, Austria, 2012.
12. Yu, X.; Xue, Y. Smart Grids: A Cyber–Physical Systems Perspective. *Proc. IEEE* **2016**, *104*, 1058–1070. [CrossRef]
13. El-Emam, R.S.; Subki, M. Hadid. Small modular reactors for nuclear-renewable synergies: Prospects and impediments. *Int. J. Energy Res.* **2021**, *45*, 16995–17004. [CrossRef]
14. U.S. Department of Energy (DOE). Nuclear Reactor Technologies. Available online: https://www.energy.gov/ne/nuclear-reactor-technologies (accessed on 6 April 2025).
15. Gabbar, H.A.; Esteves, O.L.A. Real-Time Simulation of a Small Modular Reactor in-the-Loop within Nuclear-Renewable Hybrid Energy Systems. *Energies* **2022**, *15*, 6588. [CrossRef]
16. Ho, A.Q.H. Nuclear Hybrid Energy Systems as a Solution for Carbon-Free Flexible Dispatch to the Grid: A Study on Design, Dynamic Modeling, and Optimization. Ph.D. Thesis, The University of Utah, Salt Lake City, UT, USA, 2023.
17. World Nuclear Association. Small Modular Reactors. Available online: https://www.world-nuclear.org/information-library/nuclear-fuel-cycle/nuclear-power-reactors/small-nuclear-power-reactors.aspx (accessed on 6 April 2025).
18. OECD Nuclear Energy Agency (NEA). *Advanced Nuclear Fuel Cycles and Radioactive Waste Management*; OECD: Paris, France, 2020. Available online: https://www.oecd.org/en/publications/advanced-nuclear-fuel-cycles-and-radioactive-waste-management_9789264024861-en.html (accessed on 6 April 2025).
19. Locatelli, G.; Bingham, C.; Mancini, M. Small modular reactors: A comprehensive overview of their economics and strategic aspects. *Prog. Nucl. Energy* **2014**, *73*, 75–85. [CrossRef]
20. Gabbar, H.A.; Abdussami, M.R.; Adham, M.I. Optimal Planning of Nuclear-Renewable Micro-Hybrid Energy System by Particle Swarm Optimization. *IEEE Access* **2020**, *8*, 181049–181073. [CrossRef]
21. Carelli, M.D.; Garrone, P.; Locatelli, G.; Mancini, M.; Mycoff, C.; Trucco, P.; Ricotti, M.E. Economic features of integral, modular, small-to-medium size reactors. *Prog. Nucl. Energy* **2010**, *52*, 403–414. [CrossRef]
22. Vinoya, C.L.; Ubando, A.T.; Culaba, A.B.; Chen, W.-H. State-of-the-Art Review of Small Modular Reactors. *Energies* **2023**, *16*, 3224. [CrossRef]
23. NEA. *Small Modular Reactors: Challenges and Opportunities*; OECD Publishing: Paris, France, 2021. Available online: https://www.oecd-nea.org/jcms/pl_57979/small-modular-reactors-challenges-and-opportunities?details=true (accessed on 6 April 2025).
24. Mignacca, B.; Locatelli, G. Economics and Finance of Small Modular Reactors: A Systematic Review and Research Agenda. *Renew. Sustain. Energy Rev.* **2020**, *118*, 109519. [CrossRef]
25. U.S. Department of Energy. *Benefits of Small Modular Reactors (SMRs)*; Office of Nuclear Energy: Washington, DC, USA, 2021. Available online: https://www.energy.gov/ne/benefits-small-modular-reactors-smrs (accessed on 6 April 2025).
26. International Atomic Energy Agency (IAEA). *What are Small Modular Reactors (SMRs)?* IAEA News Center: Vienna, Austria, 2021. Available online: https://www.iaea.org/newscenter/news/what-are-small-modular-reactors-smrs (accessed on 6 April 2025).
27. NuScale Power. The NuScale Power Module. NuScale Power. 2023. Available online: https://www.nuscalepower.com/products/nuscale-power-module (accessed on 6 April 2025).
28. GE Vernova. BWRX-300 Small Modular Reactor. GE Vernova. 2023. Available online: https://www.gevernova.com/nuclear/carbon-free-power/bwrx-300-small-modular-reactor (accessed on 6 April 2025).
29. Rolls-Royce SMR. Rolls-Royce SMR Progresses to Final Step of UK Assessment. World Nuclear News. 2024. Available online: https://www.world-nuclear-news.org/articles/rolls-royce-smr-progresses-to-final-step-of-uk-ass (accessed on 6 April 2025).
30. Terrestrial Energy. Terrestrial Energy. Terrestrial Energy. 2023. Available online: https://www.terrestrialenergy.com/ (accessed on 6 April 2025).
31. Seaborg Technologies. Seaborg Technologies | Rethinking Nuclear Energy. Seaborg Technologies. 2023. Available online: https://www.seaborg.com/ (accessed on 6 April 2025).
32. International Energy Agency (IEA). Life Cycle Upstream Emission Factors 2023 (Pilot Edition). 2023. Available online: https://iea.blob.core.windows.net/assets/69b838f4-12ad-4f51-9155-9da6435b5d53/IEA_UpstreamLifeCycleEmissionFactors_Documentation.pdf (accessed on 6 April 2025).
33. National Renewable Energy Laboratory (NREL). Life Cycle Greenhouse Gas Emissions from Solar Photovoltaics. 2021. Available online: https://www.nrel.gov/docs/fy21osti/80580.pdf (accessed on 6 April 2025).
34. U.S. Energy Information Administration (EIA). Carbon Dioxide Emissions Coefficients by Fuel. Available online: https://www.eia.gov/environment/emissions/co2_vol_mass.php (accessed on 6 April 2025).
35. World Nuclear Association. Carbon Dioxide Emissions from Electricity. Available online: https://world-nuclear.org/information-library/energy-and-the-environment/carbon-dioxide-emissions-from-electricity (accessed on 6 April 2025).

36. International Energy Agency (IEA). Emission Factors. 2022. Available online: https://iea.blob.core.windows.net/assets/631bfd9a-fea7-4ef3-8cc0-a11ab416805d/CO2KWH_Methodology.pdf (accessed on 6 April 2025).
37. International Atomic Energy Agency (IAEA). Advances in Small Modular Reactor Technology Developments. 2022 Edition. Available online: https://www.iaea.org/publications/14750/advances-in-small-modular-reactor-technology-developments (accessed on 6 April 2025).
38. Rolls-Royce SMR. Rolls-Royce SMR Cost Estimates. Available online: https://www.rolls-royce-smr.com/ (accessed on 6 April 2025).
39. International Renewable Energy Agency (IRENA). Renewable Power Generation Costs in 2023. Available online: https://www.irena.org/Publications/2024/Sep/Renewable-Power-Generation-Costs-in-2023 (accessed on 6 April 2025).
40. International Renewable Energy Agency (IRENA). Renewable Power Generation Costs in 2022. Available online: https://www.irena.org/Publications/2023/Aug/Renewable-power-generation-costs-in-2022 (accessed on 6 April 2025).
41. BloombergNEF. Lithium-Ion Battery Pack Prices Hit Record Low of $139/kWh. Available online: https://about.bnef.com/blog/lithium-ion-battery-pack-prices-hit-record-low-of-139-kwh/ (accessed on 6 April 2025).
42. Python Software Foundation. Python Official Website. Available online: https://www.python.org/ (accessed on 21 April 2025).
43. Schmitt, T.; Leptinsky, S.; Turner, M.; Zoelle, A.; White, C.W.; Hughes, S.; Homsy, S.; Woods, M.; Hoffman, H.; Shultz, T.; et al. Cost and Performance Baseline for Fossil Energy Plants Volume 1: Bituminous Coal and Natural Gas to Electricity. Available online: https://www.osti.gov/biblio/1893822 (accessed on 6 April 2025).
44. IRENA. *Innovation Outlook: Smart Charging for Electric Vehicles*; International Renewable Energy Agency: Abu Dhabi, United Arab Emirates, 2019. Available online: https://www.irena.org/publications/2019/May/Innovation-Outlook-Smart-charging (accessed on 16 April 2025).
45. IEA. *Digitalization and Energy*; International Energy Agency: Paris, France, 2017. Available online: https://www.iea.org/reports/digitalisation-and-energy (accessed on 16 April 2025).
46. Yaldız, A.; Gökçek, T.; Şengör, İ.; Erdinç, O. Optimal Sizing and Economic Analysis of Photovoltaic Distributed Generation with Battery Energy Storage System Considering Peer-to-Peer Energy Trading. *Sustain. Energy Grids Netw.* **2021**, *28*, 100540. [CrossRef]
47. International Atomic Energy Agency. *Small Modular Reactors: Advances in SMR Developments 2024*; Non-serial Publications; IAEA: Vienna, Austria, 2024. [CrossRef]
48. Wang, Y.; Chen, W.; Zhang, L.; Zhao, X.; Gao, Y.; Dinavahi, V. Small Modular Reactors: An Overview of Modeling, Control, Simulation, and Applications. *IEEE Access* **2024**, *12*, 39628–39650. [CrossRef]
49. Rosner, R.; Goldberg, S. *Small Modular Reactors—Key to Future Nuclear Power Generation in the U.S.*; University of Chicago Press: Chicago, IL, USA, 2011.
50. Oettingen, M. Assessment of the Radiotoxicity of Spent Nuclear Fuel from a Fleet of PWR Reactors. *Energies* **2021**, *14*, 3094. [CrossRef]
51. Oettingen, M. The Application of Radiochemical Measurements of PWR Spent Fuel for the Validation of Burnup Codes. *Energies* **2022**, *15*, 3041. [CrossRef]
52. Todorov, G.; Kralov, I.; Koprev, I.; Vasilev, H.; Naydenova, I. Coal Share Reduction Options for Power Generation during the Energy Transition: A Bulgarian Perspective. *Energies* **2024**, *17*, 929. [CrossRef]

energies

MDPI

Article

Parametric Evaluation of Coolant Channels for Proton-Exchange Membrane Fuel Cell Based on Multi-Pass Serpentine Flow Field

Qingsheng Liu [1,2], Xuanhong Ye [3,4], Hai Huang [3], Junjie Cheng [3], Kai Meng [1], Qinglong Yu [3], Junyi Liu [3], Waqas Ahmad [3], Zulkarnain Abbas [2,3,4], Muhammad Aurangzeb [2,3,4], Muhammad Ahmed [2,3,4,*] and Shusheng Xiong [2,3,4,5,*]

1 Jiashan Power Supply Company, State Grid Zhejiang Electric Power Co., Ltd., Jiashan 314100,China
2 Provincial Key Laboratory of New Energy Vehicles Thermal Management, Longquan 323700, China
3 College of Energy Engineering, Yuquan Campus, Zhejiang University, Hangzhou 310027, China
4 Longquan Industrial Innovation Research Institute, Longquan 323050, China
5 Jiaxing Research Institute, Zhejiang University, Jiaxing 314016, China
* Correspondence: engrahmed916@zju.edu.cn (M.A.); xiongss@zju.edu.cn (S.X.)

Abstract

Proton-exchange membrane fuel cells (PEMFCs) stand out for their exceptional efficiency, high power density, and zero emissions, as they produce merely heat and water as byproducts. Appropriate and robust thermal management is the key to ensuring the maximum efficiency of the fuel cell (FC) as its optimum operating temperature is 70~80 °C. The current study was designed for the parametric evaluation of coolant channels (CCs) based on the multi-pass serpentine flow field (MPSFF) to investigate the relationship between channel geometry and thermal performance in PEM fuel cells, offering novel insights into optimal design configurations for improved thermal management. Six 3D computational models of PEMFCs with varying numbers of coolant channels were created and evaluated using COMSOL 6.2. The acquired results suggested that longer channel lengths with more serpentine turns cause the maximum number of hot spots around turns and offer a maximal pressure drop, whereas increasing the number of channels results in a uniform thermal distribution and leads to a minimal pressure drop. The findings indicate that systematic variations in geometrical configurations of MPSFFs can significantly enhance thermal uniformity and minimize the pressure drop, offering valuable insights for improving the efficiency of PEMFCs.

Keywords: new energy vehicle; vehicle thermal management; polymer electrolyte membrane fuel cell; coolant channels; multi-pass serpentine flow field

1. Introduction and Literature Review

Global demographic expansion, exponentially increasing energy demands, and anthropogenic climate challenges, including rising emissions and air pollution, demand a paradigm shift from carbon-intensive energy systems to clean, green, and sustainable energy solutions. As the global population is expected to surpass 8.9 billion by 2035, this will require additional energy resources to meet the needs of almost 720 million more individuals [1]. The current power generation system is the biggest contributor to global pollution, as the share of renewable energy in the electrical sector is only 30%. However, it remains challenging to expand the scope of renewable energy in other sectors, such as heating and transportation [2]. Therefore, this emphasizes the need for a clean, green, and sustainable energy solution to not only meet increasing energy demands but also contribute to achieving the goal of net-zero emissions by the middle of the century, set by the major economies of the world.

In recent years, PEMFCs have emerged as a highly promising and sustainable technology for clean energy generation as they offer a much higher energy density than conventional energy resources (gasoline/diesel) and even traditional renewable energy resources (solar and wind), which makes them particularly suitable for transportation applications. Furthermore, low-temperature PEMFCs offer many other benefits, including their quick startup, continuous operation, quick refueling, low operating temperatures, and minimal environmental impact as they primarily produce only water as a byproduct. Their scalability, versatility, and compact nature make them a promising solution for the decarbonization of the transportation sector [3]. Therefore, the adaptation of PEMFCs can play a pivotal role in bridging the renewable energy adaptation gap, particularly in heating and, especially, in the transportation sector, due to their compact size and high energy density.

A PEMFC is an electrochemical device that converts the chemical energy of reactants (hydrogen and oxygen) directly into electricity after a reaction and produces only heat and water as byproducts [4]. When operating, the fuel (hydrogen) and the oxidizer (oxygen) are supplied externally, and the system involves an exchange of both energy and matter with the environment, as hydrogen fuel must be replenished, and the byproducts (water) must be discharged. In principle, as long as the necessary reactants are continuously supplied and the reaction products are effectively removed, the FC can generate power without interruption, making it a highly efficient and sustainable energy solution [5,6].

The choice of a low-temperature PEMFC is due to its various advantages, which make it suitable for automotive, portable, and backup power applications. Its low operating temperature (60 to 80 °C) offers several benefits, including quick startup, as it is easy to reach this operating temperature in cold regions, typically taking from seconds to minutes. Another benefit is its rapid response to changing power demands. Further, the use of a solid polymer electrolyte membrane (PEM) and a compact catalyst layer enables very compact stack designs, which results in a high power density [7,8].

The smooth operation of a PEMFC is affiliated with the proper thermal management of the system. For a single cell, the overall heat generation rate (W/cm^2) is given by the following equation:

$$q = (E_{tn} - E_{cell}) \times i \tag{1}$$

where E_{tn} refers to the thermal voltage (which gives the hypothetical voltage of the FC when the change in enthalpy is converted into electricity), i shows the current density (A/cm^2), and E_{cell} is the operating voltage. Therefore, the rate of heat generation is directly proportional to the current density and cell voltage, as given in Equation (1). As a result, the heat generation exceeds the power generation at higher current densities in PEMFCs. This causes critical challenges in thermal management, especially for vehicle applications in which a higher current density is required.

The performance of the stack is greatly impacted by temperature. When the stack temperature is low, the chemical reaction's activation energy inside the cell is larger, and the activation loss increases. Lower temperatures also lead to lower conductivity and increased ohmic polarization. At the same time, the expansion rate of the gas decreases at low temperatures, and the concentration polarization increases. Similarly, at very high temperatures, the water content of the proton-exchange membrane (PEM) decreases, and the conductivity and transport characteristics of the membrane decrease, which increases the ohmic loss and concentration loss and decreases the output performance of the stack [9–12].

Furthermore, excessively high temperatures can dry out the membrane, and over time, this may lead to damage in both the membrane and the catalyst layers [13]. Very low temperatures can reduce catalyst activity and increase the risk of water buildup, leading to flooding inside the cell [14,15]. Temperature impacts nearly all the transport properties of an FC, and studies suggest that thermal management systems can account for up to 8% of the total stack design in some cases [16,17]. Therefore, the proper thermal management of PEMFCs is necessary, and the main objective of thermal management is to maintain PEMFCs within an optimal temperature range (70–80 °C) to ensure maximum efficiency [18].

However, effectively cooling a PEMFC stack in real-world applications is quite challenging. In practice, multiple single cells are combined to form a stack, which makes uniform temperature control even more difficult [19]. As a result, the temperature across the cells in the stack can become uneven. Addressing thermal management is the key to improving PEMFC stack performance and durability [20]. Various thermal management techniques can be used, such as edge cooling, separate airflow cooling, air cooling, cooling with the cathode air supply, phase change cooling, and liquid cooling. Each method has its own advantages and limitations, making the choice dependent on specific system requirements [18]. Due to its high heat removal capability and simple operation, liquid cooling is preferred in all FC electric vehicles (FCEVs). Typically, coolant channels are engraved in the bipolar plates, and the coolant is circulated in those coolant channels with the help of liquid pumps [21]. Serpentine designs are widely recognized as the industry standard because of their excellent performance and their ability to maintain a uniform thermal distribution. They are often used as a benchmark when evaluating or developing new flow field configurations [22,23].

Despite significant advancements in the design and analysis of SFFs for PEMFCs' thermal management, there remain critical gaps in our understanding of the interplay between the impact of varying channel counts and channel widths of multi-path serpentine flow fields (MPSFFs). Previous studies have predominantly focused on limited configurations and simplified geometries, often neglecting the combined effects of varying channel counts and channel widths on the fluid dynamics and heat transfer efficiency of SFFs. Existing studies often isolate design parameters (e.g., varying channel counts, varying channel widths, or comparing different types of FFs). A few research studies have considered the combined effect of varying channel counts and widths by holistically addressing their coupled impact on thermo-electrochemical performance. For instance, prior works typically explored fixed channel widths while varying channel counts (or vice versa).

For instance, Yihong Gu et al. [24] demonstrated that coupling LBM-LES effectively models high-speed gear lubrication with baffles, revealing optimal baffle designs that enhance fluid enthalpy and vorticity while maintaining stable oil film formation at radial speeds of 8 m/s. This approach provides insights into turbulence control and energy dissipation, which are critical for aerospace and industrial applications. Similarly, Yunfeng Tan et al. [25] reviewed advancements in SAF, highlighting the integration of ultrasonic and magnetic fields to enhance turbulence and abrasive particle kinetics. Techniques like CFD-DEM coupling and PIV measurements have elucidated particle–wall interactions and cavitation effects, improving surface finish uniformity and efficiency. For example, gas–liquid–solid three-phase flows with microbubbles were shown to increase fluid velocity by 20%, significantly boosting polishing performance for hard materials like optical glass and silicon wafers.

S.M. Rahgoshay et al. [10] analyzed the index of uniform temperature (IUT), pressure drops, and temperature gradients of two different types of flow fields (SFF and Parallel FF) and found that SFF outperformed Parallel FF by 24%. Yang Wang et al. [26] investigated six different PEMFC models by integrating six different CFFs (single serpentine channel, triple serpentine channels, quintuple serpentine channels, and two configurations with mixed serpentine channels (I–V and V–I)) to obtain a better temperature distribution and found that V–I yielded the best performance. Xin Xu et al. [27] compared two novel CFFs (A and B) with the traditional SFF through numerical modelling. They found the A novel CFF was the best CFF in terms of cooling performance.

Xuanyu Liu et al. [28] compared three different CFFs (parallel flow field (PFF), parallel serpentine flow field (PSFF) along the Y-axis, and PSFF along the X-axis) in terms of thermal performance and found that the PSFF along the X-axis contributes to the effective thermal management of PEMFCs. Similarly, Ben Chen et al. [29] analyzed the impact of the different coolant directions and found that when the coolant flow direction is the same as that of the reactants, an improved IUT is achieved and the coolant temperature difference is less than 6 K. Zhang et al. [30] also analyzed a metallic PEMFC structure by integration with a wavy flow field. Their study demonstrated that the thermal distribution is more uniform when the CFF is has the same direction as that of the cathode.

Moreover, while the remarkable advantages of SFFs are well documented, the pressure drop remains a significant challenge. Numerous researchers have attempted to address this issue using various techniques, but there remains a research gap. The study by Xuanyu Liu et al. [28] highlighted an important observation regarding pressure drops in their research, in which they found that while the PSFF along the X-axis provided improved thermal management, it also resulted in a pressure loss that was consistently 1.06 times higher than that of the PFF. This emphasizes the challenge of optimizing the trade-off between cooling efficiency and pressure loss in MPSFFs. Specifically, the research indicated that while reducing the number of channels in a manifold is beneficial for enhancing the uniformity of flow, it inevitably leads to an increase in the pressure drop within the CFFs. This underscores the need for further analyses to balance improved cooling performance and acceptable pressure losses.

Seung Man Baek et al. [31] also assessed the performance of parallel MPSFFs in designing a flow field for a larger cooling plate measuring 18 cm $\times$ 18 cm. Their findings indicated that parallel MPSFFs demonstrated a superior cooling efficiency compared to conventional parallel SFFs, while the pressure drop characteristics remained similar when the number of parallel channels was unchanged. The study attributed the improved temperature uniformity in parallel MPSFFs to their highly intertwined structure.

Similarly, G Amarnath et al. [32] conducted a CFD analysis of a PEMFC using ANSYS Fluent 17.2, comparing two different flow configurations: the dual serpentine flow field (DSFF) and the split serpentine counter flow field (SSCFF). The DSFF consisted of two distinct single serpentine flow fields within the active region of the FC, whereas the SSCFF was designed with a counterflow arrangement to enhance thermal management. The study revealed that the SSCFF configuration demonstrated a superior performance in terms of temperature regulation and uniformity across the FC system compared to the DSFF.

Extensive research on the thermal performance of PEMFCs has provided a thorough evaluation of various coolant flow field (CFF) configurations, particularly serpentine flow fields (SFFs) used with bipolar plates or cooling plates. These studies have rigorously examined aspects such as temperature distribution, heat dissipation efficiency, and overall thermal behavior and have proposed that the SFFs greatly enhance all these parameters. Variations in SFFs' design configurations can augment the effective coolant fluid distribution and could help in keeping PEMFCs under an ideal operating temperature [2]. While these investigations emphasize the importance of effective thermal regulation, they also consistently highlight the often overlooked need to address pressure drops.

To bridge this gap, the present study focuses on advanced computational fluid dynamics (CFD) modeling of multi-pass serpentine flow fields (MPSFFs). It aims to systematically analyze channel geometry to both optimize thermal management and minimize pressure losses within the flow field. This dual optimization is expected to significantly enhance the efficiency and operational performance of PEMFC systems.

While existing studies have explored various aspects of CFFs, few have investigated the impact of geometric variations in MPSFFs, particularly the number of channels. Increasing the number of channels has shown promise in improving both the thermal uniformity and reducing pressure drops. He Lu et al. [33] studied the impact of increasing the number of channels on the reactant distribution in PEMCs and indicated that increasing the number of channels of SFFs makes the length of individual channels shorter, which improves the fluid supply capacity and distribution uniformity. Further, they concluded that the increased number of channels has a shunt-like effect on the fluid, which significantly minimizes the pressure drop.

However, the effectiveness of SFFs in maintaining a uniform thermal distribution and reliable coolant flow has led to them being a widely adopted design in proton-exchange membrane fuel cells (PEMFCs). Despite their popularity, recent studies have highlighted the potential for further performance improvements through careful geometric optimization of the flow field. Specifically, variations in the number of the channels can significantly enhance the thermal distribution and reduce pressure drops within the flow path. Existing research has demonstrated that even modest adjustments to the channel geometry can lead to better heat management and more efficient coolant flow. By systematically evaluating the impact of the number of channels, it is possible to optimize the thermal profile and minimize the pressure drop, which are the two critical factors that directly influence the efficiency and longevity of PEMFCs. Given the limitations of traditional SFFs and the significant gap in the literature regarding geometric optimization, this study focuses on the dual optimization of multi-pass serpentine flow fields (MPSFFs). The aim is to enhance thermal management while simultaneously reducing pressure drops, addressing these key challenges to improve the overall performance of PEMFCs.

Furthermore, the most compelling aspects of geometric variation in MPSFFs are their simplicity and cost-effectiveness. Unlike complex multi-material or multi-layer designs, adjusting the number and width of channels is a straightforward modification that can be implemented without the need for expensive materials or intricate manufacturing processes. This makes it an attractive approach for improving PEMFC performance, especially in commercial applications for which cost and efficiency are critical factors.

This study uses numerical simulations to evaluate the cooling efficacy and pressure drop of CFFs based on MPSFFs in a PEMFC. For this purpose, six three-dimensional (3D) computational models were developed using COMSOL Multiphysics 6.2. This investigation aimed to optimize the cooling capabilities of MPSFFs and to minimize the pressure drop, which plays a critical role in maintaining the operating temperature of PEMFCs within an ideal range. The novelty of this study is the dual optimization of the coolant flow field (CFF) configuration based on a multi-path serpentine flow field (MPSFF) strategy to enhance the thermal distribution uniformity while minimizing the pressure drop in PEMFCs and the proposed geometrical designs, which are not only structurally effective but also inherently simple, making them highly practical and cost-efficient for real-world manufacturing and integration.

2. Model Description

2.1. Geometrical Modeling

Six computational models with distinct configurations of CFFs, all having a fixed channel width of 1.5 mm and varying numbers of channels, ranging from two to twelve channels (two, four, six, eight, ten, and twelve) were designed, which are depicted in Figure 1. The 3D layout of the structural model of all the components including the

CFFs, gas channels, GDL, CL, and membrane is shown in Figure 2. Detailed geometrical specifications for all these components except for those of the coolant channels are provided in Table 1, and the geometrical parameters of the coolant channels are presented in Table 2.

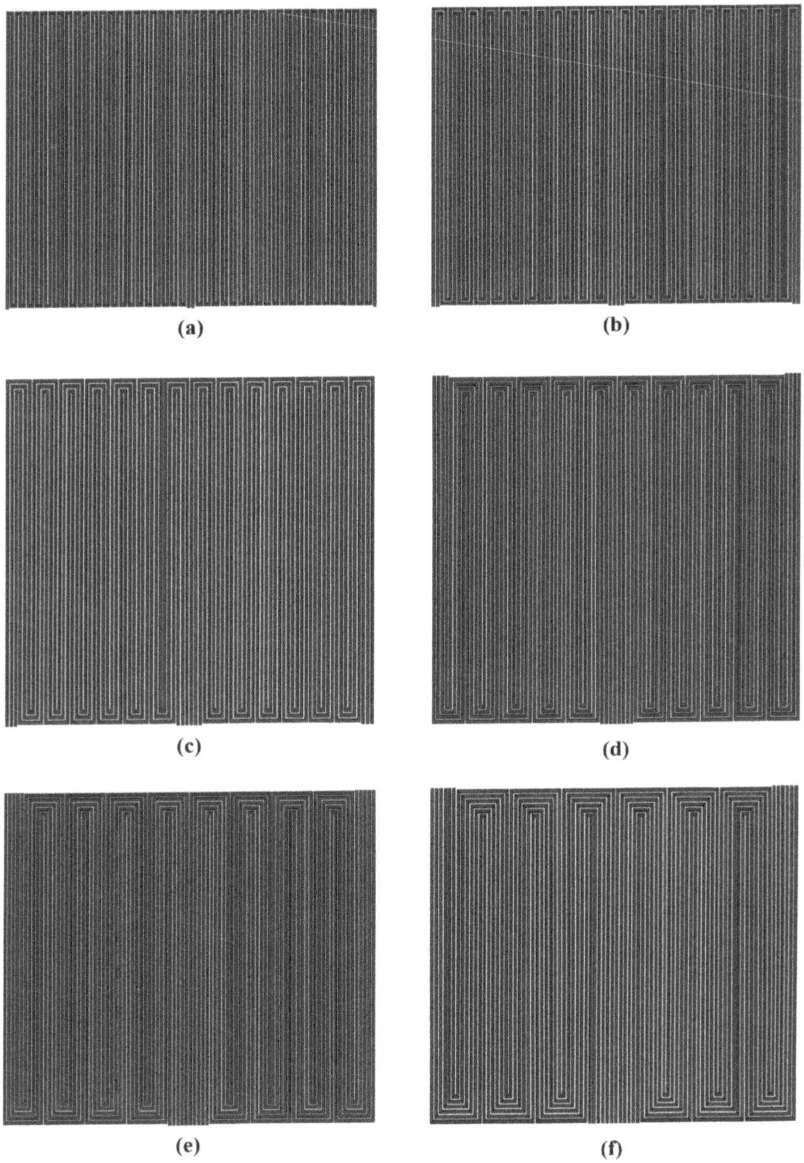

Figure 1. Coolant flow fields—(**a**) two-path SFF, (**b**) four-path SFF, (**c**) six-path SFF, (**d**) eight-path SFF, (**e**) ten-path SFF, (**f**) twelve-path SFF.

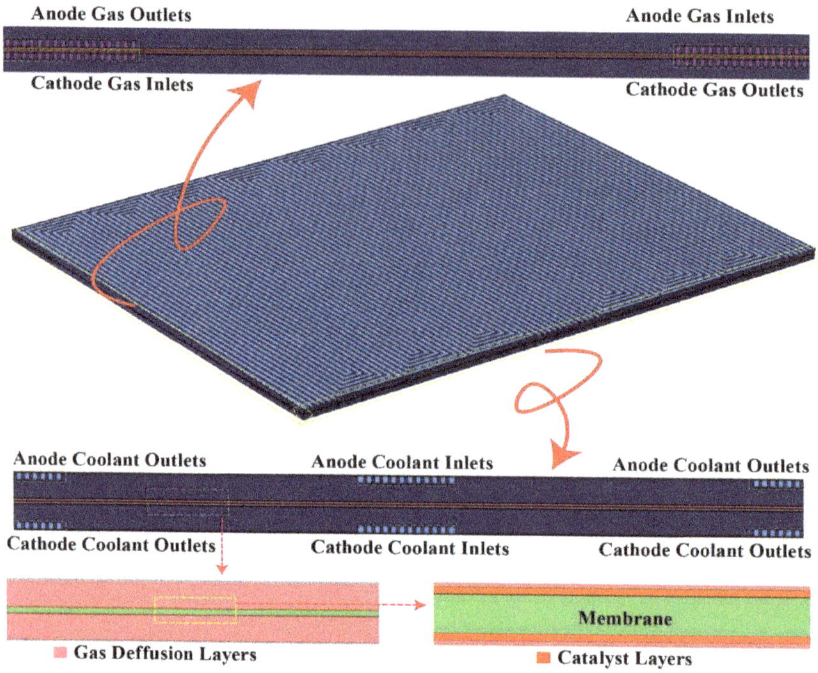

Figure 2. 3D layout of the structural model.

Table 1. Geometrical parameters of the PEMFC model.

Details	Units	Values
Active area of PEMFC	mm^2	50,000
Length of PEMFC	mm	200
Width of PEMFC	mm	250
Bipolar plate thickness (anode/cathode)	mm	2.8
Gas diffusion layer thickness (anode/cathode)	µm	240
Catalyst layer thickness (anode/cathode)	µm	10
Membrane thickness	µm	50
Number of gas channels	--	17
Height of gas channels	mm	1
Width of gas channels	mm	1
Height of cooling channels	mm	0.8

Table 2. Detailed geometrical parameters of the coolant channels.

CW (mm)	CH (mm)	L (mm)	NS	NC	HD	AR
1.5	0.8	8860	44	2	1.0435	0.533
1.5	0.8	4436	22	4	1.0435	0.533
1.5	0.8	3020	14	6	1.0435	0.533
1.5	0.8	2116	10	8	1.0435	0.533
1.5	0.8	1812	08	10	1.0435	0.533
1.5	0.8	1610	07	12	1.0435	0.533

Note: Abbreviations: CW stands for channel width; CH stands for channel height; L stands for length; NS stands for number of serpentine turns; NC stands for no. of channels; HD stands for hydraulic diameter; AR stands for aspect ratio.

Experimental data from previous fuel cell (FC) studies were comprehensively analyzed at the initial stage of this research to establish accurate and reliable baseline parameters for designing the structure of the above model. Subsequently, six FC models were designed by integrating the MPSFFs with them to provide a comprehensive analysis. The gas flow fields utilized in this study are consistent across all of the six models, comprising 17-path serpentine gas flow channels. Each channel is designed with a uniform width and height of 1 mm, ensuring geometric consistency throughout the configurations. This specific cross-sectional design was chosen because it aligns with findings from previous research. For instance, S. Toghyani [34] demonstrated that reducing the cross-sectional area of gas flow channels can significantly enhance the performance of PEMFCs. The reduction in channel width, in particular, leads to an improved gas distribution, enhanced reactant availability, and better water management within the cell, thereby optimizing the overall cell efficiency. So, the deliberate selection of these dimensions reflects an evidence-based approach to achieving a PEMFC system with a superior thermal and electrochemical performance.

2.2. Computational Model

The computational models presented in this study were designed to simulate fluid flow and thermal characteristics in PEMFCs with different CFF configurations. To effectively fulfill the objectives of this study, the computational model was developed with a comprehensive and detailed representation of all essential components of the PEMFC. These components include the gas diffusion layers (GDLs), catalyst layers (CLs), polymer electrolyte membrane (PEM), gas flow fields, and coolant flow fields (CFFs). Such an integrated modeling approach is crucial for accurately capturing the complex interactions of electrochemical reactions, heat transfer, and fluid flow within the cell. For this study, the active area was defined as 50,000 mm^2, corresponding to dimensions of 200 mm in width and 250 mm in length. This specification ensures a highly realistic representation of practical PEMFC configurations, offering a robust foundation for the accurate evaluation of the thermal performance of various CFF geometries. All the numerical models were solved by using COMSOL 6.2, and all the physics parameters utilized are listed in Table 3.

To solve all the computational models, an XFUSION (Zhengzhou, China) computer was used, which consisted of an Intel Xeon Platinum 8374C processor, with 72 physical cores, 84 logical cores, a processor base frequency of 2.70 GHz, and a max turbo frequency of 4.40 GHz. The system is equipped with a total of 512 GB of RAM and 10 TB of ROM, running on Windows Server 2019. The time taken by each model to converge was about 7 h. To optimize the solver time and enhance the computational efficiency, we employed the biconjugate gradient stabilization (BiCGStab) solver, which is known for its effectiveness in solving large and complex numerical simulations. We configured the solver with a relative tolerance of 0.001 to balance performance and accuracy. In addition, we incorporated the F-cycle multigrid preconditioner with 60 levels to accelerate the convergence process. The choice of 60 levels was made to optimize the convergence further. Each model required approximately 75 GB of physical memory (RAM). The combined use of the BiCGStab solver, a relative tolerance of 0.001, and the F-cycle multigrid preconditioner allowed us to strike an optimal balance between computational efficiency and solution accuracy. Despite the substantial memory demand (75 GB of physical memory), these settings helped reduce the solver time while ensuring precision.

The study assessed thermal performance metrics such as the temperature uniformity across the cell, peak operating temperature at the membrane, maximum temperature of CFFs, average temperature of CFFs, average temperature of the membrane, index of uniform temperature, and pressure drop in CFFs. To access the IUT, the following expression was employed:

$$\mathrm{IUT} = \frac{\int_A \left| \mathrm{T} - \mathrm{T_{avg}} \right| \mathrm{dA}}{\int_A \mathrm{dA}} \tag{2}$$

$$T_{avg} = \frac{\int_A T dA}{\int_A dA} \quad (3)$$

where T is the local temperature at a specific point on the surface, T_{avg} is the average temperature of the surface, and A is the total surface area.

Table 3. Detailed physics parameters of the model [35].

Details	Units	Values
GDL porosities (anode/cathode)		0.5
GDL permeabilities (anode/cathode)	m^2	2.7×10^{-12}
GDL electrical conductivity	$S \cdot m^{-1}$	17,000
GDL density	$kg \cdot m^{-3}$	437.5
Effective porosity of the catalyst layers in anode and cathode		0.3
Inlet temperature of gases (anode/cathode)	K	343
Inlet temperature of the coolant fluid (anode/cathode)	K	343
Inlet flow velocity (anode gas)	m/s	0.1
Inlet flow velocity (cathode gas)	m/s	0.5
Inlet flow velocity (coolant)	m/s	0.05
Contact resistance	m^{-2}	4.45×10^{-6}
Reference exchanged current density at the anode side	$A \cdot m^{-2}$	10^4
Reference exchanged current density at the cathode side	$A \cdot m^{-2}$	50
Open-circuit voltage of the cell	V	1.03
Anode and cathode transfer coefficients for the hydrogen oxidation reaction		1.2
Anode and cathode transfer coefficients for the oxygen reduction reaction		1.2
Faraday constant	$C \cdot mol^{-1}$	96,457
Density of membrane	$kg \cdot m^{-3}$	1980
Density of the collector plates	$kg \cdot m^{-3}$	1800
Density of the catalyst layers	$kg \cdot m^{-3}$	2719
Specific heat of the membrane	$J \cdot kg^{-1} \cdot K^{-1}$	2000
Specific heat of the gas diffusion and catalyst layers	$J \cdot kg^{-1} \cdot K^{-1}$	871
Specific heat of the collector plates	$J \cdot kg^{-1} \cdot K^{-1}$	1040
Thermal conductivity of the dry membrane	$W \cdot m^{-1} \cdot K^{-1}$	2
Thermal conductivity of gas diffusion layer	$W \cdot m^{-1} \cdot K^{-1}$	0.25
Thermal conductivity of collector plates	$W \cdot m^{-1} \cdot K^{-1}$	46.2
Thermal conductivity of catalyst layer	$W \cdot m^{-1} \cdot K^{-1}$	10
Electrical conductivity of the dry membrane	$S \cdot m^{-1}$	10^{-16}

2.2.1. Boundary Conditions

The following boundary conditions were considered to replicate the operational characteristics of a PEMFC under realistic conditions while maintaining computational efficiency and ensuring accurate simulation results:

- The geometry was constructed in three dimensions;
- The flow was set to be laminar and compressible;
- Models were studied under a steady-state condition;
- Terminals and side boundaries were treated as thermally insulated;
- At outlets, the flow was considered fully developed;
- At walls, zero diffusive flux was considered;
- Periodic conditions were used for top and bottom boundaries;
- Inflow and outflow boundaries were used to ensure that the only heat transfer occurring across the boundaries was through the flowing coolant in FFs by convection;

- Periodic boundary conditions were used to predict the stack-level behavior of a single cell with a reduced computational cost;
- Gravitational impact was neglected.

To emulate the thermal behavior of a practical FC stack, periodic temperature boundary conditions were applied to the top and bottom boundaries of the computational domain. This approach replicates the thermal coupling between adjacent cells in a stacked configuration, in which the heat generated in one cell influences its neighbors. By enforcing thermal periodicity, the model captures longitudinal heat propagation and stack-level temperature gradients without the computational expense of simulating an entire stack.

2.2.2. Mesh Independence Test

In computational studies, one of the important aspects that affects the simulation results is mesh independence. In PEMFC studies, mesh independence is particularly of great importance as they involve coupled multiphysics phenomena, fluid dynamics electrochemical reactions, mass transfer, and heat transfer. A poorly resolved mesh can lead to improper reaction rates and incorrect predictions of results. Achieving mesh independence ensures that the numerical results are artifacts of the discretization process and accurately represent the applied physics. Therefore, a mesh-independent solution strengthens the confidence in the prediction capability of the studied models and ensures the predicted trends are physically meaningful.

The meshing process for the studied models was performed using COMSOL 6.2. To ensure grid independence, the number of elements was systematically increased from 0.7 million to 4.5 million. A noticeable deviation in current density can be observed in Figure 3 from 0.7 to 2.3 million mesh, and this deviation is above 1%, whereas when the element count exceeded 2.3 million, the current density value stabilizes and the deviation becomes minor (under 1%). Keeping in mind the computational resources and the time requirements for computing the models, the mesh count of 2.3 million elements was utilized. In addition, the deviation in current density remained under 1% from 2.3 to 4.5 million mesh elements, which made it reasonable for further analysis. The number of mesh elements and the mesh quality in all six of the models are similar due to the fact that the overall dimensions of all the models are the same, i.e., 200 mm × 25 mm × 6.15 mm, and the difference is only in regards to channel count.

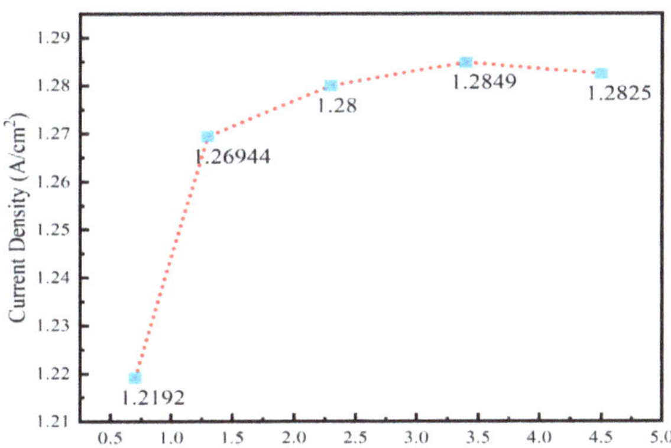

Figure 3. Mesh independence graph (current density vs number of mesh elements).

To further strengthen the mesh independence, the values for the maximum temperature were extracted to examine the impact of a different number of mesh elements. In

Figure 4, the maximum temperature can be observed with a mesh element quantity of 0.7 million and 1.3 million, whereas when the element quantity ranged from 2.3 million to 4.5 million, the variation in the value for the maximum temperature stabilizes within an acceptable limit (<1%). Therefore, the mesh element quantity of 2.3 million was utilized for further computing. These were allocated based on the complexity and resolution requirements of each component. The bipolar plates and flow channels accounted for about 0.4 million elements, the GDLs for 0.6 million, the catalyst layers (CLs) for 0.5 million, the microporous layers (MPLs) for 0.3 million, and the membrane for approximately 0.5 million elements.

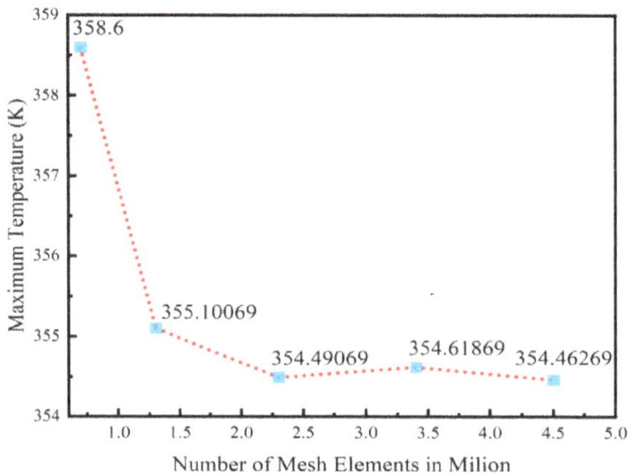

Figure 4. Mesh independence graph (maximum temperature vs number of mesh elements).

2.2.3. Model Validation

To guarantee the reliability and accuracy of the computational model, validation against experimental or numerical benchmarks is essential. However, the experimental testing of FC components at the microscopic level, such as the membrane, GDL, and catalytic layer, is often prohibitively expensive and technically challenging due to their complex geometries and operating conditions. Consequently, it is standard practice to validate simulation results using widely accepted datasets from the peer-reviewed literature. In this study, a rigorous comparative analysis was conducted by benchmarking the simulation outcomes of the model with 12-path CFFs with a 1.5 mm width against a prior established study [35].

In the present work, water was used as a coolant, and the inlet boundary condition was set to the velocity inlet with a specific inlet velocity and inlet temperature. At the outlet of the coolant channels, the boundary condition of a fully developed flow was considered. The results show consistency with the reported work. A quantitative comparison of I/V values is provided in Table 4, and the polarization curve shown in Figure 5 confirms the validity of the present study. The validation outcomes are associated with the 12-channel model, which exhibited an optimal performance across the six configurations. This robust validation ensures that the simulation aligns closely with established thermal behaviors, enhancing its reliability for practical, real-world applications.

Table 4. Quantitative comparison of I/V values.

Voltage (V)	Current (A/cm^2)		Deviation %
	SM Miri Joibary	Current Study	
1	0	0	0
0.95	0.04	0.0395	1.26
0.9	0.10294	0.10294	0
0.85	0.1828	0.18297	0.09
0.8	0.283	0.283	0
0.75	0.47	0.469	0.21
0.7	0.81	0.80501	0.62
0.65	1.06	1.06382	0.36
0.6	1.29	1.30	0.77
0.55	1.39	1.40	0.72
0.5	1.49	1.48	0.67
0.45	1.56	1.554	0.39
0.4	1.63	1.60397	1.6
0.35	1.69	1.66963	1.4
0.3	1.75	1.71272	2.2

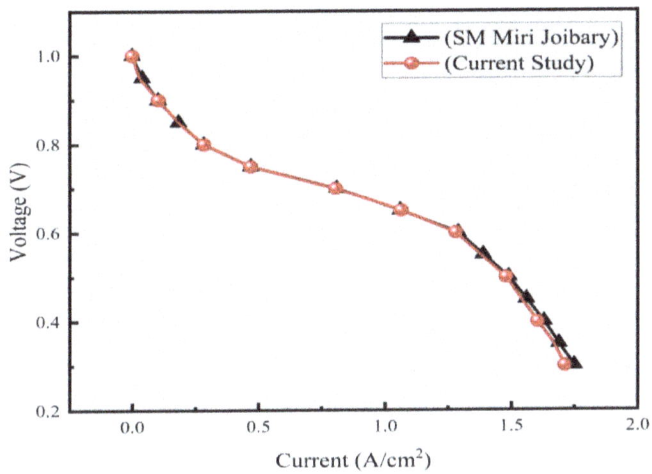

Figure 5. Polarization curves representing model validation.

2.2.4. Governing Equations [35]

A PEMFC is a multiphysics system that contains different coupled phenomena during its operation, including the mass transport of species (H_2, O_2, H_2O, and N_2) associated with fluid flow, heat transfer, and electrochemical reactions. The numerical modeling of those phenomena involves the physical and chemical processes generally occurring in a PEMFC and was solved numerically using a set of conservation equations and transport equations, such as mass, energy, and momentum conservation equations, electrochemical reaction equations, and mass transfer equations.

Mass Conservation (Continuity Equation)

This ensures that reactants and products properly account for the flow of reactant gases (hydrogen and oxygen) and the production of water within the FC.

The gas continuity equation is as follows:

$$\frac{\partial\left(\epsilon_g \rho_g\right)}{\partial t} + \nabla \cdot \left(\epsilon_g \rho_g V_g\right) = R_{reaction} - R_{phase} \tag{4}$$

where ϵ_g: the porosity of the porous medium (e.g., GDL or CL), ρ_g: the density of the gas mixture, v_g: the velocity vector of the gas phase, $R_{reaction}$: the source term representing the mass consumed or generated due to chemical reactions (e.g., hydrogen and oxygen consumption and water production), and R_{phase}: the phase change term representing the conversion between water vapor and liquid water (condensation or evaporation).

$$R_{reaction} = -\frac{M_{O_2}}{4F} j_{cathode} + \frac{M_{H_2O}}{2F} j_{cathode} (\text{Cathode Catalyst Layer}) \tag{5}$$

where M_{O_2} : the molar mass of oxygen, M_{H_2O} : the molar mass of water, F: the Faraday constant (96,485 C/mol) representing the charge of one mole of electrons, and $j_{cathode}$: the local current density at the cathode.

$$R_{reaction} = -\frac{M_{H_2}}{2F} j_{anode} (\text{Anode Catalyst Layer}) \tag{6}$$

$$R_{reaction} = 0 \ (\text{Other Region})(\text{Anode Catalyst Layer}) \tag{7}$$

where M_{H_2} : the molar mass of hydrogen, j_{anode} : the local current density at the anode, F: the Faraday constant (96,485 C/mol) representing the charge of one mole of electrons, single term: represents hydrogen consumption during the anode reaction, and $R_{reaction}$ is specific to each layer, capturing the consumption of reactants.

The liquid continuity equation is as follows:

$$\nabla \cdot (\rho_l v_l) = R_{phase} (\text{Liquid Water Transport}) \tag{8}$$

where ρ_l: the density of liquid water, v_l: the velocity of liquid water flow, and R_{phase}: the source term representing the phase change between water vapor and liquid water (e.g., condensation or evaporation).

$$\nabla \cdot (\rho_{cool} v_{cool}) = 0 \ (\text{Cooling Fluid Transport}) \tag{9}$$

where ρ_{cool}: the density of liquid water and v_{cool}: the velocity of liquid water flow.

The momentum conservation (Navier–Stokes) equation is as follows:

$$\nabla \cdot \left(\epsilon \rho_g V_g V_g\right) = \nabla_p + \nabla \cdot \tau (\text{Momentum Conservation for Gases}) \tag{10}$$

where ϵ: the porosity of the medium (e.g., GDL), ρ_g: the density of the gas mixture (dependent on temperature, pressure, and species concentration), V_g: the velocity vector of the gas phase, P: the pressure of the gas phase, and τ: the viscous stress tensor, representing the momentum dissipation due to viscosity.

$$\nabla \cdot (\rho_{cool} V_{cool} V_{cool}) = \nabla \cdot \tau - \frac{\mu_{cool}}{\alpha} V_{cool} (\text{Momentum Conservation Coolant}) \tag{11}$$

where ρ_{cool}: the density of the coolant (constant for incompressible flow), V_{cool}: the velocity vector of the coolant flow, μ_{cool}: the dynamic viscosity of the coolant, and α: the permeability of the medium, representing resistance in porous layers.

The energy conservation (heat transfer) equation is as follows:

$$\left(\rho c_p\right)_{eff}(v_g \cdot \nabla T) = \nabla \cdot (k_{eff} \nabla T) + S_{temp}(\text{Gas Phase Heat Transfer}) \quad (12)$$

where ρc_p: the effective heat capacity of the gas-phase medium, v_g: the velocity of the gas flow (for convective heat transfer), k_{eff}: the effective thermal conductivity of the region (e.g., GDL and gas channels), S_{temp}: the total heat generation in the region, and S_{temp} represents the heat generated due to various mechanisms within the PEMFC.

$$S_{temp} = \begin{cases} \frac{I^2}{\sigma_{cc}}(\text{Current Collectors}) \\ r_w h_L(\text{Gas Channels}) \\ \frac{I^2}{\sigma_{GDL}} + r_w h_L(\text{GDL}) \\ \frac{I^2}{\sigma_m} + r_w h_L(\text{Membrane}) \\ \eta_{anode,cathode} + I^2\left(\frac{1}{\sigma_m} + \frac{1}{\sigma_s}\right) + r_w h_L(\text{CL}) \end{cases} \quad (13)$$

where $\frac{I^2}{\sigma}$: the heat generated due to resistance in current collectors (σ_{cc}), GDL (σ_{GDL}), membrane (σ_m), and catalyst supports (σ_s); $\eta_{anode,cathode}$: the heat generated due to exothermic electrochemical reactions at the anode and cathode; I: the current density (the amount of current flowing through a unit area); r_w: the rate of water production or phase change in the given region; and h_L: the latent heat of vaporization (or condensation) of water.

$$\nabla \cdot \left(\rho_{cool} C_{p_{cool}} V_{cool} T\right) = \nabla \cdot (k_{eff} \nabla T)(\text{Coolant Heat Transfer}) \quad (14)$$

where $\rho_{cool} C_{p_{cool}}$: the density of the coolant (constant for incompressible flow), V_{cool}: the coolant velocity (driving convective heat removal), and k_{eff}: the effective thermal conductive species transport equation (diffusion and convection).

$$\nabla \cdot \left(\rho_g v_g Y_i\right) = -\nabla \cdot J_i + S_i(\text{Species Transport Equation}) \quad (15)$$

where the left-hand side represents the convective transport of species i in the gas phase, $(-\nabla \cdot J_i)$ accounts for the diffusion of species i, and S_i: the source or sink term for species i due to chemical reactions or phase changes.

$$J_i = -j = \sum_{j=1}^{N-1} \rho_g D_{ij} \nabla Y_j(\text{Diffusive Flux}) \quad (16)$$

where ρ_g: the gas-phase density and D_{ij}: the binary diffusion coefficient between species i and j.

$$D_{ij} = D_{ij}^0\left(\frac{P_0}{P}\right)\left(\frac{T}{T_0}\right)^{\gamma T}(\text{Binary Diffusion Coefficient}) \quad (17)$$

where D_{ij}^0: the reference diffusion coefficient at a standard pressure (P_0) and temperature (T_0), p: the local pressure, t: the local temperature, and γT: the temperature dependence exponent.

$$\text{Souce Term } S_i = \begin{cases} \begin{Bmatrix} S_{H_2} = -\frac{M_{H_2}}{2F} j_{anode} \\ S_{H_2O} = \frac{n_a M_{H_2O}}{F} \nabla I_m - r_w \end{Bmatrix} (\text{Anode CL}) \\ \begin{Bmatrix} S_{O_2} = -\frac{M_{O_2}}{4F} j_{cathode} \\ S_{H_2O} = \frac{M_{H_2O}}{2F} j_{cathode} + \frac{n_a M_{H_2O}}{F} \nabla I_m - r_w \end{Bmatrix} (\text{Cathode CL}) \\ \left\{ S_{H_2O} = \frac{n_a M_{H_2O}}{F} \nabla I_m - r_w \right\} (\text{Membrane}) \\ \left\{ r_w = \frac{n_a M_{H_2O}}{F} j_{cathode} \right\} (\text{Water Phase Change Rate}) \end{cases} \quad (18)$$

where M_{H_2}: the molar mass of hydrogen, M_{h_2o}: the molar mass of water, F: the Faraday constant, j_{anode} : the current density at the anode, ∇I_m: the proton current density gradient, r_w: the water phase change rate, M_{O_2}: the molar mass of oxygen, and $j_{cathode}$: the current density at the cathode. Water transport in the membrane is driven by proton drag (∇I_m), the water phase change is driven by (r_w), and S_i represents the production or consumption of species i.

The charge conservation equation is as follows:

$$\nabla \cdot (\sigma_m \nabla \phi_m) = S_{em}(\text{Membrane Phase Potential}) \quad (19)$$

$$S_{em} = \begin{cases} R_a & (\text{Anode Catalyst Layer (ACL)}) \\ -R_c & (\text{Cathode Catalyst Layer (CCL)}) \end{cases} \quad (20)$$

where σ_m: the proton conductivity of the membrane (s/m), ϕ_m: the electric potential in the membrane (v), S_{em}: the source term for the membrane potential, representing proton transfer due to electrochemical reactions, and (R_a) and (R_c): the reaction rates, which are derived from Faraday's law.

$$\nabla \cdot (\sigma_s \nabla \phi_s) = S_{es} \quad (21)$$

$$S_{es} = \begin{cases} -R_a & (\text{Anode Catalyst Layer}) \\ -R_c & (\text{Cathode Catalyst Layer}) \end{cases} \quad (22)$$

where σ_s: the electrical conductivity of the solid (s/m), ϕ_s: the electric potential in the solid phase (v), S_{es}: the source term for the solid-phase potential, representing the electron transfer due to electrochemical reactions, and (R_a) and (R_c): the reaction rates, which are derived from Faraday's law.

Electrochemical Kinetics (Butler–Volmer Equation)

This equation governs the rate of the electrochemical reactions at the electrodes, considering both the anodic and cathodic reactions.

The anode Butler–Volmer equation is as follows:

$$j_{anode} = j_{anode}^{ref} \left(\frac{C_{H_2}}{C_{H_2}^{ref}} \right)^{\gamma_{anode}} \left[\exp\left(\frac{\alpha_{anode} F \eta_{anode}}{RT} \right) - \exp\left(-\frac{\alpha_{cathode} F \eta_{anode}}{RT} \right) \right] \quad (23)$$

Here, overpotential at the anode (η_{anode}) is defined as follows:

$$\eta_{anode} = \phi_s^{anode} - \phi_m - E_{eq}^{anode} \quad (24)$$

where j_{anode} : the local current density at the anode (a/m^2), j_{anode}^{ref} : the reference exchange current density at standard conditions, C_{H_2}: the local hydrogen concentration, $C_{H_2}^{ref.}$: the reference hydrogen concentration, γ_{anode} : the reaction order with respect to hydrogen,

α_{anode} , $\alpha_{cathode}$: the anodic and cathodic transfer coefficients, respectively, ϕ_s^{anode} : the solid-phase potential at the anode, ϕ_m: the membrane-phase potential, and E_{eq}^{anode} : the equilibrium potential of the hydrogen oxidation reaction.

The cathode Butler–Volmer equation is as follows:

$$j_{cathode} = j_{cathode}^{ref} \left(\frac{C_{O_2}}{C_{O_2}^{ref}} \right)^{\gamma_{cathode}} \left[\exp\left(\frac{\alpha_{cathode}\, F\eta_{cathode}}{RT} \right) - \exp\left(-\frac{\alpha_{anode}\, F\eta_{cathode}}{RT} \right) \right] \quad (25)$$

Here, overpotential at the anode ($\eta_{cathode}$) is defined as follows:

$$\eta_{cathode} = \phi_s^{cathode} - \phi_m - E_{eq}^{cathode} \quad (26)$$

where $j_{cathode}$: the local current density at the cathode (a/m^2), $j_{cathode}^{ref}$: the reference exchange current density at standard conditions, C_{O_2}: the local oxygen concentration, $C_{O_2}^{ref.}$: the reference oxygen concentration, $\gamma_{cathode}$: the reaction order with respect to oxygen, $\alpha_{cathode}$, α_{anode} : the cathodic and anodic transfer coefficients, respectively, $\phi_s^{cathode.}$: the solid-phase potential at the cathode, ϕ_m: the membrane-phase potential, and $E_{eq}^{cathode}$: the equilibrium potential of the oxygen reduction reaction.

3. Results and Discussion

The efficiency of a PEMFC is influenced by various factors, such as the operating temperature, species concentration in the GDL, catalyst layer (CL), membrane, membrane hydration, and the risk of drying out. All these factors could be controlled by effective thermal management. In this study, the thermal performance of MPSFFs is evaluated by varying the number of channels across various configurations, ranging from 2 to 12, while maintaining the channel width constant at 1.5 mm. All the models were evaluated to assess the impact of channel variations on the temperature distribution in CFFs and the overall efficiency of PEMFCs. Cell performance was analyzed for voltages ranging from 0.3 V to 1 V; however, the results were reported at 0.6 V to maintain consistency with comparable studies.

3.1. Thermal Distribution

The primary objective of this study was to improve the thermal management capability of MPSFFs in PEMFC applications. To achieve this goal, a comprehensive thermal analysis of the critical components of PEMFC, like anode coolant channels, the membrane midplane, and cathode coolant channels, was conducted. This approach is widely regarded as best practice for thermal optimization studies of PEMFC systems. Therefore, temperature plots were extracted at the above-mentioned locations from the computed models. As liquid coolant channels were utilized to enhance the performance of the FC, the heat distribution in the coolant channels could effectively reflect the cooling performance of CFFs.

This study is based on varying the number of channels of MPSFFs, and it is a fact that increasing the number of channels results in a decreased length of the channels, which ultimately shortens the distance covered by the fluid from the inlet to outlet. Figure 6 shows the temperature distribution on the membrane midplane, where it is evident that as the number of channels increases from two to twelve, the average temperature of the plane is reduced sufficiently. The temperature at the membrane midplane in geometries with two- to six-paths yields a large number of hotspots, which shows a nonuniform temperature distribution. This uneven temperature distribution in the models with two-, four-, and six-path CFFs could be attributed to the greater number of serpentine curves in the channels, which hinders the coolant in these FFs. Planes with eight to twelve paths yielded a much better thermal performance as fewer and less intense hotspots could be seen in planes with these geometries. This is due to the shorter channel length and fewer serpentine curves. The highest thermal uniformity was observed in the membrane midplane with twelve-path CFF geometry.

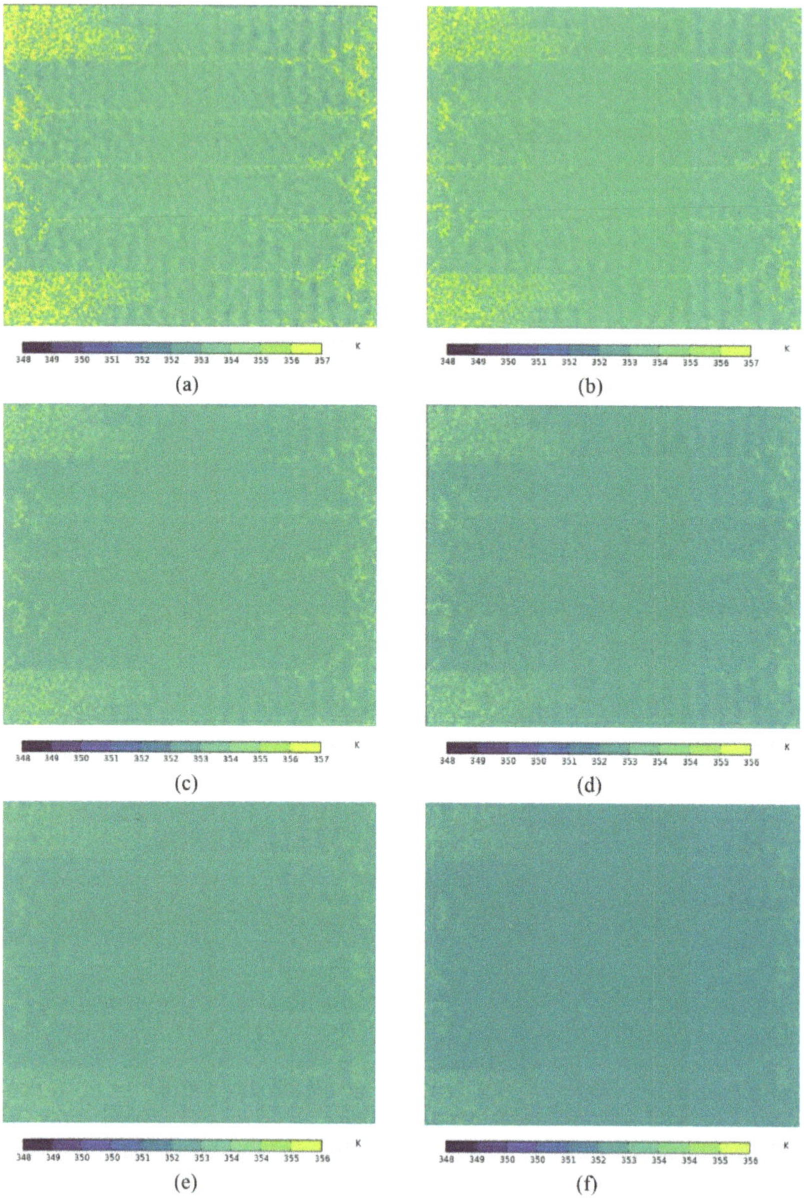

Figure 6. Temperature distribution at membrane midplane—(**a**) two-path SFF, (**b**) four-path SFF, (**c**) six-path SFF, (**d**) eight-path SFF, (**e**) ten-path SFF, (**f**) twelve-path SFF.

It is clearly noticeable in Figures 7 and 8 that the temperature varies from the inlet to outlet in the cathode and anode coolant channels. Considering the overall performance of the coolant channels, it is observed that the FFs with more serpentine turns and a greater number of curves (two- to six-path FFs) experience higher temperatures near the outlets. It is also observed that these FFs failed to extract uniform heat from the system and yielded more hotspots, which could be attributed to their longer channels and the greater number of serpentine curves. In contrast, geometries with a higher number of channels (six to twelve) presented a better cooling performance, as fewer hotspots were observed in these FFs, and the thermal extraction by these FFs was remarkable. The 12-path FF geometry was observed to have the best cooling performance in terms of its uniform heat distribution throughout.

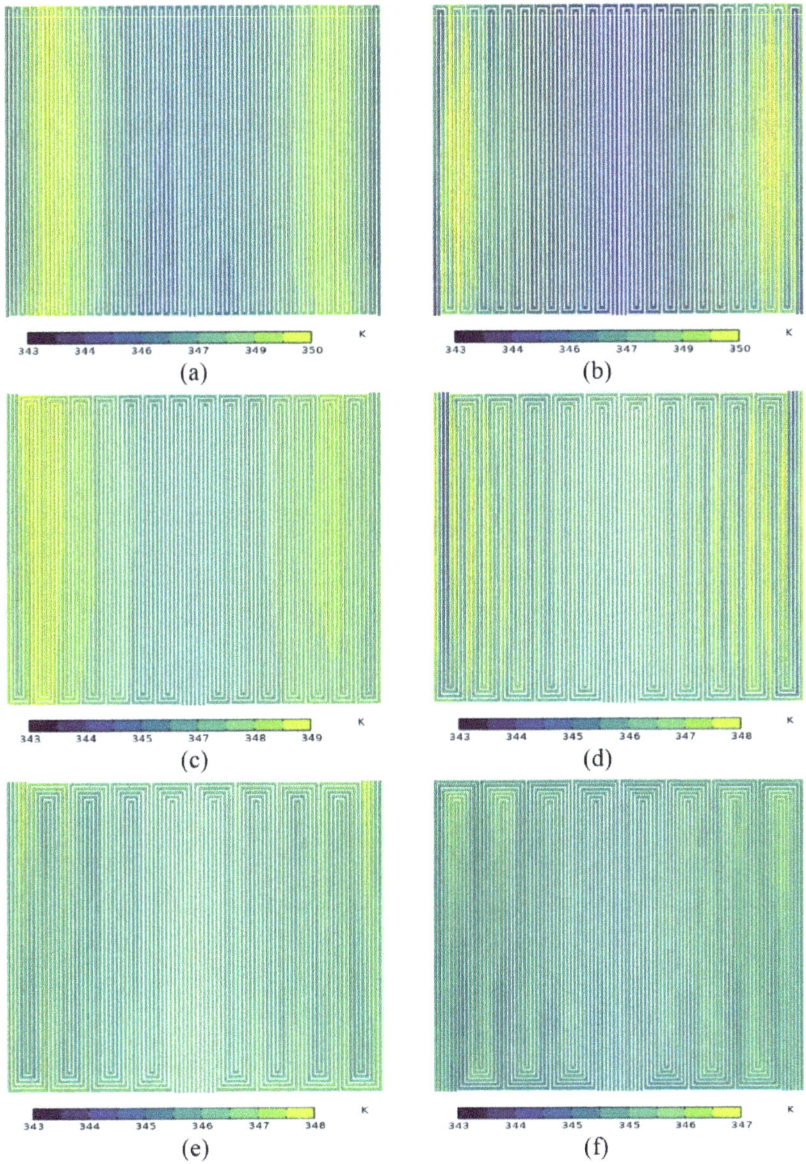

Figure 7. Temperature distribution in cathode coolant channels—(**a**) two-path SFF, (**b**) four-path SFF, (**c**) six-path SFF, (**d**) eight-path SFF, (**e**) ten-path SFF, (**f**) twelve-path SFF.

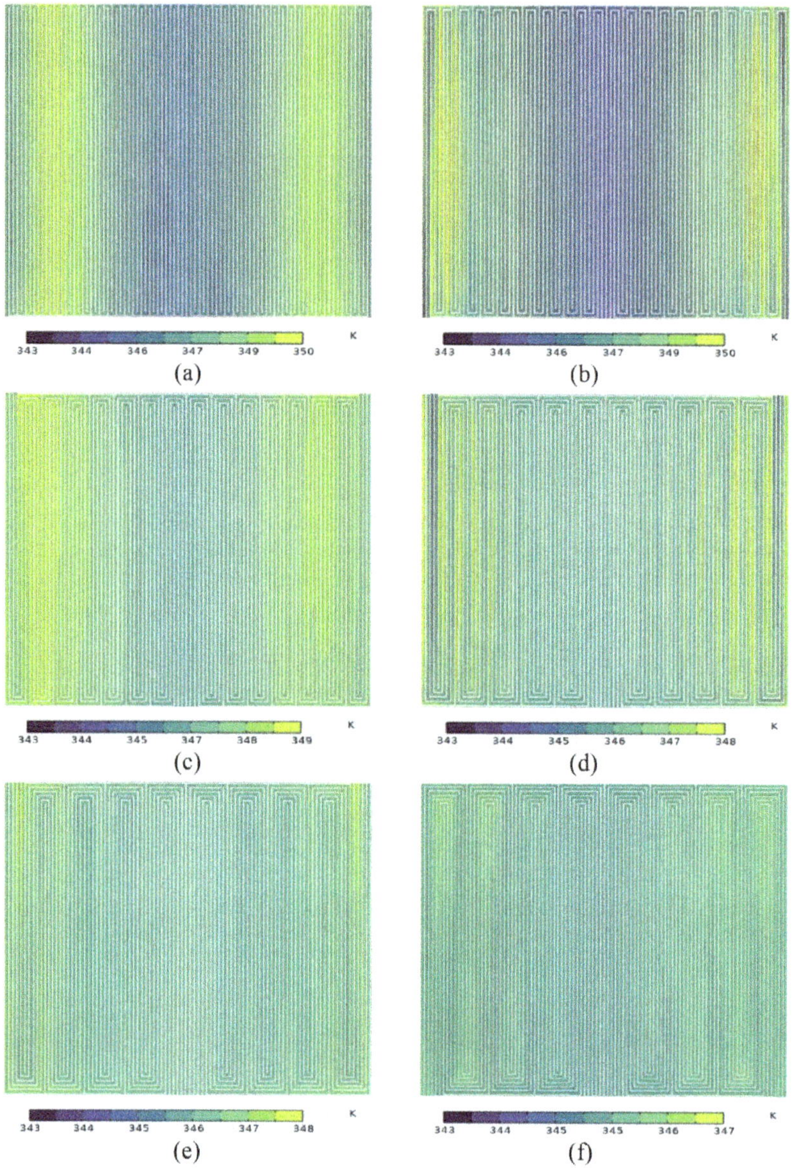

Figure 8. Temperature distribution in anode coolant channels—(**a**) two-path SFF, (**b**) four-path SFF, (**c**) six-path SFF, (**d**) eight-path SFF, (**e**) ten-path SFF, (**f**) twelve-path SFF.

Figure 9 represents the average temperature distribution at the membrane midplane, anode coolant channel, and cathode coolant channels of various models. The highest average temperature was observed in the two-path CFF, whereas the lowest average temperature was observed in the 12-path FF. Further, a consistent temperature differential of nearly 1 K was measured between the anode- and cathode-side CFFs across all tested models, with the cathode side exhibiting elevated temperatures compared to the anode side. This thermal asymmetry can be linked to the distinct electrochemical reactions occurring at each electrode, in which exothermic processes at the cathode dominate, while the anode experiences comparatively lower levels of heat generation. The primary source of heat generation in PEMFCs is the oxygen reduction reaction (ORR), which releases heat as the oxygen molecules are reduced to water. Further, the average temperature difference of the membrane midplane of the models is interesting: the temperature difference between two-, four-, and six-path FFs is less than one degree, whereas as the number of channels increases from six to twelve, the temperature of the membrane midplane drops with a difference of more than one degree, which shows the 10- and 12-path FFs have a better thermal performance.

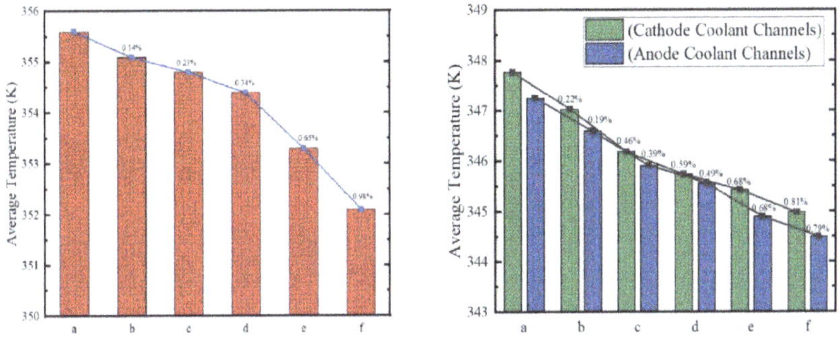

Figure 9. (**A**)Average temperature distribution at membrane midplane, (**B**) average temperature distribution at cathode and anode coolant channel—(a) two-path SFF, (b) four-path SFF, (c) six-path SFF, (d) eight-path SFF, (e) ten-path SFF, (f) twelve-path SFF.

The maximum temperature measurements also serve as a crucial indicator of the inefficiency or inadequacy of a flow channel design. It is evident in Figure 10 that the two-path CFF exhibits the maximum temperature, reaching approximately 350 K at the cathode-side channels and around 349 K at the anode-side channels. In contrast, the 12-path CFF demonstrates the lowest maximum temperature. This temperature disparity can be attributed to several factors inherent to the channel design. The two-path CFF exhibits the highest temperature because it has fewer paths, resulting in a longer coolant residence time in each channel. Longer channels with more serpentine curves and turns allow the coolant to remain inside the channels longer. Consequently, this extended flow time allows the coolant to absorb more heat from the FC system, resulting in a higher temperature near turns. Additionally, the longer path means that the fluid has more time to equilibrate with the temperature of the FC, leading to greater thermal transfer during the fluid's passage through the system.

However, a lower maximum temperature was observed in the 12-path CFF. This does not indicate a failure in heat extraction, but it is due to the balanced heat distribution inside the cell. It has also been discussed above that improved thermal management was achieved by the twelve-path CFF, followed by the eight-path and six-path CFF. The twelve-path configuration resulted in more balanced heat exchange, ensuring that the coolant can distribute the extracted heat more evenly across the system. The eight-, ten-, and twelve-path CFFs allow for more temperature uniformity, leading to more balanced thermal profiles, ensuring that the system operates within an optimal temperature range. It is crucial to emphasize that the role of the CFFs in PEMFC applications is not to extract the maximum possible heat from the system, but to keep the FC temperature within an appropriate operational range, typically between 70 °C and 80 °C. As highlighted in the literature, temperature uniformity is just as important, if not more so, than heat extraction efficiency. Maintaining a stable thermal distribution ensures that the FC operates efficiently, minimizes the risk of thermal stresses on the components, and contributes to the longevity and reliability of the system. Therefore, the 12-path CFF in this study offers the best overall thermal performance not necessarily by extracting more heat, but by ensuring an optimal heat distribution and maintaining the system within its ideal temperature range.

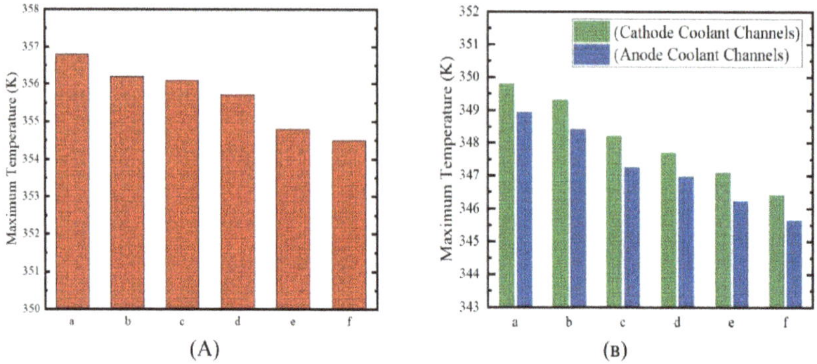

Figure 10. (**A**) Maximum temperature distribution at membrane midplane, (**B**) maximum temperature distribution at cathode and anode coolant channel—(a) two-path SFF, (b) four-path SFF, (c) six-path SFF, (d) eight-path SFF, (e) ten-path SFF, (f) twelve-path SFF.

The temperature distribution in the membrane is the most critical issue as it directly influences the performance and durability of the FC. The maximum temperature in the membrane midplane refers to the highest recorded temperature in the membrane midplane. This is an important parameter that shows the localized hotspots in the membrane. Figure 10 reflects the maximum temperature at the membrane midplanes of various models. The maximum temperature in this region is offered by the two-path FF, followed by four and six-path configurations. These FFs reflect the most critical behavior at the membrane midplane. This critical thermal behavior greatly influences the current density in the membrane midplane. Moreover, this thermal pattern also affects the temperature uniformity index in these models. The 12-path geometry depicts the appropriate maximum temperature patterns, followed by the ten- and eight-path geometries.

3.2. Temperature Uniformity Index

The IUT quantifies the degree of temperature variation across the FC, which directly impacts several performance aspects. Areas with excessive heat (hot spots) may experience an overconsumption of hydrogen, while cooler regions may face fuel starvation. A high IUT indicates that these imbalances are present, leading to inefficient energy production and a reduction in performance. Monitoring the IUT helps to maintain stable operating conditions and avoid thermal imbalance during operation. Similarly the IUTs of membrane midplanes in various models were calculated and are reflected in Figure 11. The highest temperature uniformity was achieved with the twelve-path CFF geometry, followed by the ten-path and eight-path geometries. Overall, in all the models, the IUT was noted as being in the acceptable range, reflecting the proper thermal distribution capability of MPSFF. This is because the cooling fluid starts the cooling process at the center of the FC, which is crucial. After all, this area often experiences higher temperatures due to the concentration of electrochemical reactions. This geometrical arrangement ensures that the most critical regions receive adequate thermal management by initiating cooling at the center. Similarly, as the cooling fluid reaches the outlets towards the edges, it enhances the reaction rates in areas with high hydrogen and oxygen mass ratios. This is because optimal temperatures facilitate faster electrochemical reactions, leading to the increased efficiency and performance of the FC.

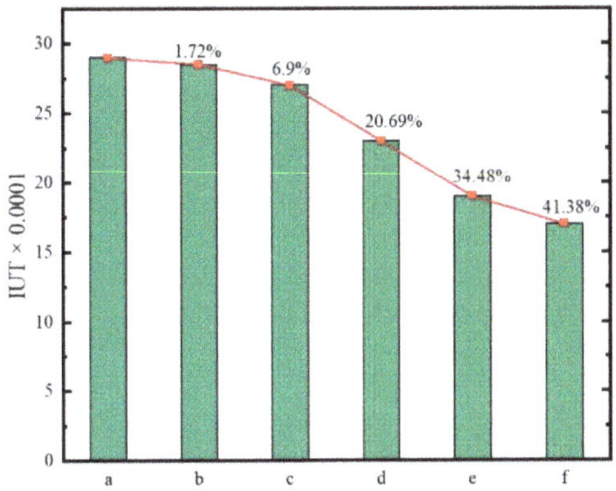

Figure 11. Index of uniform temperature at membrane midplane—(a) two-path SFF, (b) four-path SFF, (c) six-path SFF, (d) eight-path SFF, (e) ten-path SFF, (f) twelve-path SFF.

This approach promotes efficient thermal management by ensuring that the regions with higher reaction rates (typically where the hydrogen and oxygen mass ratios are the greatest) are adequately cooled. By optimizing cooling in these areas, reaction rates increase, leading to better fuel utilization and improved performance. As the chemical species are consumed more effectively in the well-cooled regions, the temperature becomes more uniform across the FC. This uniformity results from the more balanced reaction rates throughout the FC, where the energy input and reaction output are harmonized. With a more even temperature distribution, less cooling is required in regions further from the inlet, as these areas operate at more stable, lower temperatures. Consequently, this efficient cooling strategy not only optimizes the overall FC performance but also reduces the cooling energy consumption, making the system more energy-efficient and enhancing its operational stability over time.

3.3. *Hydrogen Concentrations*

The concentration of hydrogen in the membrane midplane plays a crucial role in examining the overall performance of PEMFCs. It refers to how evenly the hydrogen is distributed across the membrane. Since the membrane is responsible for transporting the protons from the anode to the cathode, a uniform hydrogen distribution is key for the uniform transportation of protons across the membrane. This critical factor influences multiple aspects of the FC, including reaction kinetics, proton conductivity, membrane hydration, temperature distribution, and overall efficiency. Areas with excessive hydrogen result in oversaturation.

In contrast, areas with a lower hydrogen concentration would suffer from the underutilization of the electrochemical sites; both factors contribute to reducing the overall current density and power output. All these phenomena are greatly affected by the uneven temperature distribution, as hotter regions would cause a faster hydrogen flow, while cooler regions may result in fuel starvation. Further, poor cooling would cause water evaporation (drying out the membrane), whereas excessive cooling would cause water accumulation (flooding). Therefore, effective thermal management not only prevents fuel starvation and overconsumption but also keeps the membrane in an optimally hydrated state, which helps to safeguard it from both drying out and flooding.

The hydrogen distribution at the membrane midplane is depicted in Figure 12. It could be observed that the hydrogen concentration is highest at the inlet regions (e.g., edges or entry zones of the channels), consistent with the depletion of the reactant as the hydrogen flows toward the center. This concentration reduction toward the FC center is caused by the interplay of flow dynamics, reactant consumption, and electrochemical reaction distribution, as different patterns were observed in different models. A clear relation between temperature and hydrogen concentration was observed. The variation in concentration patterns across different models suggests that the geometry of CFFs plays a significant role in the hydrogen distribution. The clear relationship between temperature and hydrogen concentration likely indicates that higher temperatures result in a reduced hydrogen concentration gradient. This phenomenon resulted from the enhanced hydrogen diffusion and accelerated reaction kinetics at elevated temperatures, promoting a more uniform hydrogen distribution across the membrane.

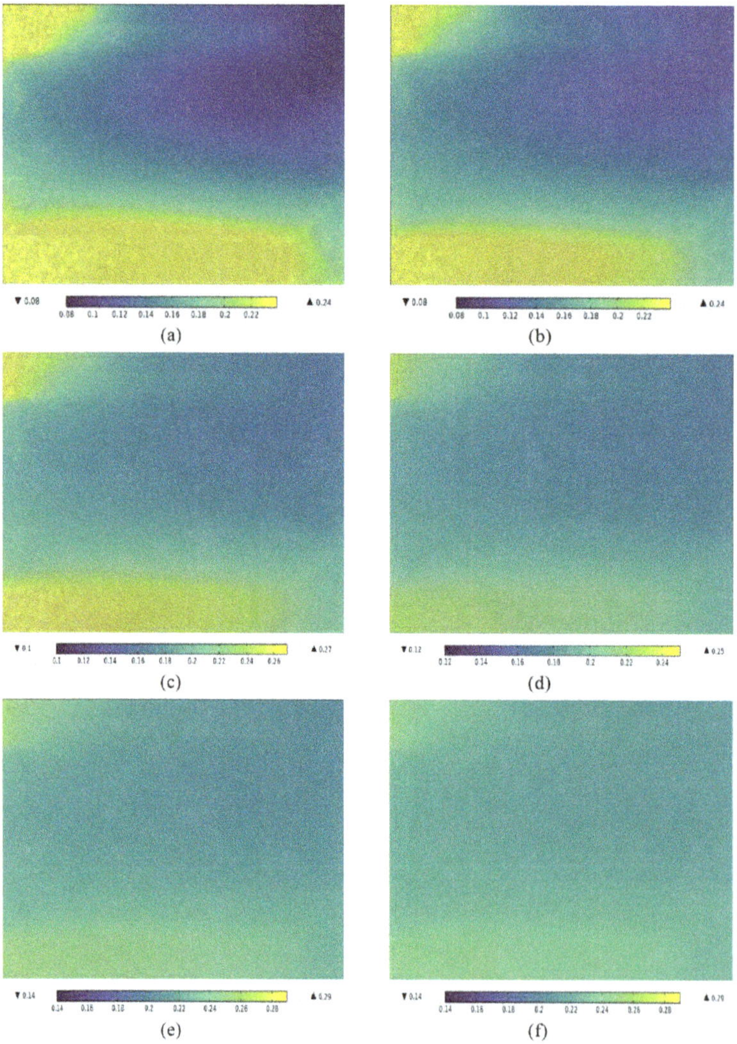

Figure 12. Hydrogen concentration at membrane midplane—(**a**) two-path SFF, (**b**) four-path SFF, (**c**) six-path SFF, (**d**) eight-path SFF, (**e**) ten-path SFF, (**f**) twelve-path SFF.

3.4. Pressure Drop

In serpentine coolant flow channels for PEMFC applications, a pressure drop is an important factor that influences the overall performance of the FC, as the serpentine channels have a series of serpentine turns that offer remarkable resistance to the fluid. The length of the SFF also plays a critical role in pressure reduction. Longer channels result in a greater level of resistance to the flow and, thus, a greater pressure drop. There are many other factors affecting the pressure, but in this study, the effect of channel geometry was considered as a prime factor. Considering that the pressure drop is also influenced by the number of paths in the serpentine design, various models with different numbers of paths were produced. As it is also the fact that having fewer number of passes, increases the length of the channel, leading to a higher resistance to fluid. Therefore, this study has evaluated various models with different numbers of passes to investigate the impact of varying the number of channels.

The total pressure drop in the coolant flow field is reflected in Figure 13. The pressure drops in CFFs follows the normal trend of SFFs and significant reduction in the pressure drop was observed with increasing number of channels. This optimization in the pressure drop could be attributed to the reduced length and serpentine turns of channels, as increasing the number of channels results in a reduced length and fewer serpentine turns of channels, ultimately minimizing the residence period of the coolant inside the CFF. This shorter residence time of the coolant inside the CFF facilitates the extraction of a balanced amount of heat from the cell, ultimately increasing the thermal uniformity. Similarly, the maximum pressure drop was observed in the FF with the maximum length and the highest number of serpentine turns. In contrast, the minimum pressure drop was reported in the FF with the fewest serpentine turns and the minimum length. Therefore, this increased pressure drop results in the increased residence of the coolant inside the CFF, which also ultimately increases the heat extraction by the coolant near turns, causing a non-uniform temperature distribution, as seen in Figures 7a–c and 8a–c. Evaluating the overall pressure drop in the CFF, the 12-path CFF offered the lowest pressure drop of the fluid.

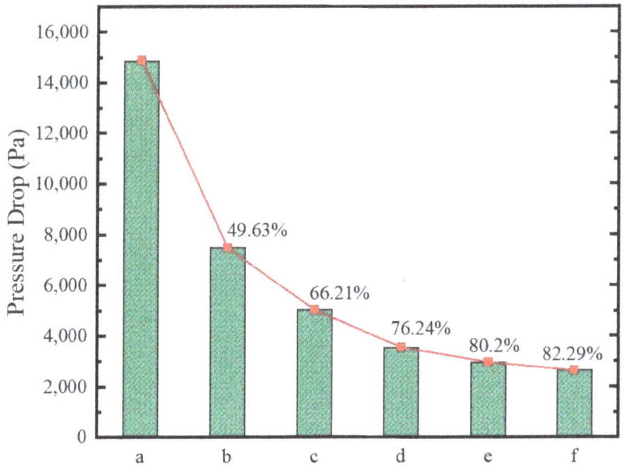

Figure 13. Total pressure drops in coolant flow channels—(a) two-path SFF, (b) four-path SFF, (c) six-path SFF, (d) eight-path SFF, (e) ten-path SFF, (f) twelve-path SFF.

The key mechanisms influencing the thermal and flow behavior in PEMFCs with MPSFF are the convective heat transfer, pressure drop, and the distribution of the flow within the channels. These factors interact in complex ways to influence the temperature distribution, coolant flow dynamics, and ultimately, the performance of the fuel cell.

3.5. Key Mechanisms Influencing Thermal and Flow Behaviors

3.5.1. Convective Heat Transfer

The increase in the number of channels generally improves the convective heat transfer by increasing the surface area available for heat exchange. This effect is particularly important for larger active areas, as a higher channel count leads to a more efficient thermal distribution, helping to avoid the formation of localized hot spots. The path length also plays a significant role, as longer channels promote a more uniform heat distribution, which reduces the thermal gradient across the membrane.

3.5.2. Flow Distribution and Pressure Drop

Path length and channel geometry impact the flow distribution, with longer channels typically resulting in a more uniform flow, while shorter paths may lead to higher velocity fluctuations and potential areas of stagnant flow. This has a direct effect on heat dissipation, as uneven flow can cause localized heating or a reduced cooling efficiency. Additionally, the pressure drop in the flow field increases with longer channels, which can affect the overall fuel cell efficiency by requiring greater pumping power.

3.5.3. Membrane Water Management

The temperature distribution across the membrane is also closely tied to membrane hydration levels, which affect ionic conductivity. As the coolant flow is adjusted, both the channel count and path length can influence water removal and hydration at different locations within the membrane. Effective temperature management, therefore, not only minimizes thermal stress but also helps in maintaining optimal hydration levels for ionic conduction.

4. Conclusions and Future Recommendations

This study investigated the impact of MPSFF on the thermal management and pressure drop of PEMFCs. It was concluded that introducing a coolant at the center led to effective thermal management, preventing hotspots, maintaining optimal temperatures, and ensuring peak reaction rates. As the reaction concentration is usually high at the center, effective cooling in high-reactant-concentration regions prevents thermal inhibition, stabilizing temperatures to maintain efficient charge transfer, reactant utilization, and an enhanced power output.

This study systematically analyzed the thermal behavior of coolant channels with two- to twelve-paths, while maintaining a constant channel width of 1.5 mm. Models with two-to-six-path CFFs demonstrated notable high-temperature hotspots, primarily caused by the extended length of the cooling channels and the excessive number of serpentine bends. However, models with eight-to-twelve-path CFFs had fewer hotspots and a uniform temperature distribution. At the membrane midplane, the highest average temperature was observed for the model with the two-path CFF, having an average membrane temperature around 356 K. In contrast, the model with 12-path CFFs reflected the 0.98% lowest average temperature (352.1) in the region. Similarly, a temperature difference of up to 1 K was observed between the anode and cathode CFFs, which could be due to the ORR on the cathode side. The highest maximum temperature was also observed in two-path CFFs, which could be attributed to the extended coolant residence time inside the FF due to longer channels with the maximum number of serpentine turns. In contrast, the value of the maximum temperature observed in the 12-path CFF was 0.81% lower, which outperformed all other CFF configurations, which could be due to the same reason that the maximum temperature in the two-path CFF is elevated.

The study concluded that the minimum number of coolant channels (i.e., two paths) causes longer channel lengths and a greater number of serpentine turns, resulting in the maximum number of hot spots around turns, whereas maximizing the number of channels (12 paths) results in the shorter length of the channels, causing a uniform thermal distribution. In addition, the maximum uniformity of the thermal distribution was reflected in 12-path CFFs with the lowest average temperature of around 352.3 K at the membrane midplane. However, this investigation is constrained by fixed channel dimensions, static operational parameters, and limited material variations. To advance this research, future work should incorporate dynamic operating conditions, such as modulated coolant flow rates, and evaluate advanced cooling media, e.g., nanofluids. Additionally, conventional CFD tools face computational limitations in exhaustively exploring geometric design spaces. Thus, a data-driven approach leveraging machine learning models, trained on high-fidelity CFD datasets, is proposed to efficiently predict optimized CFF architectures.

Building on the current findings, the optimized CFF configuration warrants further investigation through parametric variations in channel width and height, guided by a systematic evaluation of hydraulic diameter and aspect ratio. Such an approach could refine the trade-off between thermal uniformity and pressure drops. A future work will specifically examine width-adjusted channel designs derived from the present optimized geometry, targeting enhanced pressure loss mitigation. This methodology holds significant potential to redefine thermal–fluid performance boundaries in MPSFFs, enabling next-generation designs that harmonize thermal regulation and hydrodynamic efficiency.

Author Contributions: Conceptualization, Q.L., X.Y., H.H., Q.Y., Z.A., M.A. (Muhammad Ahmed) and S.X.; Methodology, Q.L., H.H., J.C., K.M., Q.Y., J.L., W.A. and M.A. (Muhammad Aurangzeb); Software, Q.L., X.Y., J.C., W.A., Z.A. and M.A. (Muhammad Aurangzeb); Validation, Q.L., H.H., K.M., J.L., Z.A. and M.A. (Muhammad Aurangzeb); Formal analysis, Q.L., X.Y., J.C., Q.Y., W.A., Z.A., M.A. (Muhammad Aurangzeb) and M.A. (Muhammad Ahmed); Investigation, Q.L., X.Y., H.H., K.M., J.L., W.A., M.A. (Muhammad Aurangzeb) and S.X.; Resources, M.A. (Muhammad Ahmed) and S.X.; Data curation, Q.L. and J.L.; Writing—original draft, Q.L., X.Y., H.H., J.C., K.M., Q.Y., J.L., W.A., M.A. (Muhammad Aurangzeb), M.A. (Muhammad Ahmed) and S.X.; Writing—review & editing, Q.L., X.Y., H.H., J.C., K.M., Q.Y., J.L., W.A., Z.A., M.A. (Muhammad Aurangzeb), M.A. (Muhammad Ahmed) and S.X.; Visualization, J.C., K.M., Q.Y. and M.A. (Muhammad Aurangzeb); Supervision, M.A. (Muhammad Ahmed) and S.X.; Project administration, M.A. (Muhammad Ahmed) and S.X.; Funding acquisition, M.A. (Muhammad Ahmed) and S.X. All authors have read and agreed to the published version of the manuscript.

Funding: This research was funded by Zhejiang Province Spearhead and Leading Goose Research and Development Key Program "Research and Development of Intellectualized Key Technologies for Thermal Management System Components of New Energy Vehicles–Research and Development of Intellectualized Key Technologies for Thermal Management System Compressors of New Energy Vehicles" (grant no. 2024C01229) and the APC was funded by Longquan Industrial Innovation Research Institute, Longquan 323050, China.

Data Availability Statement: The original contributions presented in the study are included in the article, further inquiries can be directed to the corresponding author.

Acknowledgments: This work was supported by National Key Research and Development Program Project (grant no. 2022YFB2502400; "High-power density fuel cell stack and engine technology for passenger vehicles") and Zhejiang Province's "Spearhead and Leading Goose + X" Research and Development Breakthrough Plan (grant no. 2023C01239; "Ultra-long endurance hydrogen-electric hybrid drone").

Conflicts of Interest: Authors Qingsheng Liu, Kai Meng were employed by the company Jiashan Power Supply Company, State Grid Zhejiang Electric Power Co., Ltd. The remaining authors declare that the research was conducted in the absence of any commercial or financial relationships that could be construed as a potential conflict of interest.

Nomenclature

Abbreviations	
PEMFC	Proton-Exchange Membrane Fuel Cell
CC	Coolant Channel
CFFs	Coolant Flow Fields
UNEP	United Nations Environment Program
WHO	World Health Organization
EV	Electric Vehicle
PEM	Proton-Exchange Membrane
FCEV	Fuel Cell Electric Vehicle
FF	Flow Field
MPSFF	Multi-Path Serpentine Flow Field
FC	Fuel Cell
SFF	Serpentine Flow Field
BiCGStab	Biconjugate Gradient Stabilization
RAM	Random Access Memory
ROM	Read-Only Memory
IUT	Index of Uniform Temperature
GDL	Gas Diffusion Layer
CL	Catalyst Layer
CFD	Computational Fluid Dynamics
ORR	Oxygen Reduction Reaction
Greek Letters	
ϵ	Porosity (typically used for GDL or CL)
ρ	Density
μ	Dynamic Viscosity
σ	Electrical or Thermal Conductivity
α	Permeability (of the medium)
Φ	Potential (usually electric potential)
Symbols	
T	Temperature (Local or Average)
P	Pressure
I	Current Density
R	Resistance
V	Voltage
F	Faraday's Constant
M	Molar Mass
t	Time
c	Concentration
Q	Heat or Heat Transfer

References

1. Department of Economic and Social Affairs-Population Division. *2024 Revision of World Population Prospects*; United Nations: New York, NY, USA, 2024. Available online: https://population.un.org/wpp/ (accessed on 23 May 2025).
2. Ahmed, M.; Shusheng, X. Comprehensive review of serpentine flow fields for effective thermal management of proton exchange membrane fuel cel. *Heliyon* **2024**, *10*, e39793. [CrossRef] [PubMed]
3. Afshari, E.; Ziaei-Rad, M.; Jahantigh, N. Analytical and numerical study on cooling flow field designs performance of PEM fuel cell with variable heat flux. *Mod. Phys. Lett. B* **2016**, *30*, 1650155. [CrossRef]

4. Ajanovic, A.; Haas, R. Economic prospects and policy framework for hydrogen as fuel in the transport sector. *Energy Policy* **2018**, *123*, 280–288. [CrossRef]
5. Jie, W.; Cunguang, F. Research on Hydrogen Fuel Cell Controllers Based on Thermal Management. *Intern. Combust. Engine Accessories* **2023**, *23*, 27–29. [CrossRef]
6. Teimouri, A.; Kabeh, K.Z.; Changizian, S.; Ahmadi, P.; Mortazavi, M. Comparative lifecycle assessment of hydrogen fuel cell, electric, CNG, and gasoline-powered vehicles under real driving conditions. *Int. J. Hydrogen Energy* **2022**, *47*, 37990–38002. [CrossRef]
7. Qasem, N.A.A. A recent overview of proton exchange membrane fuel cells: Fundamentals, applications, and advances. *Appl. Therm. Eng.* **2024**, *252*, 123746. [CrossRef]
8. Zucconi, A.; Hack, J.; Stocker, R.; Suter, T.A.M.; Rettie, A.J.E.; Brett, D.J.L. Challenges and opportunities for characterisation of high-temperature polymer electrolyte membrane fuel cells: A review. *J. Mater. Chem. A* **2024**, *12*, 8014–8064. [CrossRef]
9. Chen, Q.; Zhang, G.; Zhang, X.; Sun, C.; Jiao, K.; Wang, Y. Thermal management of polymer electrolyte membrane fuel cells: A review of cooling methods, material properties, and durability. *Appl. Energy* **2021**, *286*, 116496. [CrossRef]
10. Rahgoshay, S.M.; Ranjbar, A.A.; Ramiar, A.; Alizadeh, E. Thermal investigation of a PEM fuel cell with cooling flow field. *Energy* **2017**, *134*, 61–73. [CrossRef]
11. Bao, C.; Ouyang, M.; Yi, B. Analysis of the water and thermal management in proton exchange membrane fuel cell systems. *Int. J. Hydrogen Energy* **2006**, *31*, 1040–1057. [CrossRef]
12. Jeo, D.H. Numerical study of serpentine flow-field cooling plates on PEM fuel cells performance. *Int. J. Energy Res.* **2013**, *37*, 510–521. [CrossRef]
13. Choi, E.J.; Park, J.Y.; Kim, M.S. A comparison of temperature distribution in PEMFC with single-phase water cooling and two-phase HFE-7100 cooling methods by numerical study. *Int. J. Hydrogen Energy* **2018**, *43*, 13406–13419. [CrossRef]
14. Gong, D.; Xu, S.; Gao, Y. Investigation of Water and Heat Transfer Mechanism in PEMFCs Based on a Two-Phase Non-Isothermal Model. *Energies* **2023**, *16*, 697. [CrossRef]
15. Nolan, J.; Kolodziej, J. Modeling of an automotive fuel cell thermal system. *J. Power Sources* **2010**, *195*, 4743–4752. [CrossRef]
16. Wang, J.; Wang, H.; Fan, Y. Techno-Economic Challenges of Fuel Cell Commercialization. *Engineering* **2018**, *4*, 352–360. [CrossRef]
17. Zhang, G.; Jiao, K. Multi-phase models for water and thermal management of proton exchange membrane fuel cell: A review. *J. Power Sources* **2018**, *391*, 120–133. [CrossRef]
18. Li, Q.; Liu, Z.; Sun, Y.; Yang, S.; Deng, C. A Review on Temperature Control of Proton Exchange Membrane Fuel Cells. *Processes* **2021**, *9*, 235. [CrossRef]
19. Veeranjaneyulu, K.; Joshi, S.; Devalla, V.; Kiran, K.S.; Khushal. Recent advancements of PEMFC in transport applications. *AIP Conf. Proc.* **2023**, *2492*, 050019. [CrossRef]
20. Deng, Q.; Meng, K.; Chen, W.; Yang, G.; Zhang, N.; Chen, B. Study on the heat and mass transfer mechanisms of liquid-cooled PEMFC stacks based on non-isothermal model. *Energy Convers. Manag.* **2024**, *313*, 118628. [CrossRef]
21. Carrère, P.; Prat, M. Impact of non-uniform wettability in the condensation and condensation-liquid water intrusion regimes in the cathode gas diffusion layer of proton exchange membrane fuel cell. *Int. J. Therm. Sci.* **2019**, *145*, 106045. [CrossRef]
22. Jeon, D.H.; Greenway, S.; Shimpalee, S.; Zee, J.W.V. The effect of serpentine flow-field designs on PEM fuel cell performance. *Int. J. Hydrogen Energy* **2008**, *33*, 1052–1066. [CrossRef]
23. Zhang, G.; Guo, L.; Ma, B.; Liu, H. Comparison of current distributions in proton exchange membrane fuel cells with interdigitated and serpentine flow fields. *J. Power Sources* **2009**, *188*, 213–219. [CrossRef]
24. Gu, Y.; Li, L.; Zheng, G. Study on the Dynamic Characteristics of the Gear Lubrication Flow Field with Baffles and Optimization Design Strategies. *Lubricants* **2025**, *13*, 143. [CrossRef]
25. Tan, Y.; Ni, Y.; Xu, W.; Xie, Y.; Li, L.; Tan, D. Key technologies and development trends of the soft abrasive flow finishing method. *J. Zhejiang Univ.-Sci. A* **2023**, *24*, 1043–1064. [CrossRef]
26. Wang, Y.; Sun, Z.Y.; Yang, L. Optimizing temperature distribution in a PEMFC stack: A computational study on cooling plate and coolant dynamics. *Int. J. Hydrogen Energy* **2025**, *97*, 88–103. [CrossRef]
27. Xu, X.; Zhang, L.; Wang, S.; Han, D.; You, S.; Zhou, J. Numerical and experimental analyses of a novel type PEMFC coolant channel. *Int. J. Hydrogen Energy* **2024**, *49*, 652–673. [CrossRef]
28. Liu, X.; Bai, M.; Zhou, Z.; Poramapojana, P.; Li, Y.; Gao, L.; Li, Y.; Song, Y. Three-dimensional multi-phase numerical study for the effect of coolant flow field designs on water and thermal management for the large-scale PEMFCs. *Int. J. Hydrogen Energy* **2023**, *48*, 23681–23705. [CrossRef]
29. Chen, B.; Deng, Q.; Yang, G.; Zhou, Y.; Chen, W.; Cai, Y.; Tu, Z. Numerical study on heat transfer characteristics and performance evaluation of PEMFC based on multiphase electrochemical model coupled with cooling channel. *Energy* **2023**, *285*, 128933. [CrossRef]
30. Yong, Z.; Shirong, H.; Xiaohui, J.; Yuntao, Y.; Mu, X.; Xi, Y. Performance study on a large-scale proton exchange membrane fuel cell with cooling. *Int. J. Hydrogen Energy* **2022**, *47*, 10381–10394. [CrossRef]

31. Baek, S.M.; Yu, S.H.; Nam, J.H.; Kim, C.-J. A numerical study on uniform cooling of large-scale PEMFCs with different coolant flow field designs. *Appl. Therm. Eng.* **2011**, *31*, 1427–1434. [CrossRef]
32. Amarnath, G.; Babu, A.V. Comparative computational fluid dynamic analysis between split and dual serpentine flow field for proton exchange membrane fuel cells. *Chem. Pap.* **2023**, *77*, 5033–5045. [CrossRef]
33. Lu, H.; Pei, X.; Yan, F.; Yao, J.; Feng, S. Numerical Analysis of the Effect of a Radial Serpentine Flow Field on PEMFC Performance. *J. Energy Eng.* **2025**, *151*, 04024046. [CrossRef]
34. Toghyani, S.; Nafchi, F.M.; Afshari, E.; Hasanpour, K.; Baniasadi, E.; Atyabi, S.A. Thermal and electrochemical performance analysis of a proton exchange membrane fuel cell under assembly pressure on gas diffusion layer. *Int. J. Hydrogen Energy* **2018**, *43*, 4534–4545. [CrossRef]
35. Joibary, S.M.M.; Rahgoshay, S.M.; Rahimi-Esbo, M.; Firouzjaee, K.D. Numerical investigation of the influence of different cooling flow channels on the thermal and water saturation distribution in a real dimensional polymer electrolyte membrane fuel cell. *Int. J. Hydrogen Energy* **2023**, *48*, 2762–2787. [CrossRef]

energies

MDPI

Article

Enhancing the Adsorption Performance of HKUST-1 by Adding NH_4F During Room-Temperature Synthesis for Desulfurization of Fuel Oil

Jiawei Fu [1], Xinchun Liu [1], Yuqing Kong [1], Ruyu Zhao [1], Yinyong Sun [1,*] and Ahmed S. Abou-Elyazed [2]

1 MIIT Key Laboratory of Critical Materials Technology for New Energy Conversion and Storage, School of Chemistry and Chemical Engineering, Harbin Institute of Technology, Harbin 150001, China
2 Institute of Intelligent Manufacturing Technology, Shenzhen Polytechnic University, Shenzhen 518055, China
* Correspondence: yysun@hit.edu.cn

Abstract

Adsorption desulfurization of fuel oil is regarded as one of the most promising technologies for obtaining clean fuel because it can remove refractory sulfur compounds at ambient temperature and pressure. Studies indicate that HKUST-1, as an important type of metal–organic framework (MOF), is a potential candidate for adsorption desulfurization of fuel oil. In this work, we report that defective HKUST-1 can be rapidly synthesized at room temperature with the aid of NH_4F and exhibit superior adsorption desulfurization performance compared to conventional HKUST-1 by the solvothermal method. Moreover, the influence of adsorption parameters on the desulfurization performance of HKUST-1 prepared with the aid of NH_4F was investigated. We used 50 mg of HKUST-1-5 synthesized with 5 wt% added NH_4F to adsorb 5 g of model oil with a sulfur concentration of 1000 ppm at 25 °C for 1 h, and the adsorption capacity of the adsorbent reached 23.8 mgS/g, 46.8 mgS/g and 36.8 mgS/g for benzothiophene (BT), dibenzothiophene (DBT) and 4,6-dimethyldibenzothiophene (4,6-DMDBT), respectively, which are higher values than those of conventional HKUST-1. Such performance can be mainly attributed to its relatively small particle size and the presence of more unsaturated Cu sites. The results of regeneration experiments show that HKUST-1-5 still maintains excellent adsorption performance after four cycles. These findings highlight the great potential of this material as an efficient adsorbent for adsorption desulfurization of fuel oil.

Keywords: MOFs; room-temperature synthesis; adsorption desulfurization; HKUST-1; dibenzothiophene

1. Introduction

With the acceleration of industrial development, the consumption of fossil fuels is on the rise. In the process of fuel processing, sulfur compounds in fuel oil would cause not only corrosion of equipment and poison the catalyst used in the reaction [1,2], but also serious environmental problems and damage to the respiratory system due to the combustion of the subsequent processed products [3]. Therefore, it is important for protecting the environment and human health to develop relevant technologies to realize the deep removal of sulfur compounds in fuel oil.

Currently, the main desulfurization technology in industry is hydrodesulfurization (HDS) [4], which needs higher operating temperature (300–400 °C), pressure (20–60 bar) and expensive hydrogen. Comparatively, adsorption desulfurization technology without hydrogen consumption is simple to operate and can remove refractory sulfur compounds that are difficult to desulfurize by HDS technology [5]. It is regarded as one of the most promising technologies to obtain ultra-low-sulfur fuel. Therefore, the development of highly efficient adsorbents has been of great interest.

Metal–organic frameworks (MOFs), with advantages such as large specific surface area, adjustable pore structure and easy functionalization, are considered as a kind of ideal adsorbent for adsorption desulfurization of fuel oil [6–10]. Matzger et al. [11] reported for the first time that MOFs as adsorbents were utilized in adsorption desulfurization of model fuel oil. It was found that the adsorption capacity of MOFs was higher than that of zeolites. Achmann et al. [12] studied four kinds of MOFs for the adsorption desulfurization of different types of real fuel oils. The results indicated that HKUST-1 had highest adsorption capacity among these MOFs. Peralta et al. [13] compared the selectivity and reproducibility of HKUST-1 and CPO-27 in adsorption desulfurization. It was noted that HKUST-1 was more selective and reproducible than CPO-27. In addition, Duan et al. [14] synthesized several HKUST-1 materials by various methods. The results showed that HKUST-1 prepared by the hydrothermal method possessed the largest adsorption capacity for the removal of thiophene.

In this contribution, defective HKUST-1 was rapidly synthesized with the addition of NH_4F at room temperature and exhibited superior adsorption desulfurization performance. The obtained materials were characterized by various techniques. Their adsorption desulfurization performance was evaluated by the adsorption of sulfur-containing compounds like dibenzothiophene (DBT) in model fuel oil. Based on the experimental results, a possible adsorption mechanism was proposed.

2. Materials and Methods

2.1. Materials

N, N-Dimethylformamide (DMF, >99.5%), ethanol (>99.7%), methanol (>99%), copper nitrate trihydrate ($Cu(NO_3)_2 \cdot 3H_2O$, >99%) and ammonium fluoride (NH_4F, >99%) were obtained from China National Pharmaceutical Group Limited. *n*-Octane(99.5%), 1,3,5-benzenetricarboxylic acid (H_3BTC, 98%), benzothiophene (BT, 98%), dibenzothiophene (DBT, >98%) and 4,6-dimethyldibenzothiophene (4,6-DMDBT, 99%) were purchased from Beijing InnoChem Science & Technology Co., Ltd., China.

2.2. Solvothermal Synthesis of HKUST-1-S

HKUST-1-S was prepared by using previously established methodologies [15]: 1.00 g H_3BTC was dissolved in 30 mL mixed solution of ethanol/DMF with a volume ratio of 1:1 to form solution A, and 2.077g $Cu(NO_3)_2 \cdot 3H_2O$ was dissolved in 15 mL deionized water to form a transparent light blue solution, referred to as solution B. Under the condition of strong stirring, solution B was poured into solution A at room temperature and kept stirring for 10 min. The obtained solution was transferred into an autoclave with a PTFE lining and heated at 120 °C for 12 h. After that, the autoclave was naturally cooled to room temperature. The resultant product was washed with anhydrous ethanol at 60 °C for 2 h, dried at 70 °C for 3 h, and activated at 150 °C for 12 h under vacuum. The final product was named HKUST-1-S.

2.3. Room-Temperature Synthesis of HKUST-1-x with the Addition of NH_4F

1.00 g H_3BTC was dissolved in 30 mL mixed solution of ethanol/DMF with a volume ratio of 1:1 to form solution A, and 2.077 g of $Cu(NO_3)_2 \cdot 3H_2O$ was dissolved in 15 mL of NH_4F aqueous solution with different mass fractions (2.5%, 5%, 10%, 15%) to form a transparent light blue solution, referred to as solution B. Under strong stirring, solution

B was poured into solution A at room temperature and keep stirring for 30 min. The resultant sample was centrifugally separated and then washed with deionized water trice and methanol twice at room temperature. Suction filtration and rotary evaporation were sequentially carried out on the separated liquid to recover NH_4F. After drying in an oven at 70 °C for 3 h, the obtained sample was activated at 150 °C for 12 h under vacuum. The final product was named HKUST-1-x, where x represents the mass fraction of NH_4F. The mass fraction of ammonium fluoride ranged from 2.5% to 15%, and the synthesized HKUST-1-x was 1.69 g, 1.52 g, 1.54 g and 1.49 g in turn.

2.4. Batch Adsorption Test

Model fuel oil with different sulfur concentrations (the sulfur atomic mass fraction in the compound) was prepared by mixing different amounts of sulfur compounds such as BT, DBT and 4,6-DMDBT with *n*-octane, respectively. The adsorbent was degassed at 150 °C in vacuum for 2 h in advance. Then, 50 mg of adsorbent was quickly added to a flask containing 5 g of model fuel oil. The flask was placed in a water bath at 25 °C. A few samples of liquid were taken at different times and then analyzed by an Agilent 7890A gas chromatography (GC) apparatus. The adsorption capacity was calculated by the following formula:

$$Q_t = \frac{V(C_0 - C_t)}{m} \tag{1}$$

Q_t, V, C_0, C_t, and m represent the adsorption capacity at different times (mgS/g), the volume of model oil (mL), initial sulfur concentration (mgS/mL), sulfur concentration at different times (mgS/mL), and mass of adsorbent (g), respectively.

2.5. Cycle Test

The adsorbent was degassed at 150 °C in vacuum for 2 h in advance. Then, 100 mg of adsorbent was quickly added to a flask containing 10 g of model fuel oil. The flask was placed in a water bath at 25 °C. After the adsorption experiment, the simulated oil and adsorbent were separated by centrifugation. The resultant adsorbent was washed with anhydrous ethanol 3 times and dried at 70 °C for 3 h. Finally, the sample was activated at 150 °C under vacuum for 12 h for the next experiment.

2.6. Characterization

Powder X-ray diffraction (XRD) patterns were recorded on a Bruker D8 advance diffractometer in Bragg–Brentano geometry equipped with a Ge-focusing primary monochromator (Cu-Kα radiation, λ = 0.15406 nm) at 40 kV and 40 mA with a scanning speed of 9° min^{-1} and a step size of 0.02°. Scanning electron microscopy (SEM) images were obtained using a SUPRA 55 instrument with an acceleration voltage of 15 kV. Fourier transform infrared (FT-IR) spectra were collected in the range of 400–4000cm^{-1} on the Thermo Fisher Scientific Nicolet iS20 spectrometer. The N_2 adsorption–desorption isotherms were obtained at 77 K on a BSD-PS1 Static Volumetric Specific Surface and Aperture Analyzer. Before nitrogen adsorption, the samples were pretreated at 423 K for 2 h under vacuum to remove the impurities. The total surface area was calculated via the Brunauer–Emmett–Teller (BET) equation. An X-ray photoelectron spectrometer (XPS) was used to analyze the chemical composition of all the samples on the Thermo Scientific K-Alpha spectrometer. Thermogravimetric analysis (TGA) was performed on a HITACHI STA200 instrument under air atmosphere at a ramp rate of 10 °C/min. The NH_3-TPD test was carried out on a chemisorption instrument with the model Micromeritics AutochemIII2930.

3. Results and Discussion

3.1. Structural Characterization

The power XRD patterns of HKUST-1-x are shown in Figure 1a. As seen, all the samples showed well-resolved diffraction peaks, which are in good agreement with the simulated ones belonging to the crystal structure of HKUST-1. The diffraction peaks of 2θ = 7.2°, 9.4° and 11.8° correspond to (200), (220) and (222) crystal planes, respectively [16,17]. Obviously, the addition amount of NH_4F during synthesis had a great influence on the crystallinity of HKUST-1. Notably, when the mass fraction of NH_4F is less than or equal to 5%, the intensity of diffraction peaks is high, meaning that most of the crystal planes can be well observed. When the mass fraction of NH_4F is more than 5%, the intensity of diffraction peaks becomes weak. It is possible that excessive NH_4F would lead to a drop in the pH of the reaction system, which is harmful for the crystallization of products [18].

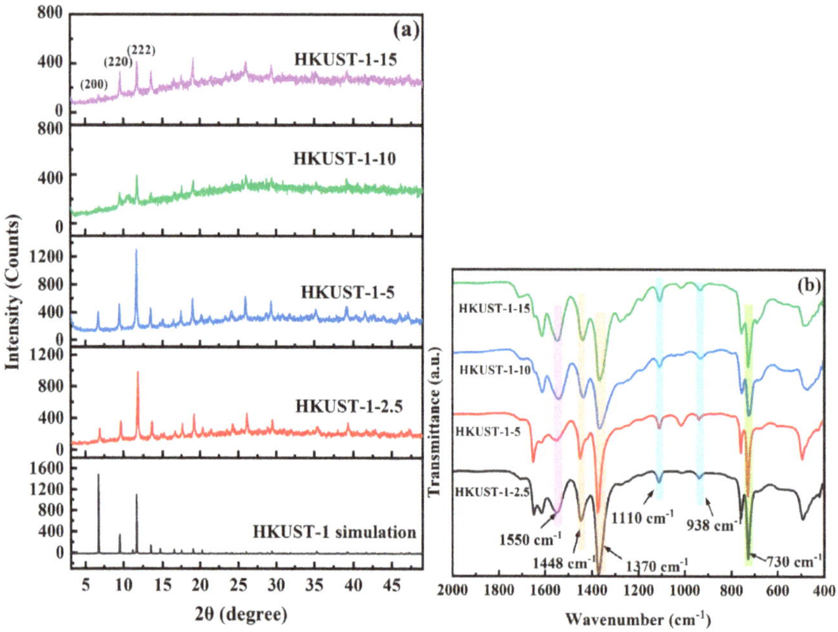

Figure 1. (**a**) XRD patterns; (**b**) FT-IR spectra of HKUST-1 prepared with different addition amounts of NH_4F.

Figure 1b shows the FT-IR spectra of HKUST-1-x. The absorption peak at 730 cm^{-1} corresponds to the C-H bending vibration of the benzene ring in trimellitic acid [19]. The absorption peak at 1550 cm^{-1} is associated with the asymmetric stretching vibration of -COO- in the structure [20]. Additionally, the absorption peaks at 1371 cm^{-1} and 1447 cm^{-1} are ascribed to the symmetric stretching vibration of the -COO- group [21]. The vibration of Cu-O absorption peaks at 1110 cm^{-1} and 938 cm^{-1} directly confirm the successful coordination of metal ions with ligands [22]. Based on the above results, it can be concluded that HKUST-1 has been rapidly synthesized at ambient temperature and pressure with the aid of NH_4F. Compared with the traditional solvothermal method, the NH_4F-assisted synthesis does not require a long heating and crystallization process. Under the same feeding conditions, the production efficiency of HKUST-1 by the solvothermal method and the NH4F-assisted synthesis method is 0.18 g/h and 3.04 g/h, respectively. Therefore, the NH4F-assisted method has lower energy consumption and cost.

Figure 2 shows the N_2 adsorption–desorption isotherms and corresponding pore size distribution curves. As seen in Figure 2a, all the samples exhibited typical type I isotherm, indicating the existence of micropores [23,24]. Their average micropore size is between 0.6 and 0.8 nm (Figure 2b). The detailed sorption data are shown in Table 1. Apparently, HKUST-1-5 gave the highest BET specific surface area (1507 m^2/g) among these studied materials, which is in agreement with the XRD results. We speculated that a low NH_4F concentration is not beneficial for the rapid crystallization of HKUST-1. If a high NH_4F concentration is used, the pH value of the system would decrease and thus result in the formation of amorphous product.

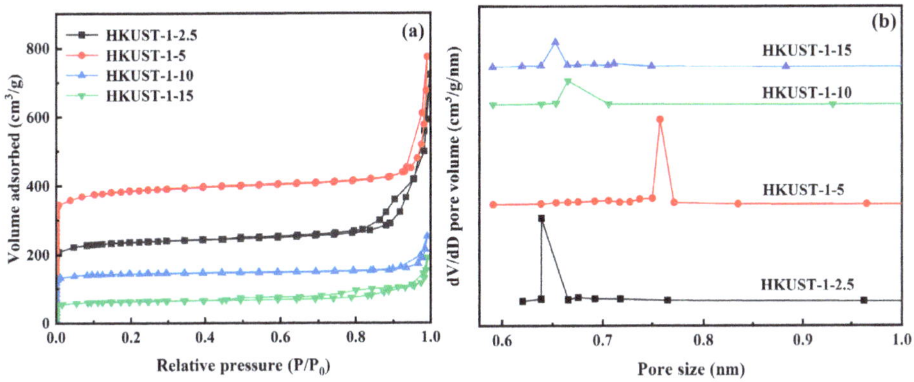

Figure 2. (**a**) N_2 sorption and desorption isotherms; (**b**) pore size distribution curves of HKUST-1 prepared with different addition amounts of NH_4F.

Table 1. Nitrogen sorption data of HKUST-1-x materials.

Samples	BET Surface Area (m^2/g)	Micropore Volume (mL/g)	Average Micropore Size (nm)
HKUST-1-2.5	927	0.36	0.68
HKUST-1-5	1508	0.58	0.73
HKUST-1-10	574	0.22	0.68
HKUST-1-15	245	0.09	0.66

Figure 3 presents the SEM images of HKUST-1-x. It is noted that HKUST-1 prepared with a low mass fraction of NH_4F (HKUST-1-2.5 and HKUST-1-5) possessed small particle size (100~200 nm) and a morphology of irregular polyhedrons. Differently, the surface of HKUST-1-2.5 is smooth. The surface of HKUST-1-5 is rough and there are some obvious dimples. Comparatively, HKUST-1 samples prepared with a high mass fraction of NH_4F (HKUST-1-10 and HKUST-1-15) have large particle size (2–8 μm) and a rodlike morphology. These results indicate that the addition amount of NH_4F had a great influence on the particle size and morphology of HKUST-1. Such effect can be also observed in the synthesis of HKUST-1 with the addition of certain metal ions like K^+ or Na^+ [25].

Figure 3. SEM images of HKUST-1 prepared with different addition amounts of NH_4F: (**a**) 2.5%; (**b**) 5%; (**c**) 10%; (**d**) 15%.

3.2. Evaluation of Adsorption Desulfurization Performance

The adsorption desulfurization performance of HKUST-1-x was firstly evaluated by the removal of DBT in model fuel oil. As seen in Figure 4, HKUST-1-5 exhibited the best adsorption performance among these HKUST-1-x materials. Its adsorption capacity for DBT could reach 46.8 mgS/g in 20 min at room temperature. Comparatively, the adsorption capacity of HKUST-1-S prepared by the solvothermal method is 40.9 mgS/g under the same conditions. Such adsorption performance is still outstanding compared with that of some reported MOFs (Table 2).

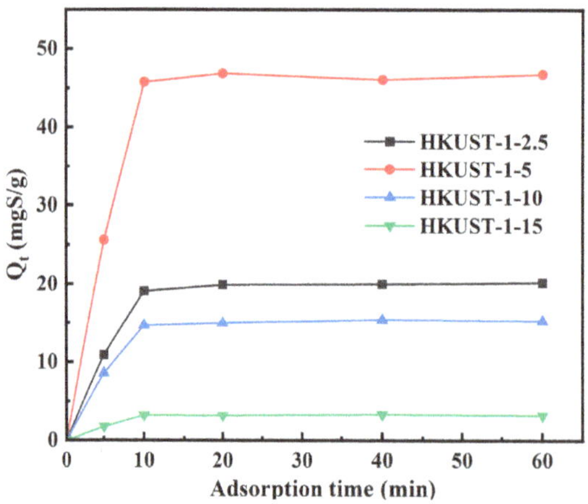

Figure 4. The adsorption desulfurization performance of HKUST-1-x prepared with different addition amounts of NH_4F for the removal of DBT in model fuel oil.

Table 2. Comparison of adsorption desulfurization performance of HKUST-1-5 with some reported MOFs.

Adsorbents	Sulfur Compounds	Sulfur Concentration (ppm)	Q_e (mgS/g)	References
MIL-53(Al)	BT	1000	4.5	
MIL-53(Cr)	BT	1000	9.5	[26]
MIL-47	BT	1000	21.5	
MIL-53(Cr)-G	DBT	1000	86.4	[27]
MIL-100(Fe)	BT	1000	14.3	[28]
MIL-101(Cr)	DBT	1000	6.5	[29]
Co-ZIF-67-M	DBT	500	2.2	[30]
UiO-66	BT	100	19.8	[31]
	BT	1000	6.6	[32]
ZIF-8	BT	1500	40	
UMCM-150	DBT	1500	83	
	4,6-DMDBT	600	41	[7]
	BT	1500	25	
HKUST-1	DBT	1500	45	
	4,6-DMDBT	600	16	
	BT	1000	16.6	
HKUST-1-S	DBT	1000	40.9	
	4,6-DMDBT	1000	22.4	This work
	BT	1000	23.8	
HKUST-1-5	DBT	1000	46.8	
	4,6-DMDBT	1000	36.8	

To further assess the adsorption performance of HKUST-1-5, a comprehensive comparison was made with HKUST-1-S. Firstly, the influence of sulfur compounds on adsorption desulfurization of HKUST-1-5 was investigated. As shown in Figure 5, the adsorption equilibrium of both adsorbents can be reached in 20 min for three kinds of simulated fuel oils. Overall, the adsorption performance of HKUST-1-5 for the removal of all sulfur compounds was better than that of HKUST-1-S. Nevertheless, the adsorption capacity varies with the different sulfur compounds. The order of adsorption capacity is DBT > 4,6-DMDBT > BT. Such difference might be mainly caused by the different electron cloud density distributions around S atoms in sulfur compounds and steric hindrance effects [33,34]. Interestingly, as for the removal of bulky 4,6-DMDBT that is difficult to desulfurize by the HDS technique, HKUST-1-5 gave a adsorption capacity of 36.8 mgS/g, which is much higher than that (22.4 mgS/g) of HKUST-1-S. This result suggests that HKUST-1-5 could be a good adsorbent for deep desulfurization of fuel oil.

A pseudo-first-order reaction model and pseudo-second-order reaction model were used to fit the adsorption data (Figures S1 and S2). The important parameters are shown in Tables 3 and 4. All correlation coefficients are less than 0.90 based on the pseudo-first-order reaction model and higher than 0.99 according to the pseudo-second-order reaction model, revealing that the adsorption desulfurization process on HKUST-1 should belong to pseudo-second-order kinetics. In addition, the measured adsorption equilibrium capacities for BT, DBT and 4,6-DMDBT are close to the Q_e values calculated by the pseudo-second-order kinetic model. This suggests that the rate-limiting step of the adsorption process should be a chemical interaction process. Moreover, K_2 of HKUST-1-5 is higher than that of HKUST-1-S, which is consistent with the results for adsorption capacity.

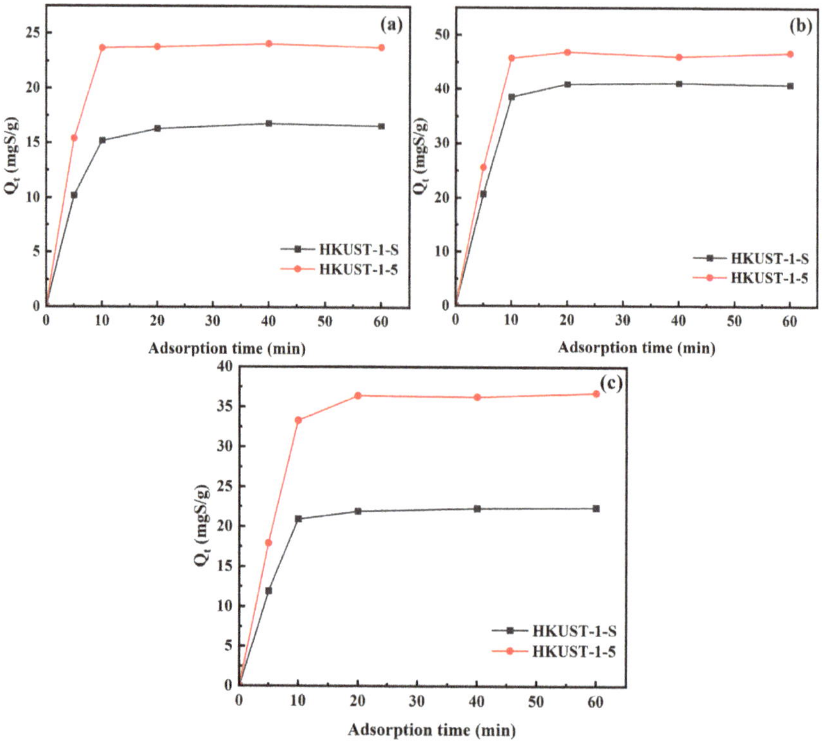

Figure 5. Adsorption desulfurization performance of HKUST-1-S and HKUST-1-5 for various sulfur compounds: (**a**) BT; (**b**) DBT; (**c**) 4,6-DMDBT.

Table 3. Pseudo-first-order kinetic parameters of adsorption desulfurization on HKUST-1-S and HKUST-1-5.

Adsorbents	Sulfur Compounds					
	BT		DBT		4,6-DMDBT	
	K_1	R^2	K_1	R^2	K_1	R^2
HKUST-1-S	0.0896	0.8612	0.1077	0.6082	0.0056	0.8910
HKUST-1-5	0.0679	0.4103	0.1281	0.8037	0.0936	0.6381

Table 4. Pseudo-second-order kinetic parameters of adsorption desulfurization on HKUST-1-S and HKUST-1-5.

Adsorbents	Sulfur Compounds					
	BT		DBT		4,6-DMDBT	
	K_2	R^2	K_2	R^2	K_2	R^2
HKUST-1-S	0.0274	0.9982	0.0084	0.9936	0.0056	0.9959
HKUST-1-5	0.0297	0.9977	0.0093	0.9953	0.0077	0.9939

In order to confirm the adsorption model, the adsorption isotherms are shown in Figure 6. For three kinds of sulfur compounds, the initial sulfur content was tuned from 250 ppm to 1000 ppm. As seen, the adsorption capacity of both adsorbents displayed a similar upward trend with the increase in initial concentration. The Langmuir model and Freundlich model were used to fit the desulfurization isotherms (Figures S3 and S4). The important parameters are shown in Tables 5 and 6. The obtained R^2 value based on the

Freundlich model is larger than that from the Langmuir model. Obviously, the Freundlich model can better simulate the adsorption desulfurization process of both materials, indicating that the adsorption of three sulfur compounds on HKUST-1 should belong to multi-layer chemical adsorption [35].

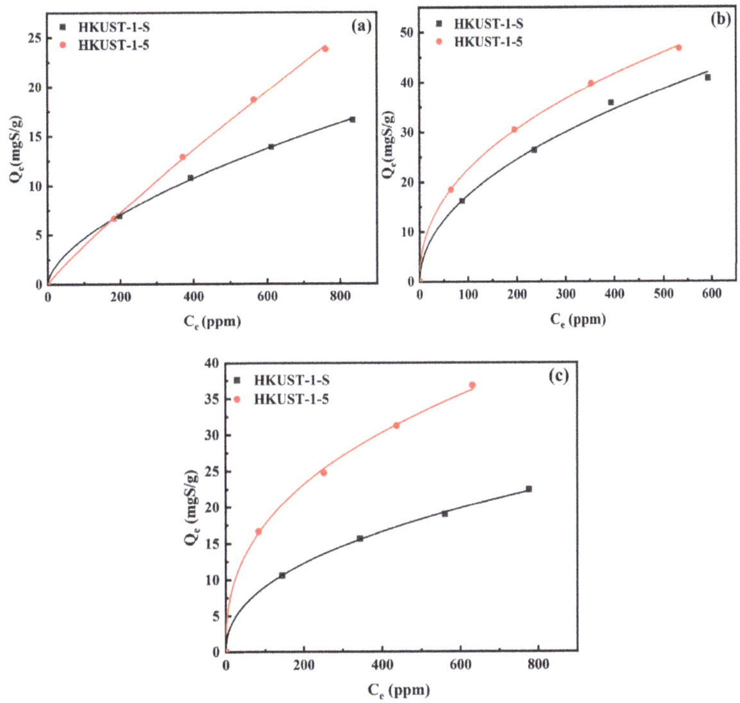

Figure 6. Adsorption desulfurization isotherms of HKUST-1-S and HKUST-1-5 for various sulfur compounds: (**a**) BT; (**b**) DBT; (**c**) 4,6-DMDBT.

Table 5. Langmuir model parameters of adsorption desulfurization.

Adsorbents	Sulfur Compounds					
	BT		DBT		4,6-DMDBT	
	Q_{max} (mgS/g)	R^2	Q_{max} (mgS/g)	R^2	Q_{max} (mgS/g)	R^2
HKUST-1-S	29.1	0.9931	56.8	0.9887	29.9	0.9854
HKUST-1-5	125	0.9969	60.2	0.9885	45.7	0.9798

Table 6. Freundlich model parameters of adsorption desulfurization.

Adsorbents	Sulfur Compounds								
	BT			DBT			4,6-DMDBT		
	K_F	n	R^2	K_F	n	R^2	K_F	n	R^2
HKUST-1-S	0.298	1.168	0.9989	1.791	2.024	0.9945	1.201	2.281	0.9989
HKUST-1-5	0.064	1.120	0.9993	2.939	2.259	0.9995	2.964	2.576	0.9975

To evaluate the adsorption performance of HKUST-1-5 and HKUST-1-S in complex oil, 10 wt% benzene was added to the model oil. The experimental results are shown in Figure 7. The adsorption capacity of both materials would greatly decrease with the addition of benzene. However, HKUST-1-5 still exhibited better adsorption performance than HKUST-1-S, which shows that HKUST-1-5 has stronger adsorption selectivity for sulfur-containing compounds and is more suitable for adsorption desulfurization of real fuel.

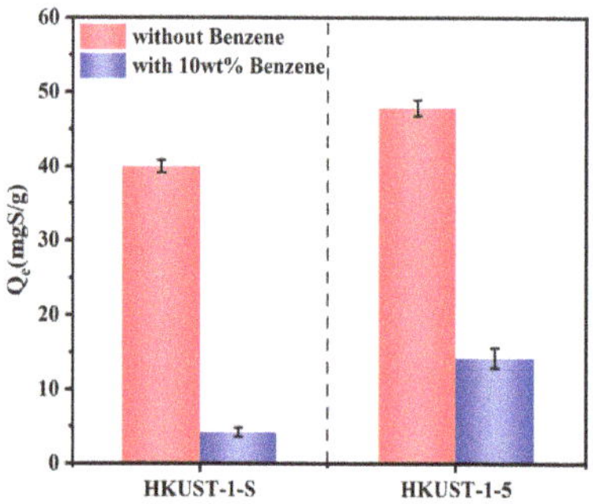

Figure 7. The adsorption desulfurization performance of HKUST-1-S and HKUST-1-5 in the presence or absence of 10% benzene in model oil.

The stability of adsorbents is of practical importance during adsorption. In this study, a cycle experiment on HKUST-1-5 was carried out. HKUST-1-5 materials before and after adsorption were characterized (Figure 8, Table 7). In the beginning, the adsorption capacity for DBT was 49.4 mgS/g. After four consecutive cycles of adsorption, the adsorption capacity for DBT can still stay above 45 mgS/g, indicating that HKUST-1-5 may be reused for several cycles. In addition, the XRD patterns of used HKUST-1-5 are consistent with those of fresh HKUST-1-5, demonstrating that the crystal structure of HKUST-1-5 is stable. Furthermore, the fact that the used HKUST-1-5 has a similar specific surface area to its fresh counterpart can support this point. Notably, the average micropore size and volume are slightly decreased after reuse, which might be attributed to the case that partial adsorption active sites in the pores are covered by adsorbates [14]. Also, it may hinder the effective contact between adsorbates and adsorption active sites, resulting in a decrease in adsorption performance.

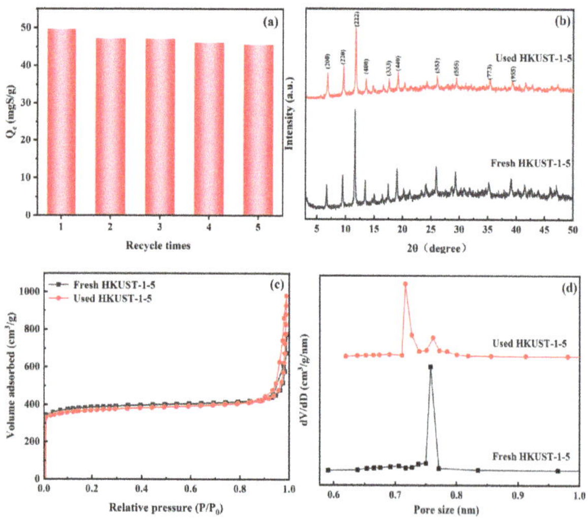

Figure 8. (**a**) Reusability of HKUST-1-5; (**b**) XRD patterns; (**c**) N_2 adsorption–desorption isotherms; (**d**) pore size distribution curves of HKUST-1-5 before and after adsorption.

Table 7. Nitrogen sorption data of fresh and used HKUST-1-5.

Adsorbents	BET Surface Area (m^2/g)	Micropore Volume (mL/g)	Average Pore Size (nm)
Fresh HKUST-1-5	1508	0.58	0.73
Used HKUST-1-5	1450	0.51	0.71

3.3. Relationship Between Adsorption Desulfurization Performance and Structure

To clarify the relationship between adsorption performance and structure, XPS of HKUST-1-5 and HKUST-1-S was measured. As shown in Figure 9, the results confirm that both adsorbents contain elements such as carbon (C), oxygen (O), and copper (Cu). The peaks of 284.8 eV, 286 eV and 288.65 eV are attributed to C=C, C=O and O=C-O respectively [36,37]. The relative content of C=O in HKUST-1-5 is higher than that in HKUST-1-S. In the high-resolution O 1s spectrum, the peaks at 231.64 eV, 232.46 eV and 233.68 eV belong to C-O, C=O and adsorbed oxygen, respectively [38]. However, all peaks corresponding to HKUST-1-5 shifted toward high binding energy, indicating that the chemical environment of O was different and the electron cloud density around O decreased. Moreover, the high-resolution XPS spectra of copper only exhibited the existence of Cu (II) associated with Cu-O bonds in the HKUST-1 framework and both adsorbents possessed similar proportions of Cu species. This result revealed that the difference in the adsorption desulfurization performance for both adsorbents had no close relationship with the valence of Cu species.

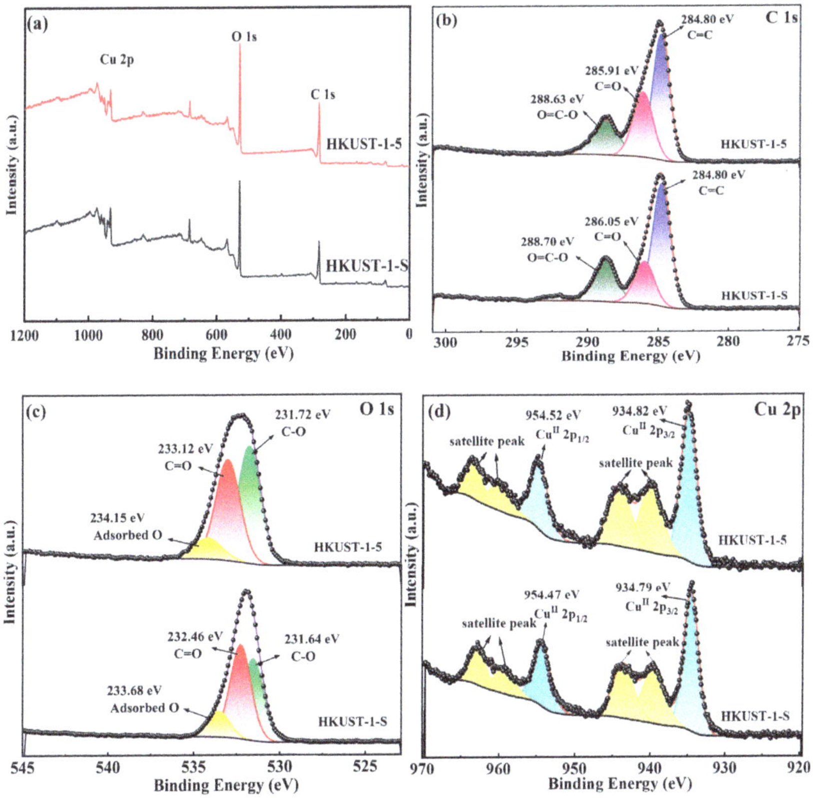

Figure 9. XPS spectra of HKUST-1-5 and HKUST-1-S: (**a**) full spectrum; (**b**) C 1s; (**c**) O 1s; (**d**) Cu 2p.

From the SEM results, it can be seen that there a big difference in the particle size of both adsorbents (Figure S5). HKUST-1-5 had much smaller average particle size (around 0.15 μm) than HKUST-1-S (about 16 μm). This suggests that the surface on HKUST-1-5 might expose more adsorption sites. Moreover, the average pore size of HKUST-1-5 is slightly larger than that of HKUST-1-S (Figure S6, Table S1). The large average pore size can make adsorbates enter the pore channels of HKUST-1 easily and thus enhance the adsorption performance.

Thermogravimetric analysis is an effective method for characterizing ligand defects in MOFs [39]. In order to determine the defects and their thermal stability, TG curves of HKUST-1-5 and HKUST-1-S were measured (Figure 10). The whole weight loss process of both samples can be divided into three stages. The first weight loss is attributed to the removal of water or other gases trapped in micropores. The second one corresponds to the evaporation of physically adsorbed water and the removal of μ-OH in clusters [40]. The third one originates from the removal of organic ligands [41]. The mass fractions of organic ligands in HKUST-1-S and HKUST-1-5 are 50.57% and 39.23%, respectively. Meanwhile, the mass fraction of the remaining CuO is 38.43% and 42.06%, respectively. The molar ratios of Cu to organic ligand in HKUST-1-S and HKUST-1-5 are 3:1.516 and 3:1.075, respectively. It is known that the molar ratio of Cu to organic ligand in perfect HKUST-1 crystals is 3:2 [42]. Based on these data, the calculated ligand loss rates for HKUST-1-S and HKUST-1-5 are 24.2% and 44.2%, respectively. This means that HKUST-1-5 has more Cu unsaturated coordination sites (defects), which can be helpful for improving the adsorption desulfurization performance.

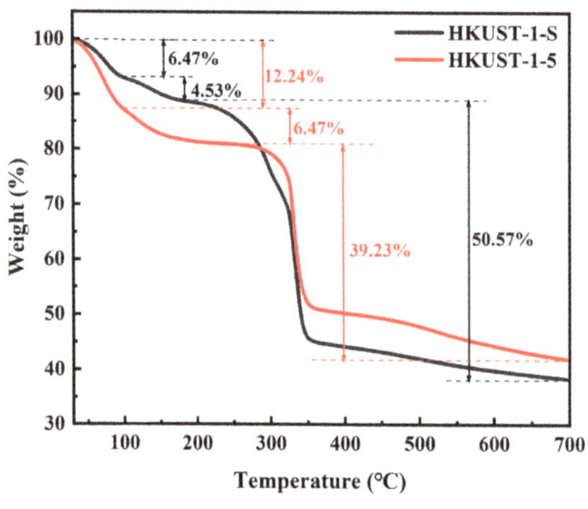

Figure 10. TG curves of HKUST-1-5 and HKUST-1-S.

To further confirm the defects in both samples, the NH_3-TPD technique was used [43]. As obtained from Figure 11, the number of Lewis acid sites in HKUST-1-5 and HKUST-1-S was 6.59 mmol/g and 6.19 mmol/g, respectively. This means that HKUST-1-5 possessed more Cu unsaturated coordination sites than HKUST-1-S, which is consistent with the results from TG analysis.

FT-IR spectra of HKUST-1-5 before and after DBT adsorption were performed. As shown in Figure 12, a new absorption peak appears at 1252 cm^{-1} after DBT adsorption, which is attributed to the C-S stretching vibration in the thiophene ring from DBT [44]. This result indicates that there is a strong interaction between HKUST-1-5 and DBT.

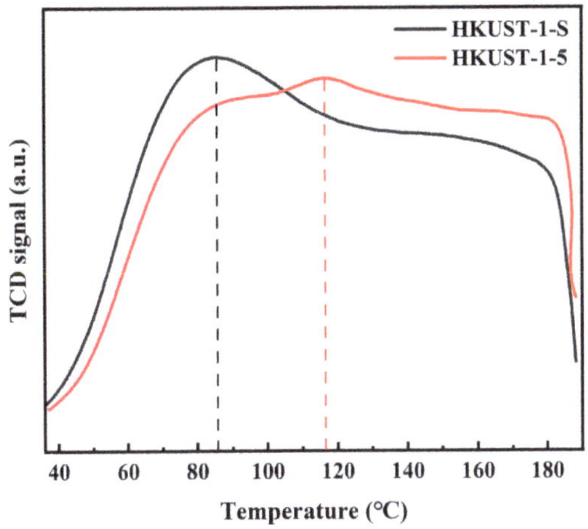

Figure 11. NH_3-TPD curves of HKUST-1-5 and HKUST-1-S. To validate the adsorption mechanism.

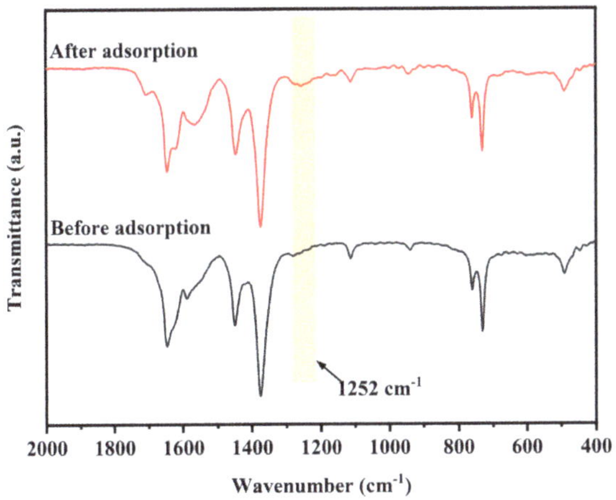

Figure 12. FT-IR spectra of HKUST-1-5 before and after DBT adsorption.

Based on the above results, it is concluded that the main reason why HKUST-1-5 exhibited enhanced adsorption desulfurization performance could be attributed to the existence of more defects and its smaller particle size. For this, a plausible adsorption process is shown in Figure 13. In this process, sulfur compounds like DBT were possibly adsorbed by two modes. One mode involves the unsaturated Cu sites in the structure of HKUST-1 behaving as electron acceptors and sulfide acting as an electron donor. The sulfur compounds can be adsorbed by the formation of Cu–S coordination bonds. As an alternative mode, these unsaturated Cu sites possessed acidic properties and sulfur compounds had weak basicity. Consequently, the adsorption desulfurization process may run through Lewis acid–base interactions [45]. There are also weak interactions such as π-π stacking, π-π interactions and Van der Waals forces between adsorbate molecules. These cumulative interactions with the strong chemical adsorption at the coordinatively unsaturated Cu sites acting as the primary anchor collectively justify the "multi-layer chemical adsorption" behavior.

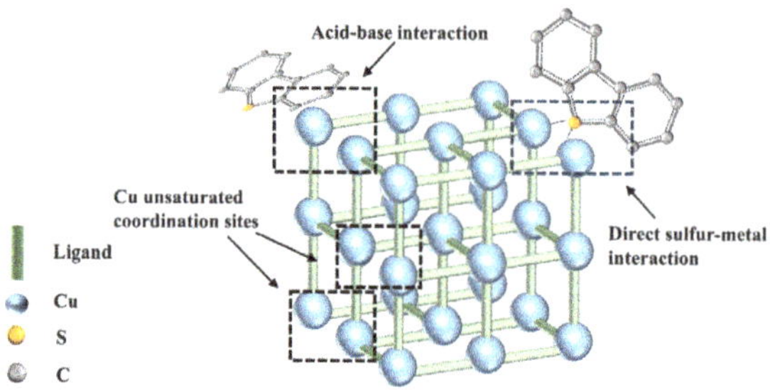

Figure 13. Schematic diagram showing adsorption desulfurization process of HKUST-1-5.

4. Conclusions

In summary, a series of HKUST-1 samples have been rapidly prepared with different addition amounts of NH_4F at room temperature. Under the conditions of that 50 mg HKUST-1-5 synthesized with the addition amount of 5wt% NH_4F was used to adsorb 5 g of model oil with a sulfur concentration of 1000 ppm at 25 °C for 1 h, the adsorption capacity of the adsorbent reached 23.8 mgS/g, 46.8 mgS/g and 36.8 mgS/g for benzothiophene (BT), dibenzothiophene (DBT) and 4,6-dimethyldibenzothiophene (4,6-DMDBT), respectively, which exhibited enhanced adsorption desulfurization performance compared with conventional HKUST-1 prepared by the solvothermal method. The characterization results from various techniques suggested that such superior adsorption performance could be attributed to the existence of more defects and its smaller particle size. In addition, the recycling experiments demonstrate that HKUST-1-5 can maintain good adsorption desulfurization performance after several cycles. This work will pave a way to design and synthesize new highly efficient adsorbents for desulfurization of fuel oil.

Supplementary Materials: The following supporting information can be downloaded at: https://www.mdpi.com/article/10.3390/en18205344/s1, Figure S1: Pseudo-first-order kinetic fitting of adsorption desulfurization between HKUST-1-S and HKUST-1-5: (a) BT, (b) DBT, (c) 4,6-DMDBT; Figure S2: Pseudo-second-order kinetic fitting of adsorption desulfurization between HKUST-1-S and HKUST-1-5: (a) BT, (b) DBT, (c) 4,6-DMDBT; Figure S3: Langmuir model fitting of adsorption of different sulfur compounds by HKUST-1-S and HKUST-1-5: (a) BT, (b) DBT, (c) 4,6-DMDBT; Figure S4: Freundlich model fitting of adsorption of different sulfur compounds by HKUST-1-S and HKUST-1-5: (a) BT, (b) DBT, (c) 4,6-DMDBT; Figure S5: The SEM images of (a) HKUST-1-5 and (b) HKUST-1-S; Figure S6: (a) The N_2 sorption and desorption isotherms and (b) the corresponding pore size distributions of HKUST-1-5 and HKUST-1-S; Figure S7: Adsorption desulfurization performance of HKUST-1-S and HKUST-1-5 for the removal of DBT; Figure S8: Non-linear pseudo-second-order kinetic fitting of adsorption desulfurization for HKUST-1-S and HKUST-1-5; Table S1: Comparison of pore structure parameters for HKUST-1-5 and HKUST-1-S.

Author Contributions: Writing—Original Draft, Experiments, Methodology, Investigation, J.F.; Writing—Review and Editing, Data Curation, X.L.; Writing—Review and Editing, Resources, Y.K.; Writing—Review and Editing, R.Z.; Writing—Review and Editing, Conceptualization, Methodology, Funding Acquisition, Supervision, Y.S. Writing—Review and Editing, Supervision, A.S.A.-E. All authors have read and agreed to the published version of the manuscript.

Funding: This research was funded by the National Natural Science Foundation of China (No. 22172042).

Data Availability Statement: The original contributions presented in this study are included in the article or Supplementary Material. Further inquiries can be directed to the corresponding author.

Acknowledgments: The authors acknowledge the financial support from the National Natural Science Foundation of China (No. 22172042).

Conflicts of Interest: The authors declare no conflicts of interest.

References

1. Niquille-Röthlisberger, A.; Prins, R. Hydrodesulfurization of 4,6-Dimethyldibenzothiophene and Dibenzothiophene over Alumina-Supported Pt, Pd, and Pt-Pd Catalysts. *J. Catal.* **2006**, *242*, 207–216. [CrossRef]
2. Moustafa, T.M.; Froment, G.F. Kinetic Modeling of Coke Formation and Deactivation in the Catalytic Cracking of Vacuum Gas Oil. *Ind. Eng. Chem. Res.* **2003**, *42*, 14–25. [CrossRef]
3. Rajendran, A.; Cui, T.Y.; Fan, H.X.; Yang, Z.F.; Feng, J.; Li, W.Y. A Comprehensive Review on Oxidative Desulfurization Catalysts Targeting Clean Energy and Environment. *J. Mater. Chem. A* **2020**, *8*, 2246–2285. [CrossRef]
4. Zhou, S.; Liu, Y.; Zhu, J.; Liu, F.; Cheng, K.; Cheng, H.; Zhu, W. Modulating Hydrogen Spillover Effect to Boost Ultradeep Hydrodesulfurization of 4,6-Dimethyldibenzothiophene over Pt-Ni_2P/Al_2O_3. *Sep. Purif. Technol.* **2024**, *351*, 128095. [CrossRef]
5. Ye, G.; Wang, H.; Zeng, X.; Wang, L.; Wang, J. Defect-Rich Bimetallic UiO-66(Hf-Zr): Solvent-free Rapid Synthesis and Robust Ambient-temperature Oxidative Desulfurization Performance. *Appl. Catal. B—Environ. Energy* **2021**, *299*, 120659. [CrossRef]
6. Nilavu, M.C.; Leelasree, T.; Aggarwal, H.; Rajesh, N. Cobalt-Based Metal–Organic Framework for Desulfurization of Thiophene as a Model Fuel. *Sustain. Energy Fuels* **2024**, *8*, 1679–1690. [CrossRef]
7. Zhang, Y.Y.; Lin, Q.J.; Gao, D.L.; Liu, Y.X.; Qin, N.; Xue, Y.X.; Yang, Q. The Introduction of Nitrogen-Containing Heterocyclic Compounds (NHC) and Cu^{2+} in Co-MOF Further Improves the Removal of Dibenzothiophene from Model Fuels. *Sep. Purif. Technol.* **2024**, *335*, 126096. [CrossRef]
8. Liu, J.; Fu, Z.; Tan, P.; Li, M.; Liu, X.Q.; Sun, L.B. Constructing Light-Insensitive Ag (I) Sites by in situ Integration of Nanoheaters for Efficient Adsorptive Desulfurization. *Chem. Eng. J.* **2024**, *484*, 149717. [CrossRef]
9. Mansour, E.A.; Taha, M.; Mahmoud, R.K.; Shehata, N.; Abdelhameed, R.M. A Combined Experimental and Computational Studies on Thiophene Adsorption from Liquid Fuels over ZIF-67@ZnFe LDH Composites. *Mater. Chem. Phys.* **2025**, *334*, 130499. [CrossRef]
10. Yu, Q.; Zhan, T.; Xia, Z.; Lu, G.; Ma, N.; Dai, W. Well-Construction of Hollow-Pocket Shaped Zn-based Metal-Organic Framework for Boosting the Capture Ability towards Thiophene Sulfur. *Particuology* **2025**, *99*, 116–127. [CrossRef]
11. Cychosz, K.A.; Wong-Foy, A.G.; Matzger, A.J. Liquid Phase Adsorption by Microporous Coordination Polymers: Removal of Organosulfur Compounds. *J. Am. Chem. Soc.* **2008**, *130*, 6938–6939. [CrossRef]
12. Achmann, S.; Hagen, G.; Hämmerle, M.; Malkowsky, I.M.; Kiener, C.; Moos, R. Sulfur Removal from Low-sulfur Gasoline and Diesel Fuel by Metal-Organic Frameworks. *Chem. Eng. Technol.* **2010**, *33*, 275–280. [CrossRef]
13. Peralta, D.; Chaplais, G.; Simon-Masseron, A.; Barthelet, K.; Pirngruber, G.D. Metal-Organic Framework Materials for Desulfurization by Adsorption. *Energy Fuels* **2012**, *26*, 4953–4960. [CrossRef]
14. Duan, M.; Wang, Y.; Fan, H.; Su, Z.; Yang, C.; Kou, J.; Shangguan, J. Revealing the Contribution of Cu(II) and Cu(I) Inherent in MOF-199 for Efficient Thiophene Removal at Room Temperature. *Sep. Purif. Technol.* **2024**, *349*, 127922. [CrossRef]
15. Li, Y.; Wang, L.J.; Fan, H.L.; Shangguan, J.; Wang, H.; Mi, J. Removal of Sulfur Compounds by a Copper-based Metal Organic Framework under Ambient Conditions. *Energy Fuels* **2015**, *29*, 298–304. [CrossRef]
16. Liu, P.; Zhao, T.; Cai, K.; Chen, P.; Liu, F.; Tao, D.J. Rapid Mechanochemical Construction of HKUST-1 with Enhancing Water Stability by Hybrid Ligands Assembly Strategy for Efficient Adsorption of SF_6. *Chem. Eng. J* **2022**, *437*, 135364. [CrossRef]
17. Duan, C.; Li, F.; Luo, S.; Xiao, J.; Li, L.; Xi, H. Facile Synthesis of Hierarchical Porous Metal-Organic Frameworks with Enhanced Catalytic Activity. *Chem. Eng. J.* **2018**, *334*, 1477–1483. [CrossRef]
18. Duan, Y.; Li, H.; Shi, X.; Ji, C.; Imbrogno, J.; Zhao, D. Stability of Metal–Organic Frameworks in Organic Media with Acids and Bases. *Ind. Eng. Chem. Res.* **2025**, *64*, 5372–5382. [CrossRef]
19. Li, Y.; Xu, Y.; Huang, E.; Kong, L.; Zeng, Y. Efficient Adsorption Desulfurization via Encapsulation of CeO_2 Nanoparticles in Hierarchical Porous HKUST-1. *J. Colloid. Interface Sci.* **2025**, *689*, 137237. [CrossRef] [PubMed]
20. Qi, S.C.; Qian, X.Y.; He, Q.X.; Miao, K.J.; Jiang, Y.; Tan, P.; Liu, X.Q.; Sun, L.B. Generation of Hierarchical Porosity in Metal–Organic Frameworks by the Modulation of Cation Valence. *Angew. Chem. Int. Ed.* **2019**, *58*, 10104–10109. [CrossRef]
21. Chen, M.; Chen, J.; Liu, Y.; Liu, J.; Li, L.; Yang, B.; Ma, L. Enhanced Adsorption of Thiophene with the GO-modified Bimetallic Organic Framework Ni-MOF-199. *Colloid. Surf. A—Physicochem. Eng. Asp.* **2019**, *578*, 123553. [CrossRef]
22. Petit, C.; Burress, J.; Bandosz, T.J. The Synthesis and Characterization of Copper-based Metal–Organic Framework/Graphite Oxide Composites. *Carbon* **2011**, *49*, 563–572. [CrossRef]
23. Bazzi, L.; Boukayouht, K.; Mansouri, S.; El Hankari, S. Eco-friendly and Cost-effective Synthesis of Hierarchical Porous HKUST-1 from Thin Waste Electric Cables for Enhanced Cationic Dye Removal. *Process Saf. Environ. Prot.* **2024**, *191*, 750–759. [CrossRef]
24. Yañez-Aulestia, A.; Trejos, V.M.; Esparza-Schulz, J.M.; Ibarra, I.A.; Sánchez-González, E. Chemically Modified HKUST-1(Cu) for Gas Adsorption and Separation: Mixed-metal and Hierarchical Porosity. *ACS Appl. Mater. Interfaces* **2024**, *16*, 65581–65591. [CrossRef] [PubMed]

25. Liu, Q.; Yang, J.M.; Jin, L.N.; Sun, W.Y. Metal Ion Induced Porous HKUST-1 Nano/Microcrystals with Controllable Morphology and Size. *Crystengcomm* **2016**, *18*, 4127–4132. [CrossRef]
26. Khan, N.A.; Jun, J.W.; Jeong, J.H.; Jhung, S.H. Remarkable Adsorptive Performance of a Metal–Organic Framework, Vanadium-Benzenedicarboxylate (MIL-47), for Benzothiophene. *Chem. Commun.* **2011**, *47*, 1306–1308. [CrossRef]
27. Han, L.; Zhang, J.; Mao, Y.; Zhou, W.; Xu, W.; Sun, Y. Facile and Green Synthesis of MIL-53 (Cr) and its Excellent Adsorptive Desulfurization Performance. *Ind. Eng. Chem. Res.* **2019**, *58*, 15489–15496. [CrossRef]
28. Khan, N.A.; Jhung, S.H. Low-temperature Loading of Cu^+ Species over Porous Metal-Organic Frameworks (MOFs) and Adsorptive Desulfurization with Cu+-Loaded MOFs. *J. Hazard. Mater.* **2012**, *237*, 180–185. [CrossRef]
29. Khan, N.A.; Hasan, Z.; Jhung, S.H. Ionic Liquids Supported on Metal-Organic Frameworks: Remarkable Adsorbents for Adsorptive Desulfurization. *Chem. Eur. J.* **2014**, *20*, 376–380. [CrossRef] [PubMed]
30. Jafarinasab, M.; Akbari, A.; Omidkhah, M.; Shakeri, M. An efficient Co-based Metal–Organic Framework Nanocrystal (Co-ZIF-67) for Adsorptive Desulfurization of Dibenzothiophene: Impact of the Preparation Approach on Structure Tuning. *Energy Fuels* **2020**, *34*, 12779–12791. [CrossRef]
31. Zhang, X.F.; Wang, Z.; Feng, Y.; Zhong, Y.; Liao, J.; Wang, Y.; Yao, J. Adsorptive Desulfurization from the Model Fuels by Functionalized UiO-66(Zr). *Fuel* **2018**, *234*, 256–262. [CrossRef]
32. Khan, N.A.; Bhadra, B.N.; Jhung, S.H. Heteropoly Acid-loaded Ionic Liquid@Metal–Organic Frameworks: Effective and Reusable Adsorbents for the Desulfurization of a Liquid Model Fuel. *Chem. Eng. J.* **2018**, *334*, 2215–2221. [CrossRef]
33. Wei, S.; He, H.; Cheng, Y.; Yang, C.; Zeng, G.; Qiu, L. Performances, Kinetics and Mechanisms of Catalytic Oxidative Desulfurization from Oils. *RSC Adv.* **2016**, *6*, 103253–103269. [CrossRef]
34. Huang, G.; Sun, Z.; Yu, Z.; Liu, Y.Y.; Wang, Y.; Wang, W.; Wang, A.; Hu, Y. Supported Ni_2P Catalysts Derived from Nickel Phyllosilicate with Enhanced Hydrodesulfurization Performance. *J. Catal.* **2023**, *419*, 37–48. [CrossRef]
35. Vigdorowitsch, M.; Pchelintsev, A.; Tsygankova, L.; Tanygina, E. Freundlich Isotherm: An Adsorption Model Complete Framework. *Appl. Sci.* **2021**, *11*, 8078. [CrossRef]
36. Guo, Y.; Feng, C.; Wang, S.; Xie, Y.; Guo, C.; Liu, Z.; Wang, J. Construction of Planar-type Defect-engineered Metal–Organic Frameworks with both Mixed-valence Sites and Copper-ion Vacancies for Photocatalysis. *J. Mater. Chem. A* **2020**, *8*, 24477–24485. [CrossRef]
37. Xie, H.; Yi, D.; Shi, L.; Meng, X. High Performance of CuY Zeolite for Catalyzing Acetylene Carbonylation and the Effect of Copper Valence States on Catalyst. *Chem. Eng. J.* **2017**, *313*, 663–670. [CrossRef]
38. Lu, P.; Sun, Z.; Qi, Z.; Chen, J.; Ye, C.; Qiu, T. Molecular Engineering to Fabricating Dinuclear-crossed Secondary Building Units in Metal-Organic Frameworks for Aromatic Sulfur-containing Compounds Adsorption. *Sep. Purif. Technol.* **2023**, *327*, 124879. [CrossRef]
39. Kar, A.K.; Sarkar, R.; Manal, A.K.; Kumar, R.; Chakraborty, S.; Ahuja, R.; Srivastava, R. Unveiling and Understanding the Remarkable Enhancement in the Catalytic Activity by the Defect Creation in UiO-66 during the Catalytic Transfer Hydrodeoxygenation of Vanillin with Isopropanol. *Appl. Catal. B—Environ. Energy* **2023**, *325*, 122385. [CrossRef]
40. Abou-Elyazed, A.S.; Ftooh, A.I.; Sun, Y.; Ashry, A.G.; Shaban, A.K.; El-Nahas, A.M.; Yousif, A.M. Solvent-free Synthesis of HKUST-1 with Abundant Defect Sites and its Catalytic Performance in the Esterification Reaction of Oleic Acid. *ACS Omega* **2024**, *9*, 37662–37671. [CrossRef]
41. Tian, L.; Song, X.; Liu, Y.; Zhang, C.; Shi, L.; Chen, Q.; Hu, T. Defect-Engineering Improves the Activity of Metal-Organic Frameworks for Catalyzing Hydroboration of Alkynes: A Combination of Experimental Investigation and Density Functional Theory Calculations. *J. Colloid Interface Sci.* **2024**, *662*, 263–275. [CrossRef] [PubMed]
42. Chui, S.S.Y.; Lo, S.M.F.; Charmant, J.P.; Orpen, A.G.; Williams, I.D. A Chemically Functionalizable Nanoporous Material $[Cu_3(TMA)_2(H_2O)_3]$ n. *Science* **1999**, *283*, 1148–1150. [CrossRef] [PubMed]
43. Xu, Y.P.; Wang, Z.Q.; Tan, H.Z.; Jing, K.Q.; Xu, Z.N.; Guo, G.C. Lewis Acid Sites in MOFs Supports Promoting the Catalytic Activity and Selectivity for CO Esterification to Dimethyl Carbonate. *Catal. Sci. Technol.* **2020**, *10*, 1699–1707. [CrossRef]
44. Castillo-Villalón, P.; Ramirez, J.; Castañeda, R. Relationship Between the Hydrodesulfurization of Thiophene, Dibenzothiophene, and 4,6-Dimethyl Dibenzothiophene and the Local Structure of Co in Co–Mo–S sites: Infrared Study of Adsorbed CO. *J. Catal.* **2012**, *294*, 54–62. [CrossRef]
45. Saha, B.; Vedachalam, S.; Dalai, A.K. Review on Recent Advances in Adsorptive Desulfurization. *Fuel Process. Technol.* **2021**, *214*, 106685. [CrossRef]

energies

MDPI

Article

The Impact of Brine Saturation and Distribution on Lean Gas Huff-n-Puff EOR Performance of Tight Oil Reservoirs: Examples from the Montney Formation (Canada)

Chengyao Song *, Amin Ghanizadeh and Christopher R. Clarkson

Department of Earth, Energy, and Environment, University of Calgary, 2500 University Drive NW, Calgary, AB T2N 1N4, Canada
* Correspondence: chengyao.song@ucalgary.ca

Abstract

Oil recovery from low-permeability ('tight') oil reservoirs remains low despite the application of modern drilling and completions technologies, which has increased interest in trialing enhanced oil recovery (EOR) schemes. Cyclic gas injection (Huff-n-Puff, HNP) is a promising approach to EOR for these reservoirs. However, the underlying mechanisms of EOR using the HNP scheme in tight reservoirs are not yet fully understood. This laboratory study investigates the performance of lean gas (80%C_1 + 20%C_2; approximating produced gas compositions from the field) HNP using low-permeability core plug samples from the Montney Formation of Canada. An objective of the study was to evaluate the effects of induced fractures, and brine saturation and distribution, on the efficiency of lean gas HNP performance. Both intact and artificially fractured core plugs were studied. The introduction of fractures into the low-permeability core plugs improved recovery factors by 17.5–18.5%. However, the presence of brine limited oil production from both intact and fractured core plugs. Notably, when brine was concentrated along the fracture surfaces, the recovery factor dropped significantly, down to just 1.2% of original oil in place (OOIP). This reduction is primarily attributed to the low solubility of methane and ethane (C_1 + C_2) in brine, which hinders the injectant's ability to diffuse into the core matrix and mobilize oil. The findings of this study will be of interest to operators evaluating the potential of cyclic gas injection in low-permeability reservoirs.

Keywords: brine distribution; brine saturation; EOR; Huff-n-Puff; lean gas; tight reservoirs

1. Introduction

With global oil and gas consumption continuing to rise and conventional hydrocarbon reserves on decline, there has been an increased focus on low-permeability ("unconventional") hydrocarbon (such as tight oil and liquid-rich gas) reservoir development. Multi-fractured horizontal wells (MFHWs) are currently the most popular technology used to exploit these resources. However, oil recovery from these reservoirs remains low, and production decline rates are often rapid [1]. To address this low recovery, there has been an increased focus on enhanced oil recovery schemes. The gas Huff-n-Puff (HNP) scheme using MFHWs, which involves cycles of gas injection ("huff"), soaking, and production ("puff"), has been tested in the laboratory and in the field, yielding favorable results [2–7]. Nevertheless, the complex reservoir characteristics of tight oil reservoirs—such as fracture complexity, reservoir heterogeneity, fluid saturation distributions, wettability distribution, etc.—affect lean gas HNP EOR results [5,8–13]. Moreover, there have been limited studies of EOR performance in tight oil reservoir where reservoir energy has been depleted following primary recovery.

As shown in Table 1, several experimental studies have focused on the gas HNP scheme in tight rock. Gas type, reservoir pressures and reservoir petrophysical properties (e.g., permeability, porosity, and wettability) are the primary factors influencing recovery performance [1,3]. Several studies also explored the impact of brine saturation on oil recovery from tight rock samples (e.g., Table 1; [3,8,14]) It has been observed that increased brine saturation consistently resulted in a decline in oil recovery. However, these investigations typically assumed a homogeneous distribution of brine within the rock, which does not accurately reflect in situ reservoir conditions [8,14,15]. It is believed that reservoir heterogeneity caused by the introduction of hydraulic fractures could lead to highly heterogeneous brine distributions [1,16]. Importantly, the impact of heterogeneous brine distributions on cyclic solvent injection processes, particularly lean gas HNP, in tight oil reservoir remains largely unexplored.

Table 1. Literature review of laboratory HNP experiments for tight reservoirs (partially adapted from Song et al. [17]).

No.	Formation	Sample Condition	Gas Type	Porosity (%)	Permeability (md)	Initial Brine Saturation (%)	Cycle No.	Final Recovery (%)	Reference
1	Duvernay, Canada	Fractured	CO_2	3.2	0.000125	0	4	45	[3]
	Duvernay, Canada	Fractured	Lean Gas	2.1	0.000125	0	4	29	
2	Montney/ Duvernay Canada	Fractured	Lean gas	6	~0.0008	20	7	~35	[8]
3	Bakken, Canada	Intact	CO_2	18.6–23.1	0.56–0.83	35.8–57.1	6	42.8–63	[1]
4	Haynesville Shale USA	Fractured	Lean gas	8	0.0006–0.0009	~11	7	35–50	[15]
5	Anonymous	Intact	CO_2	9.0	0.88	42.4	3	37	[14]
6	Ordos Basin, China	Intact	CO_2	9.6	0.012	0	3	~70	[11]
		Intact		3.8	0.00053		3	~55	
		Fractured		8.9	0.018		3	~78	
		Intact		4.3	0.00025		3	~58	
7	Anonymous, Russia	Intact	Rich gas	4.1–17.0	N/A	0	5	29–88	[18]
	Anonymous, Russia	Fractured		2.3–10.9	N/A		5	41–88	
8	Anonymous	Intact	CO_2	6.5	0.89	N/A	1	23	[19]
	Anonymous	Fractured		6.8	0.89		1	13	
9	Ordos Basin, China	Intact	CH_4	11.4	0.145	57.7	3	25	[20]
			Rich gas	11.7	0.145	60.6	3	31	
10	Anonymous	Intact	CO_2	11.9	0.001	0	6	84	[21]
			Rich gas	11.7	0.001		6	74	

To assess the impact of brine saturation and its distribution on lean gas HNP performance in depleted tight oil systems, a series of experiments were conducted using dead oil and core samples from the Montney Formation in Alberta, Canada. Initially, two HNP experiments were performed on two intact core plugs (i.e., without induced fractures), each with a different level of brine saturation (0% vs. 33.9%), to evaluate the influence of brine content on oil recovery. Following this, the same two core plugs were artificially fractured under in situ conditions, and subjected to a second set of HNP tests, allowing for a direct comparison of the effect of brine saturation in fractured versus unfractured samples. Additionally, a third core plug sample was fractured prior to the HNP testing, and formation brine was introduced directly through the fractures to simulate non-uniform brine distribution in the matrix—a condition more representative of the actual reservoir environment immediately after hydraulic fracturing. The HNP test with the third core sample aimed to examine the effect of heterogeneous brine distribution on lean gas HNP efficiency. The findings from these experiments provided valuable insights into the roles of brine saturation and distribution during solvent-based EOR schemes applied to low-permeability hydrocarbon reservoirs.

2. Mechanisms of Lean Gas HNP in Tight Oil Reservoirs

As previously discussed (e.g., Ghanizadeh et al. [3]), during cyclic gas injection (e.g., lean gas) in tight oil reservoirs, the primary EOR mechanisms include gas-induced oil swelling, interfacial tension (IFT) reduction, oil viscosity reduction, extraction of light-medium components, solution gas drive, and free gas expansion [7,22–27]. Additionally, the induced fractures of tight oil reservoir increase the contact interface between the injected lean gas and crude oil, enhancing the diffusion of gas into the oil-saturated matrix [28,29]. However, the presence of brine in the matrix also hinders oil flow by reducing oil phase relative permeability, which in turn leads to a lower oil recovery factor [30].

3. Materials

3.1. Rock Samples

Three core plug samples were collected from the Montney Formation of Alberta, Canada. These samples were extracted from the 2/3 slabbed cores, with the core plugs cut parallel to the bedding planes. Prior to performing the HNP experiments, a comprehensive suite of petrophysical and geochemical analyses were conducted on the core samples. As summarized in Tables 2 and 3, the quartz, K-feldspar, and clay content of the rock are around 42.03 wt.%, 7.82 wt.%, and around 22.64 wt.%, respectively. The porosity of the core samples is 4.2% for Sample #1, 4.4% for Sample #2, and 2.6% for Sample #3, respectively, while the permeability is 0.00042 md for Sample #1, 0.00078 md for Sample #2, and 0.00022 md for Sample #3, respectively. Samples #1 and #2 were sister core plugs obtained from the same depth, whereas Sample #3 was collected from a nearby location and exhibits similar petrophysical properties.

Table 2. Bulk mineralogy of Montney core sample in the vicinity of the analyzed cores (adapted from Song et al. [17]).

Quartz (wt.%)	K-Feldspar (wt.%)	Plag. Feldspar (wt.%)	Calcite (wt.%)	Dolomite (wt.%)	Pyrite (wt.%)	Illite (wt.%)	Chlorite (wt.%)
42.03	7.82	15.9	1.69	7.59	2.3	18.47	4.2

The core samples used in this study were selected to have a limited range in composition and petrophysical properties so that the impact of rock heterogeneity on results could be minimized, and the impact of brine saturation and distribution on the EOR performance could be more clearly ascertained. By minimizing the influence of other geological factors, this study allowed for a more reliable identification of key mechanisms impacting gas injectivity and recovery during the lean gas HNP process in tight oil reservoirs.

Table 3. Summary of the petrophysical properties of the analyzed core plug samples.

Sample ID	Depth (m)	Bulk Density (g/cm^3)	Grain Density (g/cm^3)	Helium Porosity (%)	Pulse-Decay (N_2) Gas Permeability (md) [1]
#1	2149.2	2.59	2.70	4.2	0.00042
#2	2149.2	2.61	2.73	4.4	0.00078
#3	2156.9	2.64	2.71	2.6	0.00022

[1] Slip-corrected pulse-decay (N_2) gas permeability evaluated at an effective stress of 500 psi.

3.2. Fluid Samples

The oil sample used in this research was also collected from the Montney Formation (AB, Canada). Because this study aimed to experimentally simulate depleted reservoirs, such as those with pressures below bubble point, dead oil was applied in the experiments. The crude oil was filtered through a 0.1-micron filter at a commercial laboratory to remove wax and solid impurities. The physical properties of the oil sample are detailed in Table 4. The oil density was estimated to be 0.83 g/cm^3, and oil viscosity was determined to be 2.48 cP under ambient conditions (i.e., 1 atm and 25 °C). The synthetic brine was prepared based on the chemical composition of field-produced brine, with its physical properties summarized in Table 5.

Table 4. Physical properties of the Montney (dead) oil sample used in this study.

Fluid	Density (g/cm^3) [25 °C, 1 atm]	Viscosity (cP) [25 °C, 1 atm]	Compressibility (psi^{-1}) [25 °C, 1 atm]
Montney Dead Oil	0.83	2.48	0.0000052

Table 5. Physical properties of the synthetic brine prepared based on the composition of formation produced brine.

Fluid	Density (g/cm^3) [25 °C, 1 atm]	Viscosity (mPa.s) [25 °C, 1 atm]	Compressibility (psi^{-1}) [25 °C, 1 atm]
Synthetic brine	1.04	0.998	0.000003

Compositional analysis of field-produced gas resulted in the following component percentage estimates: 80.1% C_1, 10% C_2, 5.6% C_3, 2.2% C_4, 1% N_2, 0.4% C_5, 0.3% C_6, 0.2% C_7, 0.1% CO_2 and other trace components. A simplified lean gas mixture—comprising 80% methane (C_1) and 20% ethane (C_2)—was used in this study to approximate produced gas in the field.

The results of this work will be used to inform future reservoir simulation and experimental studies performed under in situ conditions (e.g., live oil to be applied). In this study, dead oil and simplified lean gas mixture were used to reduce experimental complexity.

4. Experiments

4.1. Experimental Setup

The coreflooding setup used in this study (Figure 1) was largely adopted from previous work [3], with modifications to accommodate lean gas HNP tests with both intact and fractured core plug samples. For the intact core HNP test, gas injection and production were performed at the same end of the coreholder. In contrast, for the fractured core HNP test, fluids—including oil, water and gas—were produced from the opposite end of the coreholder to the injection end.

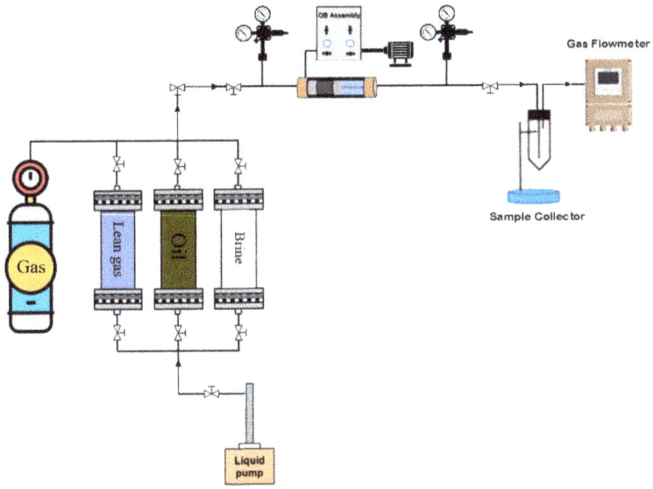

Figure 1. Schematic of lean gas HNP experimental setup used for both intact and fractured core samples (adapted from Ghanizadeh et al. [3]).

4.2. Experimental Procedures

Five experimental scenarios were designed to examine the effects of induced fractures, brine saturation and distribution on lean gas HNP EOR using the three Montney core plug samples. The experimental procedures are outlined below:

1. Helium porosity measurement—helium porosity of the three intact core plugs was determined by combing helium pycnometry (Micromeritics® Norcross, GA, USA, Accupyc II 1340™) to obtain a grain density/volume estimate and calipering to estimate bulk density/volume.
2. Gas (N_2) permeability measurement—matrix permeability values of the core plugs were determined using the gas (N_2) pulse-decay technique (Corelab® Tulsa, OK, USA, PDP-250™), under different effective stresses (500 and 1800 psi), with mean pore pressures ranging from 200 to 1000 psi to derive the slip-correct gas permeabilities.
3. Brine or oil saturation—after evacuating the core plugs for 48 hrs to remove gas, the core plugs were fully saturated with either dead oil (Sample #1 and #3) or brine (Sample #2) under 1500 psi for 5 days. Saturations were calculated from the mass change in the samples before and after saturation.
4. Liquid permeability measurement—liquid (oil or brine) pulse-decay permeability measurements [31] were conducted on the core plugs (oil for Sample #1 and #3; brine for Sample #2) under the same effective stress conditions used for the previous gas permeability testing.
5. Inducing various brine saturations—following the liquid permeability tests, brine (for sample #1 and #3) or crude oil (for sample #2) was flooded into the core samples to achieve different brine saturation levels. The brine saturations were determined from the mass difference in samples before and after saturation, combined with the measured quantities of fluids produced.
6. Lean gas HNP using intact core plugs—for each cycle of lean gas HNP, simplified lean gas was injected into the intact core plugs at a constant pressure (1300 psig) for 1 hr. The core plugs were then soaked with the injected lean gas for another hour, followed by a production period (i.e., atmosphere pressure) for 4 h. Similar HNP scheme was also employed in the previous study [3]. Three cycles of HNP were performed for Sample #1 and Sample #2.

7. Fracturing of intact core plugs—after completing the lean gas HNP experiments using the intact core plugs, the core plugs were fractured under stress within the coreholder using the procedure of Ghanizadeh et al. [3]. The axial pressure was increased in a stepwise manner, while the radial pressure was kept constant until the core plugs fractured. Sample #1 and #2 were both fractured using this procedure, while Sample #3 was unintentionally fractured during oil saturation (Step 3). Hence, for Sample #3, the intact core HNP test (Step 6) was skipped and Step 8 was implemented directly.
8. Lean gas HNP with fractured core plugs—following Step 7, lean gas HNP experiments were conducted on the fractured core plugs under similar conditions to the intact core HNP tests. Another four HNP cycles were performed for each core sample (Sample #1, #2, and #3).
9. Post-test liquid (oil) permeability measurement—after the HNP tests, liquid (oil) permeability was measured again on the fractured core plugs under stress loading (500–4000 psi) and unloading (4000–500 psi) conditions, to evaluate permeability hysteresis.

Note that Sample #1 was fully saturated with oil and contained no brine prior to the HNP testing. Sample #2 was saturated with both oil and brine (57.3% oil + 33.9% brine) prior to HNP testing. As noted, Sample #3 experienced unexpected fracturing during the oil saturation stage (Step 3). As a result, intact core HNP test (Step 6) was not performed and no additional fracturing was necessary (Step 7). Nevertheless, Sample #3 was included for fractured lean gas HNP testing and the post-test permeability measurement (Step 8–9). Following oil saturation and liquid permeability measurement (Step 3&4), brine was assumed to have entered Sample #3 through the induced fractures during Step 5, resulting in a saturation of 66.6% oil and 20% brine prior to the HNP test of Step 8.

Experiments were performed at 25 °C (room conditions). The objective of this proof-of-concept study was not to reproduce field conditions, but rather to compare the results of lean gas HNP tests for different water saturations and water distributions. Performing the experiments at room temperature helped to simplify the experimental procedures and minimize the risk of thermal-induced core damage (e.g., clay swelling or minerology dissolution). The results of this work will be used to inform future reservoir simulation and experimental studies performed under in situ conditions (e.g., reservoir temperature and live oil), which in turn will be used to support the design of potential pilot tests in the field.

5. Results

5.1. Matrix Permeability

As shown in Figure 2, the slip-corrected gas permeability for Samples #1–3 was 0.00042 md, 0.00078 md, and 0.00022 md under 500 psi effective stress, and 0.00025 md, 0.00063 md, and 0.00017 md under 1800 psi effective stress. Liquid permeability values for Sample #1 (oil) and Sample #2 (brine) are lower than gas permeability, aligning with previous studies [3]. In contrast, Sample #3 exhibited higher liquid (oil) permeability than gas permeability, which is not consistent with the literature [3,30].

The gas (N_2) permeability for Sample #3 is smaller than for Sample #1 and #2. This trend is consistent with the porosity results, suggesting that Sample #3 was not fractured during the gas permeability measurements. However, it is believed that Sample #3 was fractured during the oil saturation process, which was performed after gas permeability measurements and prior to liquid permeability measurements. Post-test analysis also revealed a significant fracture in Sample #3 (Figure 3), likely causing its elevated liquid permeability. A possible reason for fracturing is the rapid variation in effective stress at the beginning of oil saturation, when the pore pressure was increased from 0 psi to 1500 psi, while the confining pressure was maintained at 2000 psi. As a result, the effective stress changed from 2000 psi to 500 psi in a short period of time, resulting in sample fracturing. Previous research has suggested that Montney cores are relatively easy to fracture. For example, Keneti and Wong [32] reported that the tensile strength of Montney core samples parallel to the bedding was 0.3–2.8 MPa, and Vaisblat et al. [33] derived brittleness indices of Montney rocks in a range of 60–90. These values indicate the Montney cores are susceptible to fracturing caused by stress changes that are parallel to planes of weakness (laminations). Becerra Rondon [34] also highlighted that the strong heterogeneity of Montney rocks contributed significantly to fracture initiation and propagation. As such, it is believed the fracturing of Sample #3 was likely due to a combination of experimental conditions and the geomechanical properties of the sample.

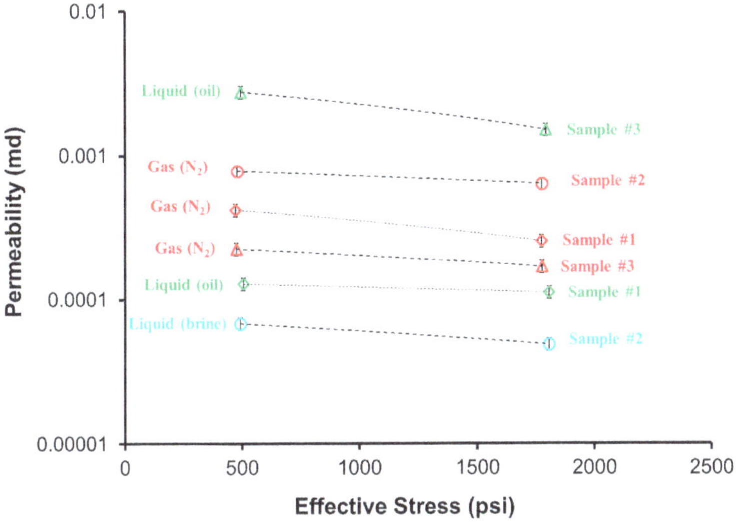

Figure 2. Stress-dependent matrix gas (N_2) permeability and liquid (oil or brine) permeability measured with the analyzed Montney core samples. Note that the oil permeability of Sample #3 is higher than expected because of sample fracturing during the oil saturation stage (after gas permeability measurements; see text for explanation).

Figure 3. Image of fracture surfaces of core plug Sample #3. This core plug was fractured during the oil saturation stage.

5.2. Lean Gas HNP Recovery

Lean gas HNP with intact core plugs. As previously mentioned, three cycles of lean gas HNP were conducted using intact core samples (Sample #1 and Sample #2). Sample #1 was fully saturated with oil prior to the HNP test, while Sample #2 was saturated with 57.3% oil and 33.9% brine. As shown in Figure 4, the maximum oil recovery from the intact Sample #1 was 4.6% versus 2.5% for intact Sample #2. The mechanisms leading to this difference will be discussed in Section 6. It is important to note that Sample #3, which was fractured during the oil saturation process, was excluded from the comparison of intact core HNP test results.

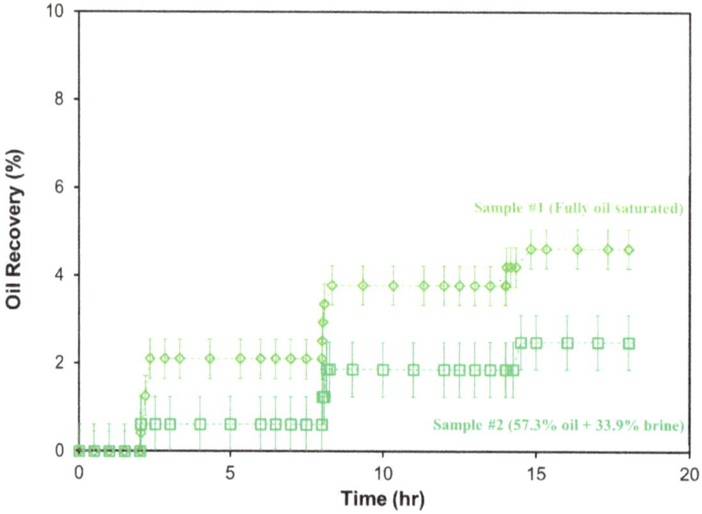

Figure 4. Oil recovery as a function of elapsed time for lean gas HNP experiments performed using intact core samples (Sample #1 and Sample #2). Results of Sample #1 were adapted from Song et al. [17].

Lean gas HNP with fractured core plugs. Following the intact core HNP experiments, Sample #1 and Sample #2 were artificially fractured under in situ stress conditions, as detailed in Section 4.2. Lean gas HNP was then conducted on each fractured core sample (Sample #1–3), for four additional cycles under similar conditions to the intact core tests, until oil production was exhausted. As shown in Figure 5, fracturing significantly enhanced oil recovery in Sample #1 and # 2, yielding additional recoveries of 18.5% and 17.5%, respectively. This led to ultimate recovery factors of 23.1% for Sample #1 and 20% for Sample #2. In contrast, Sample #3 (initially containing 66.6% oil + 20% brine) exhibited a substantially lower recovery of just 1.2%, despite its higher initial oil saturation than Sample #2 (initially containing 57.3% oil + 33.9% brine). The reasons for this discrepancy, along with the associated recovery mechanisms, will be discussed in Section 6.

It should be noted that the recovery performance with dead oil used in this study is not fully comparable to that using in situ oil during the lean gas HNP process, even for depleted reservoirs. Reservoir gases dissolved in in situ oil can lead to a lower oil viscosity and higher compressibility [35]. Moreover, the solution gas in in situ oil can expand and provide an additional driving force for oil flow when producing to lower pressure. These combined effects are expected to result in higher oil recovery during lean gas HNP with in situ oil, compared to dead oil.

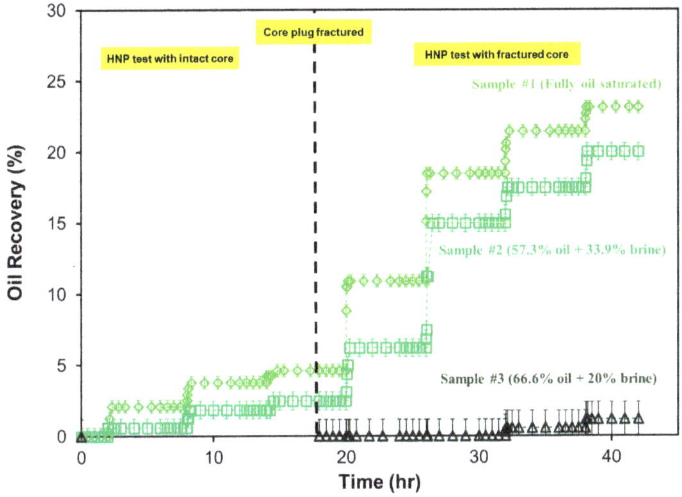

Figure 5. Oil recovery as a function of elapsed time for lean gas HNP experiments performed with intact (Sample #1 and #2) and fractured core plug samples (Sample #1–3). Results of Sample #1 were adapted from Song et al. [17].

<u>*System pressures during the lean gas HNP process.*</u> The system pressures—during injection, soaking, and production—were continuously monitored and recorded throughout the lean gas HNP experiments. As illustrated in Figure 6, a clear pressure decline was observed during the soaking period of each cycle of Sample #1. In contrast, Sample #2 and #3 exhibited minimal pressure change during the same period. This difference is primarily attributed to the variations in initial fluid saturation among the samples. Sample #1 was fully saturated with oil, allowing for more effective gas-oil interaction and gas dissolution, which contributed to the observed pressure drop. In comparison, Sample #2 was initially saturated with 57.3% oil and 33.9% brine, and Sample #3 with 66.6% oil and 20% brine. The presence of brine in Samples #2 and #3 likely reduced gas solubility and limited gas-oil mass transfer during soaking, resulting in the relatively stable system pressures.

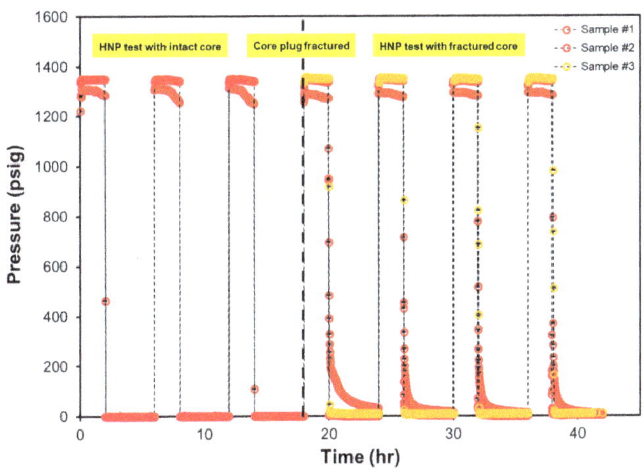

Figure 6. System pressure as a function of elapsed time for lean gas HNP experiments performed with intact (Sample #1 and #2) and fractured core plug samples (Sample #1–3). Results of Sample #1 were adapted from Song et al. [17].

Because the gas dissolution capacity of dead oil is higher than in situ oil, more lean gas is anticipated to dissolve into the core samples saturated with dead oil and create a more significant pressure decay during soaking. In this study, the use of dead oil might also contribute to the pressure performance of Sample #1.

5.3. Post-Test Permeability Evaluation with Fractured Core Plugs

Following the fractured sample HNP experiments, liquid (oil) permeability was measured as a function of stress for each fractured core sample. As shown in Figure 7, during stress loading from 500 to 4000 psi, bulk permeability decreased from 0.003 to 0.0001 md for Sample #1, from 0.6 to 0.1 md for Sample #2, and from 0.2 to 0.006 md for Sample #3. During stress unloading from 4000 to 500 psi, bulk permeability increased from 0.0001 to 0.0006 md for Sample #1, from 0.1 to 0.3 md for Sample #2, and from 0.006 to 0.01 md for Sample #3. The permeability during stress loading remained consistently higher than during unloading, indicating pronounced permeability hysteresis. This behavior reflects the dynamic changes in permeability under varying stress conditions during the HNP process.

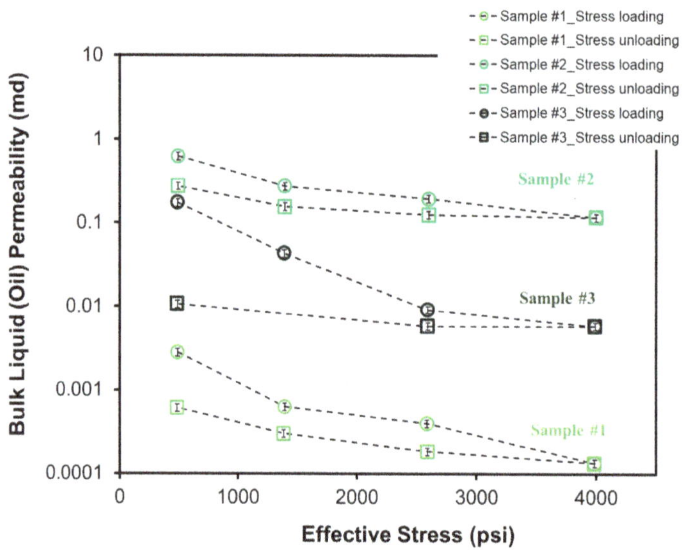

Figure 7. Liquid (oil) permeability measured during stress loading and unloading cycles for the analyzed fractured core samples.

Additionally, the measured permeability of Sample #3 was higher than that of Sample #1, but lower than Sample #2. The order contrasts with the permeability trend observed in the intact core permeability tests (Figure 2). This suggests that fractures, rather than matrix properties alone, control the permeability in tight oil reservoirs.

6. Discussion

Impact of brine saturation on lean gas HNP. As illustrated in Figures 4 and 5, Sample #1—initially fully saturated with oil—achieved a higher recovery factor than Sample #2, which was initially saturated with 57.3% oil and 33.9% brine. As discussed in Section 2, the presence of brine is known to impair the effectiveness of lean gas EOR due primarily to two mechanisms:

(1) Low solubility of the main components of the injected lean gas (80% C1 + 20% C2) in brine. For example, Zhao et al. [36] conducted methane solubility tests in brine with

salinity up to 234 g/L, which is similar to the Montney Formation brine used in this study (~210 g/L). Those tests yielded a methane solubility of only ~0.00081 mol/mol at 1450 psi and 25 °C. Kim et al. [37] reported that the solubility of ethane in brine was even lower than methane (e.g., 0.00043 mol/mol under 1450 psi and 10 °C). Solubility values of both of these components in brine are orders of magnitude lower than in liquid hydrocarbons (e.g., ~0.015 mol/mol for methane and ~0.053 mol/mol for ethane at 145 psi and 200 °C) [38]. The low solubility of lean gas components in brine reduces gas injectivity into the tight rock matrix when brine is present [39].

(2) Low oil phase relative permeability at high brine saturations. For example, Ghanizadeh et al. [30] reported that oil phase relative permeability decreases from 1 to 0. 3 as water saturation increases from 0% to 41% for a Montney core plug, which in turn limits oil mobility and overall recovery. The brine saturations selected in the current study are within a range that significantly affects oil relative permeability, allowing the impact of brine saturation on oil recovery during lean gas HNP to be evaluated.

These findings indicate that tight liquid hydrocarbon reservoirs with lower brine saturations are more likely to achieve higher oil recovery when subjected to cyclic lean gas injection.

Impact of brine distribution on lean gas HNP. As discussed above, lower brine saturation usually results in higher oil recovery. However, a notable contrast was observed between Sample #2 and Sample #3 after the core plugs were artificially fractured. Despite a higher initial brine saturation for Sample #2 (57.3% oil and 33.9% brine) than Sample #3 (66.6% oil and 20% brine), Sample #2 yielded a recovery factor of 20%, nearly 20 times higher than that of Sample #3 (around 1%). This apparent contradiction is likely attributed to the differences in brine distribution within the core samples. Prior to fracturing, Sample #2 exhibited a relatively homogeneous distribution of brine and oil throughout the core. As a result, its post-fracture fluid distribution remained largely uniform within the matrix (Figure 8a), allowing for effective contact between the injected lean gas and the oil. In contrast, fractures in Sample #3 were unintentionally induced during the oil saturation stage. During subsequent brine injection, brine preferentially flowed into the newly formed high-permeability fracture network. This resulted in the formation of a brine-rich zone adjacent to the fracture surfaces (as shown in the photograph of Figure 3 and illustrated schematically in Figure 8b). As discussed in the previous section ("Impact of brine saturation on lean gas HNP"), due to the low solubility of the injected lean gas (80% C_1 and 20% C_2) in brine [39], this brine-rich zone likely acted as a barrier to impede gas diffusion from the fractures to the matrix. Therefore, a lower oil recovery was achieved for Sample #3. Although Sample #3 was unintentionally fractured, its fracture geometry was similar to other low-permeability core samples fractured using the procedure of Ghanizadeh et al. [3]. Sample #3 represented a typical fractured tight oil system where a significant amount of water accumulates around fractures after hydraulic fracturing with a water-based fluid, or waterflooding; this allowed the impact of a heterogeneous brine distribution on lean gas HNP performance to be evaluated.

(a)

(b)

Figure 8. Illustration of two brine distribution scenarios: (**a**) relatively uniform distribution of brine and oil within the core plug matrix, achieved by saturating the core plug with brine and oil before fracturing (i.e., Sample #2); (**b**) brine accumulation near the fracture surfaces caused by saturating the core plug with brine after fracturing (i.e., Sample #3).

The findings above could be important for the design of field-scale lean gas injection in tight oil systems: for reservoirs that have recently undergone hydraulic fracturing or waterflooding, water that has accumulated around high-permeability flow paths (e.g., primary or secondary fractures) or injection wells may impair lean gas injectivity and reduce EOR performance. Appropriate reservoir conditioning and injection strategies need to be considered to mitigate this effect.

Impact of fractures on lean gas HNP. The flow of liquid hydrocarbons and gas in tight oil systems is significantly influenced by the presence of fractures [19]. To assess the recovery efficiency associated with induced fractures during the EOR process, intact core plugs (Sample #1 and #2) were artificially fractured under in situ stress conditions after completing three cycles of lean gas HNP. Four additional HNP cycles were then conducted under similar conditions (i.e., similar injection, soaking, and production pressure and durations), resulting in significant improvements in oil recovery, i.e., an additional 18.5% for Sample #1 and 17.5% for Sample #2, (Figure 5).

Post-HNP CT scans (Figure 9a–c) revealed extensive fracture networks that formed due to the combination of artificial fracturing in Sample #1 and #2, and repeat HNP cycles in Sample #1, #2 and #3. These fractures are believed to significantly improve the contact area between the injected gas and in-place oil, thereby enhancing mass transfer and fluid flow within the low-permeability matrix–fracture system. This improved connectivity is considered as a key factor controlling the observed recovery enhancement for Sample #1 and Sample #2 after fracturing [8,23,28,29].

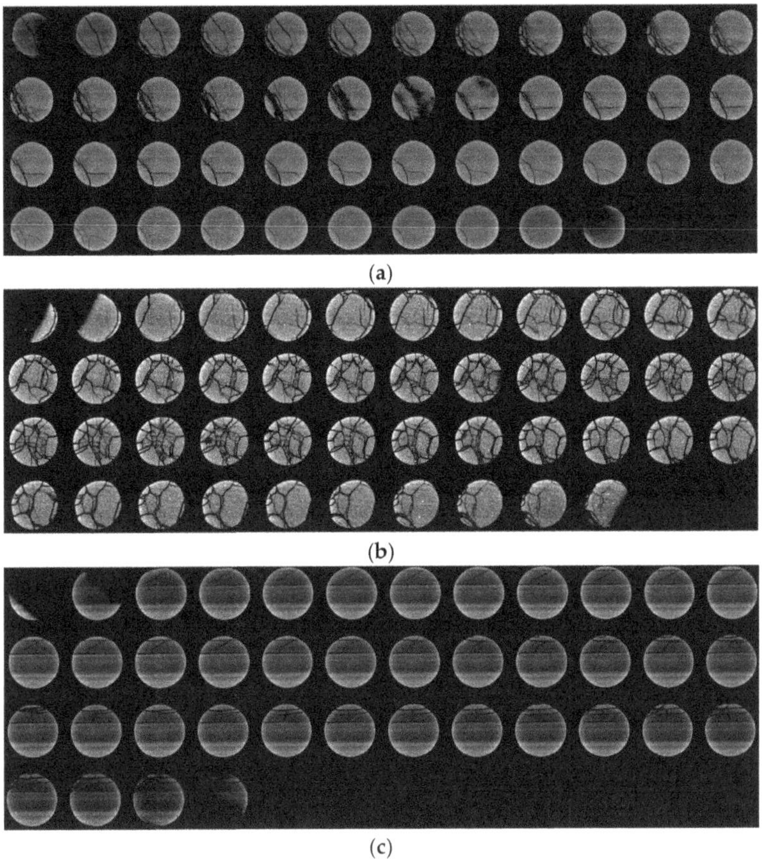

Figure 9. Post-HNP CT scan images (*X-Y* direction) for the fractured core plug samples: (**a**) Sample #1, (**b**) Sample #2, and (**c**) Sample #3. Fractures are detected from top to bottom of the core plug. Images of Sample #1 were adapted from Song et al. [17].

7. Conclusions

This laboratory study evaluated the influence of brine saturation and distribution on cyclic lean gas injection using low-permeability core plug samples collected from Montney Formation, Canada. The key conclusions of the work are summarized below:

1. Three cycles of lean gas HNP were performed on intact core samples (Sample #1 and #2), resulting in recovery factors of 4.6% and 2.5%, respectively.
2. Four additional lean gas HNP cycles were performed on the core plugs (under similar conditions to the intact core plugs) following artificial fracturing. This led to an increase in the recovery factors of 23.1% for Sample #1 and 20% for Sample #2.
3. The presence of fractures significantly improves oil recovery by increasing the contact area between injected gas and in-place oil, which enhances mass transfer and fluid flow within the matrix-fracture system.
4. Lower brine saturation within the matrix normally contributes to a higher oil recovery during the lean gas HNP process.
5. In addition to saturation, water distribution is another critical factor influencing the recovery performance. For the first time, this study quantified the negative effect of heterogeneous brine distribution on recovery performance for lean gas HNP in tight

oil reservoirs. It was determined that brine concentrated near fracture surfaces for one of the samples (Sample #3) significantly hindered lean gas diffusion into the matrix and reduced oil recovery (i.e., 1.2% only).
6. For reservoirs that have just undergone hydraulic fracturing or waterflooding, appropriate reservoir conditioning and injection strategies should be applied to mitigate the adverse effects of heterogeneous brine distribution on EOR performance.

Author Contributions: Conceptualization, C.S., A.G. and C.R.C.; Methodology, C.S., A.G. and C.R.C.; Formal analysis, C.S. and A.G.; Investigation, C.S., A.G. and C.R.C.; Writing—original draft, C.S.; Writing—review & editing, C.R.C.; Supervision, A.G. and C.R.C.; Funding acquisition, C.R.C. All authors have read and agreed to the published version of the manuscript.

Funding: This research was funded by Natural Sciences and Engineering Research Council of Canada (NSERC) through an Appliance grant [ALLRP 568627-21].

Data Availability Statement: The original contributions presented in this study are included in the article. Further inquiries can be directed to the corresponding author.

Acknowledgments: The authors gratefully thank the sponsors of the Tight Oil Consortium hosted at the Department of Earth, Energy and Environment at the University of Calgary. Chris Clarkson would like to acknowledge ARC Resources and Ovintiv for support of his Research Chair in the field of Subsurface Transitional Energy Pathways ('STEPs') at the University of Calgary, and Matt and Tara Brister for support of his Tamaratt Professorship in Transitional Energy Research. The authors further thank Natural Sciences and Engineering Research Council of Canada (NSERC) for providing funding for this work through an Alliance grant (ALLRP 568627-21).

Conflicts of Interest: The authors declare no conflict of interest.

References

1. Song, C.; Yang, D. Experimental and Numerical Evaluation of CO_2 Huff-n-puff Processes in Bakken Formation. *Fuel* **2017**, *190*, 145–162. [CrossRef]
2. Hoffman, B.T. Huff-N-Puff Gas Injection Pilot Projects in the Eagle Ford. In Proceedings of the SPE Canada Unconventional Resources Conference, Calgary, AB, Canada, 13–14 March 2018; Paper SPE-189816-MS. [CrossRef]
3. Ghanizadeh, A.; Song, C.; Hamdi, H.; Clarkson, C.R. Experimental and Computational Evaluation of Cyclic Solvent Injection in Fractured Tight Hydrocarbon Reservoirs. *Sci. Rep.* **2021**, *11*, 9497. [CrossRef]
4. Tang, W.; Sheng, J.J. Huff-n-puff Gas Injection or Gas Flooding in Tight Oil Reservoirs? *J. Pet. Sci. Eng.* **2022**, *208*, 109725. [CrossRef]
5. Milad, M.; Jinin, R.; Sidek, A.; Imqam, A.; Tarhuni, M. Huff-n-Puff Technology for Enhanced Oil Recovery in Shale/Tight Oil Reservoirs: Progress, Gaps, and Perspectives. *Energy Fuels* **2021**, *35*, 17279–17333. [CrossRef]
6. Ma, M.; Zhang, Q.; Emami-Meybodi, H. A Proxy-based Workflow for Screening and Optimizing Cyclic CO_2 Injection in Shale Reservoirs. *Fuel* **2025**, *400*, 135742. [CrossRef]
7. Emami-Meybodi, H.; Ma, M.; Zhang, F.; Rui, Z.; Rezaeyan, A.; Ghanizadeh, A.; Hamdi, H.; Clarkson, C.R. Cyclic Gas Injection in Low-Permeability Oil Reservoirs: Progress in Modeling and Experiments. *SPE J.* **2024**, *29*, 6217–6250. [CrossRef]
8. Thomas, F.B.; Piwowar, M.; Noroozi, M.; Gibb, W.; Marin, J.; Zhang, H. Experimental Measurements of Montney and Duvernay Gas-Cycling Enhanced Oil Recovery GCEOR. In Proceedings of the SPE Canada Unconventional Resources Conferences, Virtual, 29 September–2 October 2020; Paper 199994-MS. [CrossRef]
9. Torabi, F.; Asghari, K. Effect of Operating Pressure, Matrix Permeability and Connate Water Saturation on Performance of CO_2 Huff-and-puff Process in Matrix-fracture Experimental Model. *Fuel* **2010**, *89*, 2985–2990. [CrossRef]
10. Sheng, J.J. Critical Review of Field EOR Projects in Shale and Tight Reservoirs. *J. Pet. Sci. Eng.* **2017**, *159*, 654–665. [CrossRef]
11. Zhu, C.; Guo, W.; Wang, Y.; Li, Y.; Gong, H.; Xu, L.; Dong, M. Experimental Study of Enhanced Oil Recovery by CO_2 Huff-n-puff in Shales and Tight Sandstones with Fractures. *Pet. Sci.* **2021**, *18*, 852–869. [CrossRef]
12. Liu, F.; Kang, Y.; Zhang, M.; Hu, Y.; Huang, Y.; Li, L.; Pan, H. Comparative Study on the Macroscopic Exploitation and Microscopic Mobilization Characteristics of CO_2 and N_2 HNP to Enhance Shale Oil Recovery. *Energy Fuels* **2024**, *38*, 5974–5987. [CrossRef]
13. Xia, Y.; Xin, X.; Yu, G.; Wang, Y.; Lei, Z.; Zhang, L. Design of CO_2 Huff-n-Puff Parameters for Fractured Tight Oil Reservoirs Considering Geomechanical Effects. *Processes* **2024**, *12*, 2777. [CrossRef]

14. Du, D.; Pu, W.; Jin, F.; Liu, R. Experimental Study on EOR by CO_2 Huff-n-puff and CO_2 Flooding in Tight Conglomerate Reservoirs with Pore Scale. *Chem. Eng. Res. Des.* **2020**, *156*, 425–432. [CrossRef]
15. Mahzari, P.; Mitchell, T.M.; Jones, A.P.; Oelkers, E.H.; Striolo, A.; Iacoviellom, F.; Shearing, P.R.; Juri, J.E. Novel Laboratory Investigation of Huff-n-puff Gas Injection for Shale Oils under Realistic Reservoir Conditions. *Fuel* **2021**, *284*, 118950. [CrossRef]
16. Wang, L.; Tian, Y.; Yu, X.; Wang, C.; Yao, B.; Wang, S.; Winterfeld, P.H.; Wang, X.; Yang, Z.; Wang, Y.; et al. Advances in Improved/Enhanced Oil Recovery Technologies for Tight and Shale Reservoirs. *Fuel* **2017**, *210*, 425–445. [CrossRef]
17. Song, C.; Ghanizadeh, A.; Younis, A.; Clarkson, C.R. Experimental Evaluation of the Impact of Induced Fractures on Cyclic Solvent Injection Performance in Low-Permeability Hydrocarbon Reservoirs. *Fuel* **2023**, *354*, 129327. [CrossRef]
18. Shilov, E.; Dorhjie, D.B.; Mukhina, E.; Zvada, M.; Kasyanenko, A.; Cheremisin, A. Experimental and Numerical Studies of Rich Gas Huff-n-puff Injection in Tight Formation. *J. Pet. Sci. Eng.* **2022**, *208*, 109420. [CrossRef]
19. Bai, H.; Zhang, Q.; Li, Z.; Li, B.; Zhu, D.; Zhang, L.; Lv, G. Effect of Fracture on Production Characteristics and Oil Distribution during CO_2 Huff-n-puff under Tight and Low-permeability Conditions. *Fuel* **2019**, *246*, 117–125. [CrossRef]
20. Zheng, T.; Yang, Z.; Liu, X.; Luo, Y.; Xiao, Q.; Zhang, Y.; Zhao, X. Understanding Immiscible Natural Gas Huff-N-Puff Seepage Mechanism in Porous Media: A Case Study of CH4 Huff-n-Puff by Laboratory Numerical Simulations in Chang-7 Tight Core. *Nat. Resour. Res.* **2021**, *30*, 2397–2411. [CrossRef]
21. Guo, Y.; Shi, Y.; Mohanty, K. Comparison of the Effectiveness of CO_2 and Hydrocarbon Gas for Shale Huff 'n' Puff. *SPE J.* **2025**, *30*, 2942–2957. [CrossRef]
22. Li, H.; Zheng, S.; Yang, D. Enhanced Swelling Effect and Viscosity Reduction of Solvent(s)/CO_2/Heavy-oil Systems. *SPE J.* **2013**, *18*, 695–707. [CrossRef]
23. Thomas, F.B.; Noroozi, M.; Apil, R.; White, A. Considerations for Gas Cycling EOR. In Proceedings of the SPE Workshop: The Duvernay Liquid Rich Shale—What Makes It Different and How We Optimize it? Calgary, AB, Canada, 18 October 2017.
24. Ghedan, S. Global Laboratory Experience of CO_2-EOR Flooding. In Proceedings of the SPE/EAGE Reservoir Characterization and Simulation Conference, Dhabi, United Arab Emirates, 19–21 October 2009; Paper SPE 125581. [CrossRef]
25. Pu, W.; Wei, B.; Jin, F.; Li, Y.; Jia, H.; Liu, P.; Tang, Z. Experimental Investigation of CO_2 Huff-n-puff Process for Enhancing Oil Recovery in Tight Reservoirs. *Chem. Eng. Res. Des.* **2016**, *111*, 269–276. [CrossRef]
26. Assef, Y.; Kantzas, A.; Almao, P.P. Numerical Modelling of Cyclic CO_2 Injection in Unconventional Tight Oil Resources; Trivial Effects of Heterogeneity and Hysteresis in Bakken Formation. *Fuel* **2019**, *236*, 1512–1528. [CrossRef]
27. Morel, D.; Bourbiaux, B.; Latil, M.; Thiebot, B. Diffusion Effects in Gas Flooded Light Oil Fractured Reservoirs. *SPE Adv. Technol. Ser.* **1993**, *1*, 100–109. [CrossRef]
28. Silva, F.V.; Belery, P. Molecular Diffusion in Naturally Fractured Reservoirs: A Decisive Recovery Mechanism. In Proceedings of the SPE Annual Technical Conference and Exhibition, San Antonio, TX, USA, 8 October 1989; Paper SPE-19672-MS. [CrossRef]
29. Ellafi, A.; Jabbari, H. Understanding the Mechanisms of Huff-n-Puff, CO_2-EOR in Liquid-Rich Shale Plays: Bakken Case Study. In Proceedings of the SPE Canada Unconventional Resources Conference, Virtual, 29 September–2 October 2020; Paper SPE-200001-MS. [CrossRef]
30. Ghanizadeh, A.; Song, C.; Clarkson, C.R.; Younis, A. Relative Permeability of Tight Hydrocarbon Systems: An Experimental Study. *Fuel* **2021**, *294*, 119487. [CrossRef]
31. Jone, S.C. A Technique for Faster Pulse-Decay Permeability Measurements in Tight Rocks. *SPE Form. Eval.* **1997**, *12*, 19–25. [CrossRef]
32. Keneti, S.A.R.; Wong, R.C.K. Investigation of Anisotropic Behavior of Montney Shale Under Indirect Tensile Strength Test. In Proceedings of the Canadian Unconventional Resources & International Petroleum Conference, Calgary, AB, Canada, 19–21 October 2010; Paper CSUS/SPE 138104.
33. Vaisblat, N.; Shokri, A.R.; Ayranci, K.; Harris, N.; Chalaturnyk, R.J. Significance of Rock Compositional Control on Geomechanical Properties and Hydraulic Fracturing of the Montney Formation, Western Canadian Basin. In Proceedings of the Unconventional Resources Technology Conference (URTeC), Brisbane, Australia, 18–19 November 2019; Paper URTEC-198199-MS.
34. Becerra Rondon, D.M. Geological Characterization and Evaluation of Reservoir Quality in the Montney Formation in West-Central Alberta. Doctoral Thesis, University of Calgary, Calgary, AB, Canada, 2022. Available online: http://hdl.handle.net/1880/114729 (accessed on 10 October 2025).
35. Davarpanah, A.; Mirshekari, B. Experimental Study of CO_2 Solubility on the Oil Recovery Enhancement of Heavy Oil Reseroirs. *J. Therm. Anal. Calorim.* **2020**, *139*, 1161–1169. [CrossRef]
36. Zhao, Y.; Yu, J.; Shi, H.; Guo, J.; Liu, D.; Lin, J.; Song, S.; Wu, H.; Gong, J. Study of Methane Solubility Calculation Based on Modified Henry's Law and BP Neural Network. *Processes* **2024**, *12*, 1901. [CrossRef]
37. Kim, Y.S.; Lim, B.D.; Lee, J.E.; Lee, C.S. Solubilities of Carbon Dioxide, Methane, and Ethane in Sodium Chloride Solution Containing Gas Hydrate. *J. Chem. Eng. Data* **2008**, *53*, 1351–1354. [CrossRef]

38. Riazi, M.R.; Vera, J.H. Method to Calculate the Solubility of Light Gases in Petroleum and Coal Liquid Fractions on the Basis of Their P/N/A Composition. *Ind. Eng. Chem. Res.* **2005**, *44*, 186–192. [CrossRef]
39. Soltanian, M.R.; Amoois, M.A.; Cole, D.R.; Darrah, T.H.; Graham, D.E.; Pfiffner, S.M.; Phelps, T.J.; Moortgat, J. Impacts of Methane on Carbon Dioxide Storage in Brine Formations. *Groundwater* **2018**, *56*, 176–186. [CrossRef]

energies

MDPI

Article

Phase-Specific Mixture of Experts Architecture for Real-Time NO_x Prediction in Diesel Vehicles: Advancing Euro 7 Compliance

Maksymilian Mądziel

Faculty of Mechanical Engineering and Aeronautics, Rzeszow University of Technology, 35-959 Rzeszow, Poland; mmadziel@prz.edu.pl

Abstract

The implementation of Euro 7 emission standards demands advanced real-time NO_x monitoring systems for diesel vehicles. Existing unified models inadequately capture phase-dependent emission mechanisms during cold-start, urban, and highway operation. This study develops a novel Mixture of Experts (MoE) architecture with data-driven phase classification based on aftertreatment thermal dynamics. Real-world data from a Euro 6d commercial vehicle (3247 PEMS samples) were classified into three phases, cold (<70 °C coolant temperature), hot low-speed (<90 km/h), and hot high-speed (≥90 km/h), validated through t-SNE analysis (silhouette coefficient = 0.73). The key innovation integrates thermal–kinematic domain knowledge with specialized XGBoost regressors, achieving R^2 = 0.918 and a 58% RMSE reduction versus unified models (RMSE = 1.825 mg/s). The framework operates within real-time constraints (1.5 ms inference latency), integrating autoencoder-based anomaly detection (95.2% sensitivity) and Model Predictive Control (11–13% NO_x reduction). This represents the first systematic phase-specific NO_x modeling framework with validated Euro 7 OBM compliance capability, providing both methodological advances in expert allocation strategies and practical solutions for next-generation emission control systems.

Keywords: NO_x emissions; motor vehicles; mixture of experts; Euro 7; phase-specific modeling; real-time monitoring

1. Introduction

Vehicular emissions, particularly oxides of nitrogen (NO_x), remain a critical challenge in the context of urban air quality and public health [1,2]. Recent regulatory shifts, including the adoption of the Euro 7 standards, have dramatically raised the requirements for real-world tailpipe emission control [3,4]. These new standards mandate not only lower permissible emission levels but also the continuous, on-road, real-time monitoring of critical pollutants over the entire lifetime of the vehicle, resulting in unprecedented technical complexity for manufacturers, regulators, and data scientists alike [5,6].

Traditional approaches to emission compliance—relying on periodic laboratory or PEMS (Portable Emissions Measurement Systems) tests—are insufficient for capturing the full spectrum of real-world driving behaviors, transient events, and vehicle aging effects [7,8]. As a result, the automotive industry is transitioning towards robust On-Board Monitoring (OBM) systems, which must provide high-frequency, sensor-driven data streams, support over-the-air reporting, and trigger protective actions when emission exceedances are detected [9,10]. This paradigm shift is also driving the adoption of advanced data analytics and artificial intelligence (AI), as evidenced by contemporary research and frameworks leveraging machine learning for phase-specific emission prediction [11,12].

The European Union's Euro 7 emission standard represents the most stringent regulatory framework to date for controlling vehicular pollutants, mandating a 50% reduction in NO_x limits for diesel engines compared with Euro 6 and extending durability requirements to 10 years or 200,000 km [13,14]. These regulations demand the continuous real-world monitoring of NO_x, PM, and NH_3 emissions through On-Board Monitoring (OBM) systems, shifting the paradigm from periodic laboratory tests to embedded real-time surveillance in passenger and commercial vehicles [15,16]. To meet the Euro 7 requirements, predictive models must operate within strict latency constraints (<1.5 ms) while maintaining a high accuracy throughout extended operational lifespans. However, accurately forecasting NO_x formation in diesel aftertreatment systems remains challenging due to complex thermal dependencies, transient engine dynamics, and variable driving conditions, especially during cold-start and low-temperature operation phases where SCR catalyst efficiency drops substantially below optimal temperature windows [17,18]. Traditional physics-based models excel in controlled environments but fail to generalize to transient real-world cycles, and unified machine learning approaches, while demonstrating R^2 values above 0.90, do not explicitly account for distinct emission mechanisms across operational phases.

A critical gap in the current literature lies in phase-discriminative emission modeling: most studies on cold-start NO_x consider passenger vehicles without systematic criteria for phase transitions in commercial diesel applications [19,20], and SCR temperature thresholds are often arbitrary and unvalidated [21,22]. Similarly, ensemble learning methods applied to emission prediction rely on generic boosting or bagging techniques without leveraging the domain knowledge of aftertreatment thermal dynamics [23], and few approaches address the computational constraints of embedded OBM systems [24].

This study addresses these gaps by developing a novel phase-discriminative framework based on a Mixture of Experts (MoE) architecture for real-time NO_x prediction. Data were collected from a Euro 6d commercial diesel vehicle equipped with Portable Emissions Measurement Systems with 3247 samples covering cold-start, urban, and highway conditions. Data-driven criteria for phase classification—cold (coolant temperature < 70 °C, low SCR temperature), hot low-speed (<90 km/h), and hot high-speed (≥90 km/h)—were established and validated using t-SNE clustering and silhouette analysis. Advanced feature engineering captures temporal dependencies via rolling statistics and exponentially weighted moving averages. The MoE framework dynamically routes inputs to specialized XGBoost regressors optimized for each operational phase, achieving an overall R^2 of 0.918 and a 58% RMSE reduction compared with unified models. The system integrates autoencoder-based anomaly detection (95.2% sensitivity) and Model Predictive Control for proactive NO_x reduction (11–13%), all within real-time latency constraints. Integration with PTV VISSIM demonstrates applicability for traffic microsimulation and urban air quality assessment.

The contributions of this work include the first systematic phase-specific NO_x emission modeling framework, novel expert allocation strategies informed by thermal and kinematic domain knowledge, and comprehensive real-world validation for Euro 7 OBM compliance. The results provide a foundation for next-generation emissions control systems, offering both theoretical advances and practical solutions for automotive OEMs and regulatory bodies.

2. Materials and Methods

The general workflow of this study is illustrated in Figure 1. First, a vehicle equipped with a Portable Emission Measurement System (PEMS) was instrumented to collect synchronized data on NO_x, CO_2, THC, CO, coolant temperature, speed, and acceleration under various urban and highway driving conditions. Raw time-series data were preprocessed to classify each sample into one of three operational phases—cold (<70 °C coolant), hot low-speed (≥70 °C and ≤90 km/h), and hot high-speed (≥70 °C and >90 km/h)—to reflect distinct aftertreatment thermal regimes. Feature engineering involved the systematic transformation of raw time-series data into engineered variables that capture emission-relevant patterns. This process included (1) temporal indicators such as time elapsed since engine start and temperature rate of change (exponentially weighted moving averages); (2) kinematic rolling statistics computed over a 30 s windows (mean velocity, acceleration variance); (3) thermal trend metrics designed to capture aftertreatment system conditioning dynamics. These engineered features enable the predictive models to capture transient emission behaviors that depend on historical engine thermal state and driving patterns, rather than relying solely on instantaneous measurements. Three parallel modeling approaches were developed: phase-specific XGBoost regressors trained on single-phase data, a unified multi-phase XGBoost model, and a Mixture of Experts architecture that dynamically routes inputs to phase-specialist regressors. Model selection prioritized the Mixture of Experts approach based on cross-validation performance metrics (R^2, RMSE, MAE), which demonstrated superior accuracy across all operational phases. Model interpretability analysis was conducted to validate that identified features genuinely control NO_x formation rather than representing algorithm-specific artifacts. Three complementary feature attribution methods—XGBoost built-in importance, permutation-based importance, and SHAP-based importance (KernelExplainer)—were applied to each phase-specific expert. Consensus validation across these independent methods confirmed that velocity and acceleration consistently emerge as dominant predictors (ranks 1–2), while engine coolant temperature maintains stable secondary importance (rank 3). Partial dependence analysis further characterized the marginal effects of each feature on NO_x emissions, revealing phase-specific functional relationships and feature interactions. This comprehensive interpretability framework strengthens confidence that the Mixture of Experts model captures genuine physical relationships controlling NO_x formation, supporting confidence in its applicability to real-world Euro 7 compliance applications. An advanced analytics pipeline incorporated autoencoder-based anomaly detection to flag abnormal emission events and integrated the predictive model into a Model Predictive Control framework for real-time NO_x mitigation. Finally, model performance was validated through cross-validation metrics (R^2, RMSE, MAE), inference latency measurements on an ARM (Advanced RISC Machine) Cortex-A78 platform, anomaly detection sensitivity assessments, MPC-based emission reduction evaluations, and comprehensive interpretability validation demonstrating feature importance consensus across three independent attribution methods.

Data were collected from a 2020 vehicle equipped with a comprehensive real-world driving emissions monitoring system during extended road testing campaigns. The test vehicle was a Euro 6d-compliant commercial diesel vehicle representative of modern European light commercial vehicle technology, featuring a 1.56 L turbocharged diesel engine producing a maximum power of 88 kW at 3500 rpm and a maximum torque of 300 Nm at 1750 rpm. The vehicle was equipped with six-speed manual transmission and had a curb weight of 1429 kg, positioning it as a typical representative of the European LCV market segment that must comply with increasingly stringent emission regulations while maintaining practical utility for commercial applications.

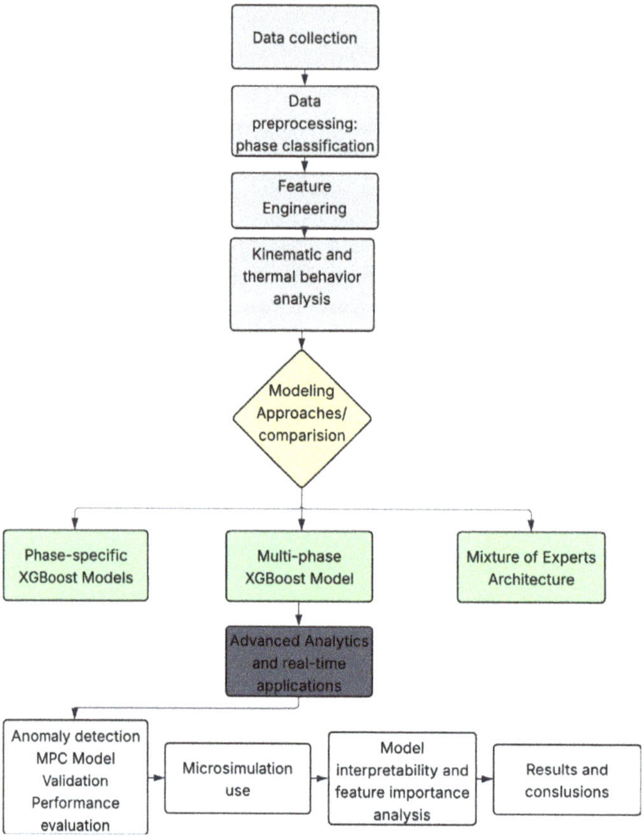

Figure 1. Flow chart of the work.

The vehicle powertrain featured advanced BlueHDi technology, representing state-of-the-art Euro 6d emission control systems that serve as a technological bridge towards future Euro 7 compliance requirements. The engine incorporated direct injection fuel delivery via a common rail with multiple injection strategies optimized for emission control and fuel economy. Most critically for this study, the vehicle was equipped with a comprehensive aftertreatment system that included a diesel oxidation catalyst (DOC), a diesel particulate filter (DPF), and a selective catalytic reduction (SCR) system with an ammonia slip catalyst (ASC). This multi-stage aftertreatment configuration represents the current industry standard for meeting the Euro 6d requirements and provides emissions characteristics that closely approximate expected Euro 7 system behavior, particularly regarding NO_x control strategies and cold-start emission patterns [25,26].

The comprehensive aftertreatment system was particularly relevant to the objectives of this study, as Euro 6d vehicles must demonstrate Real Driving Emission (RDE) compliance with conformity factors that closely approximate the Euro 7 requirements [27]. The SCR system utilized advanced urea injection control with precise dosing strategies designed to maintain a NO_x conversion efficiency exceeding 85% across a wide range of operating conditions while minimizing ammonia slip [28,29]. The integrated DPF achieved a particulate filtration efficiency exceeding 99% with intelligent regeneration management that optimized soot oxidation timing based on real-time operating conditions. This technological configuration ensures that emission characteristics closely represent the performance expectations for Euro 7 systems, making the collected data highly relevant for developing predictive models applicable to next-generation emission control requirements.

The vehicle was instrumented with a Portable Emissions Measurement System (PEMS) configured according to European RDE testing protocols and designed for continuous high-frequency data acquisition during real-world driving conditions. The PEMS installation included a comprehensive suite of gas analyzers utilizing non-dispersive infrared (NDIR) detection for CO_2 and CO measurement, electrochemical sensors for NO_x quantification, and flame ionization detection (FID) for total hydrocarbon (THC) analysis [30,31]. All emission measurements were recorded as instantaneous mass flow rates in g/s units, providing the direct quantification of pollutant formation rates rather than distance-normalized values, which enabled the more precise modeling of transient emission behavior and phase-dependent emission characteristics. A diagram showing the vehicle and the PEMS testing equipment is presented in Figure 2.

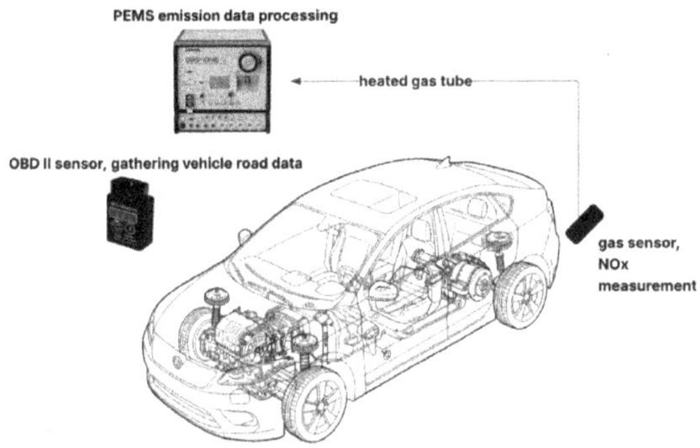

Figure 2. Experimental setup showing PEMS instrumentation, gas analyzer connections, OBD-II sensor integration, and heated gas sampling tube configuration for real-world emission data collection.

Vehicle dynamics monitoring incorporated high-precision GPS for longitude and latitude tracking, allowing the calculation of the road gradient and vehicle trajectory parameters that influence emission formation. Instantaneous vehicle velocity and acceleration were derived from GPS data with temporal smoothing to minimize measurement noise while preserving the characteristics of transient behavior essential for emission modeling. Engine operating parameters were monitored through a direct OBD-II interface connection, providing real-time access to engine coolant temperature, which served as the primary indicator for defining cold-start versus hot operation phases. The entire measurement system operated at a 1 Hz sampling frequency, generating synchronized time-series data that captured both rapid transient events and longer-term emission trends across different driving conditions and engine thermal states.

The dataset collected was systematically preprocessed to establish consistent phase classification criteria that reflect different emission formation mechanisms observed in modern diesel aftertreatment systems. The engine operating conditions were classified into three fundamental phases based on thermal and dynamic criteria that are directly correlated with the performance characteristics of the aftertreatment system. The cold phase was defined as the operation that occurs when the engine coolant temperature remained below 70 °C, representing the critical cold-start period when the aftertreatment components have not reached the optimal operating temperature and the efficiency of emission control is significantly reduced. During this phase, SCR catalyst activity is minimal due to insufficient thermal activation, and DPF regeneration capability is limited, resulting in characteristically elevated emission levels that can dominate total trip emissions despite representing only the initial 1–2 km of driving.

The hot operation was subdivided into two distinct phases based on vehicle velocity characteristics that reflect different aftertreatment system operating regimes. Hot low-speed operation was defined as an engine coolant temperature ≥ 70 °C combined with a vehicle velocity ≤ 90 km/h, representing urban driving conditions where aftertreatment systems operate under variable load conditions with frequent acceleration–braking cycles that can affect the consistency of emission control performance. This phase-based classification approach captures the fundamental differences in emission formation mechanisms that result from the thermal dynamics of the aftertreatment system and the variations in operational efficiency under different driving conditions.

Recognizing that emission formation in modern diesel vehicles exhibits strong temporal dependencies related to aftertreatment system thermal dynamics and catalytic activity, several sophisticated temporal features were engineered to capture these transient effects [32,33]. Time-dependent features included time_from_start, representing the elapsed time since engine start, which directly correlates with aftertreatment system thermal conditioning and emission control system activation timing. Rolling statistical features were calculated using 30 s moving windows to capture short-term variations in driving patterns while maintaining computational efficiency suitable for real-time applications. These features included V_mean_win (velocity moving average) and a_std_win (acceleration standard deviation), which characterize driving intensity and transient behavior patterns that directly influence exhaust temperature and aftertreatment system performance.

Advanced thermal trend analysis incorporated exponentially weighted moving averages (EWMAs) with adaptive weighting parameters to model the thermal response characteristics of the aftertreatment system. The temp_trend feature utilized EWMA calculations with $\alpha = 2/(\text{span} + 1)$ where span = 30 s, providing responsive tracking of thermal conditioning rates while suppressing measurement noise that could degrade model performance. Additional gradient-based features included road gradient calculations and jerk (acceleration rate of change) to account for topographical influences on engine load and exhaust thermal management, which directly impact SCR operating temperature and NO_x conversion efficiency. These temporal features were specifically designed to capture the dynamic relationships between driving conditions, the thermal state of the aftertreatment system, and the emission formation processes that are critical to accurate real-time emission prediction.

The modeling approach incorporated three distinct architectural strategies to systematically evaluate optimal approaches for phase-dependent emission prediction in real-time automotive applications. Phase-specific XGBoost models were developed as individual gradient boosting regressors trained separately for each operational phase, enabling specialized optimization for the distinct emission formation mechanisms observed during cold-start, urban, and highway operations. This approach takes advantage of the principle that the performance characteristics of the aftertreatment system exhibit fundamentally different behavior patterns in thermal and dynamic operating regimes, suggesting that specialized models may achieve a superior predictive accuracy compared with unified approaches.

Multi-phase unified modeling utilized a single XGBoost regressor with the operational phase encoded as categorical features through one-hot encoding, enabling the model to learn phase-dependent relationships while maintaining a unified architecture suitable for streamlined deployment. The Mixture of Experts architecture represented the most sophisticated approach, incorporating dynamic ensemble methods that combine phase-specific models with intelligent routing algorithms based on real-time engine state determination. This approach enables adaptive model selection that automatically directs prediction requests to the most appropriate specialized model based on current operating conditions, potentially achieving optimal prediction accuracy while maintaining explainable model behavior suitable for regulatory compliance applications.

The comprehensive modeling framework incorporated advanced anomaly detection and predictive control capabilities to demonstrate integrated real-time emission management functionality. A deep autoencoder network was implemented for continuous anomaly detection in emission patterns, using an encoder–decoder architecture with a progressive reduction in dimensionality from five input features through 8-node and 3-node hidden layers before reconstruction through symmetric decoder layers. The autoencoder was trained exclusively on normal operation data from hot phases to establish baseline emission pattern recognition, enabling the detection of anomalous emission behavior during cold-start periods or aftertreatment system malfunctions through reconstruction error analysis with threshold determination based on the 95th percentile validation set performance.

The integration of Model Predictive Control (MPC) demonstrated practical emission optimization capabilities through the simplified implementation of the real-time control framework. The MPC system utilized trained XGBoost regressors for NO_x emission forecasting with acceleration command optimization over candidate control input spaces, enabling proactive emission minimization through predictive driving pattern adjustment. This integrated approach demonstrates the potential to combine advanced emission modeling with real-time control strategies to achieve proactive emission management that could support Euro 7 compliance through intelligent vehicle operation optimization rather than relying solely on passive aftertreatment system performance.

2.1. Testing Protocol and Data Collection Procedure

The experimental data were collected through systematic real-world driving campaigns executed according to the following standardized procedure:

Pre-Test Configuration:

- Vehicle inspection: confirmation of nominal engine and aftertreatment system status.
- PEMS calibration: zero and span verification for all gas analyzers (NO_x, CO, CO_2, THC) as per European RDE testing protocols.
- GPS system initialization with minimum 8 satellite acquisition for spatial accuracy.
- OBD-II interface verification for stable ECU communication at 1 Hz sampling frequency.
- Cross-validation of engine coolant temperature sensor readings against OBD-II ECU values.

Test Execution:

The testing protocol comprised three consecutive driving scenarios on the same test day to minimize the variability in environmental conditions:

- Cold-Start Phase: Vehicle started at ambient temperature (18–22 °C). Initial 1–2 km driven at conservative speeds (12–25 km/h average) without aggressive maneuvers, with continuous data acquisition from engine start until coolant temperature reached 70 °C. This phase captured the critical early emission spike period when aftertreatment systems operate below optimal efficiency.
- Urban Driving Cycle: Following the cold-start phase, 30–40 min of urban driving on representative city streets with frequent acceleration/deceleration cycles, traffic light stops, and parking maneuvers. Vehicle was maintained at speeds typically 15–50 km/h with engine coolant temperature $\geq$70 °C throughout this phase.
- Highway Driving Segment: Final 25–35 min of motorway driving at sustained velocities (80–130 km/h) with minimal acceleration variability. This segment allowed the aftertreatment system to reach thermal stability (>85 °C coolant temperature) and demonstrated steady-state emission characteristics.

Data Quality Assurance:

The real-time monitoring during testing included the following:

- The continuous verification of instantaneous NO_x measurements against physical plausibility constraints.
- GPS velocity cross-validation against vehicle CAN-bus speedometer signals (required correlation coefficient > 0.95).
- Temperature sensor signal continuity assessment and thermal gradient plausibility checks.
- The identification of any communication dropouts or sensor malfunctions in real time.

Post-Test Data Processing:

Following test completion, raw data files underwent comprehensive quality assessment including the removal of identified malfunction periods, uniform resampling to 1 Hz time grid to ensure temporal synchronization across all data streams, and the visual inspection of time-series plots to flag anomalous patterns.

2.2. Model Development Workflow—Step-by-Step Implementation

The Mixture of Experts (MoE) framework combines domain knowledge of aftertreatment thermal dynamics with specialized machine learning experts to achieve phase-discriminative NO_x prediction. The architecture consists of three parallel XGBoost regressors, each trained exclusively on one operational phase (cold-start, hot low-speed urban, hot high-speed highway), coupled with a deterministic routing mechanism that dynamically selects the most appropriate expert based on the real-time engine state. This hybrid design leverages the interpretability and regulatory acceptance of gradient-boosted decision trees while maintaining the computational efficiency (<1.5 ms inference latency) required for embedded automotive deployment. The development process encompasses five sequential stages: (1) data segmentation into thermally distinct phases validated through t-SNE clustering, (2) systematic feature engineering capturing kinematic, thermal, and temporal dependencies, (3) phase-specific model optimization through grid search hyperparameter tuning, (4) ensemble integration via deterministic routing, (5) comprehensive validation including cross-validation performance, latency profiling, and interpretability analysis. This section details each stage and presents the complete model architecture.

Step 1: Data segmentation and phase validation

Raw time-series data (N_total = 3247 samples) were segmented into three operational phases using thermal–kinematic criteria:

- Cold phase (n = 205 samples): coolant temperature < 70 °C, representing cold-start operation.
- Hot low-speed phase (n = 1542 samples): coolant temperature $\geq$ 70 °C AND velocity $\leq$ 90 km/h.
- Hot high-speed phase (n = 1500 samples): coolant temperature $\geq$ 70 °C AND velocity > 90 km/h.

Phase separation quality was validated using t-SNE (t-distributed Stochastic Neighbor Embedding) dimensionality reduction with silhouette coefficient analysis (achieved: 0.73), confirming robust phase discrimination with minimal inter-phase data overlap (<5%).

Step 2: Feature engineering

Eighteen engineered features spanning five categories were generated to capture emission dynamics:

- Kinematic: V (velocity, km/h), a (acceleration, m/s^2).
- Thermal: T_coolant (°C), T_lag_30 s, T_lag_60 s (lagged temperatures).
- Temporal: time_from_start (seconds), V_mean_win, a_std_win (30 s rolling statistics).
- Advanced dynamics: jerk (j, m/s^3), road gradient (G, %), EWMA-smoothed features.

- Thermal trend: exponentially weighted moving averages with $\alpha = 2/(30 + 1)$.

Step 3: Phase-Specific Model Training

Three independent XGBoost regressors were trained on phase-specific subsets:

- Expert_Cold: trained on n = 205 cold-phase samples.
- Expert_HotLS: trained on n = 1542 hot low-speed samples.
- Expert_HotHS: trained on n = 1500 hot high-speed samples.

Hyperparameter optimization employed a grid search over n_estimators $\in$ {100, 150}, max_depth $\in$ {5, 7}, learning_rate $\in$ {0.01, 0.05}, and subsample $\in$ {0.6, 0.8}.

Step 4: Mixture of Experts integration

A dynamic routing mechanism was implemented to select the appropriate expert based on real-time engine state:

If (T_coolant < 70 °C) $\rightarrow$ Route to Expert_Cold.

Else If (T_coolant $\geq$ 70 °C AND V $\leq$ 90 km/h) $\rightarrow$ Route to Expert_HotLS.

Else If (T_coolant $\geq$ 70 °C AND V > 90 km/h) $\rightarrow$ Route to Expert_HotHS.

Routing was implemented to operate within <0.5 ms latency, enabling real-time deployment on automotive-grade processors.

Step 5: Model validation and performance assessment

Validation employed 80/20 train/test stratification by operational phase, with performance metrics including R^2 (coefficient of determination), RMSE (root mean squared error), and MAE (mean absolute error). Inference latency was profiled on an ARM Cortex-A78 platform: average expert prediction time 0.8 ms + routing overhead 0.3 ms = 1.1 ms total per sample, compliant with automotive real-time requirements (<1.5 ms constraint).

3. Results

In this chapter, a comprehensive analysis of the real-world driving emissions data is presented. The first stage of this study examines all of the exhaust compounds—including NO_x, CO_2, THC, and CO—alongside vehicle dynamics parameters, validating data quality and exploratory insights. Subsequent sections focus exclusively on NO_x modeling, as this pollutant constitutes the most critical challenge for diesel vehicles. Section 3.1 characterizes the dataset and details the phase classification methodology.

3.1. Dataset Characterization and Phase Classification

The analysis of real-world driving emissions data collected from a Euro 6d diesel commercial vehicle revealed distinct operational characteristics that validate the proposed phase-based modeling approach. The dataset comprised 3247 samples that spanned cold-start, urban, and highway driving conditions with complete PEMS measurements on all emission compounds and vehicle dynamics parameters.

Based on the initial stage of the work, the quality assessment of the data and the execution of exploratory data analysis (EDA) is presented through a correlation matrix plot. The correlation matrix plot plays a key role in the preliminary exploration of the data, as it allows the rapid identification of significant relationships between variables and helps assess data quality [34,35]. It reveals strong correlations between NO_x emissions and vehicle speed and acceleration. Simultaneously, it highlights weak or insignificant correlations—for example, the lack of a direct influence of road gradient on emissions—which permits focusing further analysis on important factors and planning appropriate data transformations or models that consider indirect effects. The correlation matrix plot for the analyzed data is presented in Figure 3.

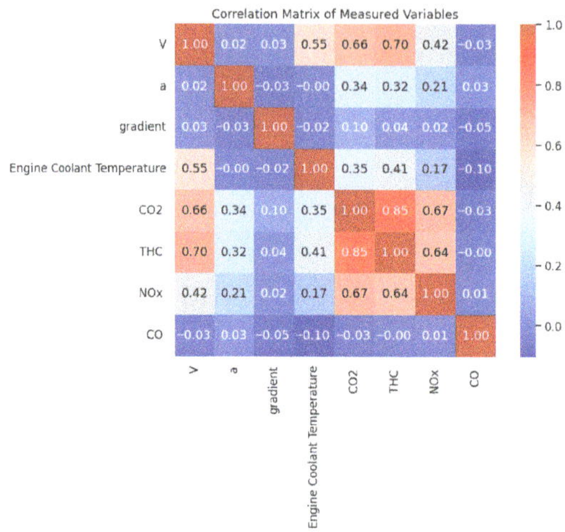

Figure 3. Correlation matrix of measured variables.

The correlation matrix analysis revealed strong inter-relationships among emission compounds and vehicle dynamics variables. NO_x emissions demonstrated significant positive correlations with vehicle velocity (r = 0.42, $p < 0.001$) and acceleration (r = 0.21, $p < 0.001$), confirming the kinematic dependence of emission formation processes. The engine temperature exhibited a moderate correlation with velocity (r = 0.55) and CO_2 emissions (r = 0.35), reflecting the thermal–load relationship in the operation of the diesel engine. Strong correlations were observed between CO_2 and THC (r = 0.85) and between THC and NO_x (r = 0.64), indicating coupled combustion and aftertreatment processes. In particular, the effects of the gradient showed a minimal direct correlation with emissions (r = 0.02), suggesting that the topographic influences are mediated by secondary effects on engine load and the thermal management of the exhaust rather than the direct impact of emissions.

Operational conditions were segmented into three phases based on the engine coolant temperature and vehicle speed, which are crucial to describe the efficiency of the exhaust aftertreatment system. This classification is based on the observed thermal and kinetic characteristics of BlueHDi systems, ensuring a clear separation of emission processes that occur under different thermal and dynamic states. This classification enables the application of phase-targeted emission models, as confirmed by the further analysis of feature distributions and phase separability demonstrated in multidimensional reduction (Figure 4).

The estimation of kernel density reveals distinct statistical characteristics of key variables in the operational phases, providing the quantitative validation of the phase classification methodology. Vehicle velocity distributions demonstrate clear phase-dependent patterns: cold operation exhibits a narrow low-speed distribution (mode = 12.4 km/h) reflecting conservative warm-up driving, while hot phases show broader distributions with hot high-speed operation extending to highway velocities (>100 km/h). The acceleration patterns reveal fundamental behavioral differences: the cold phase maintains the most constrained acceleration range ($\sigma = 0.31$ m/s^2), the high speed shows a bimodal distribution reflecting stop-and-go urban patterns, and the high speed exhibits a near-normal distribution centered around zero, consistent with steady-state highway driving. The engine temperature distributions confirm the effectiveness of the 70 °C threshold criterion, with the cold phase showing a narrow distribution below the threshold (mean = 62.8 °C, $\sigma = 8.4$ °C) and the hot phases exhibiting operating temperature ranges above 70 °C. Most significantly, NO_x emission distributions demonstrate pronounced phase-dependent characteristics: cold operation shows a right-skewed distribution with a heavy tail extending to 0.020 g/s (representing 15-fold increases over hot operation medians), while hot phases exhibit log-normal distributions with significantly lower emission rates (hot low-speed: median = 0.0018 g/s; hot high-speed: median = 0.0032 g/s). These distribution differences achieve statistical significance (Kolmogorov–Smirnov's test: $p < 0.001$ for all pairwise comparisons), confirming that phase-specific modeling approaches are statistically justified and necessary for accurate emission prediction.

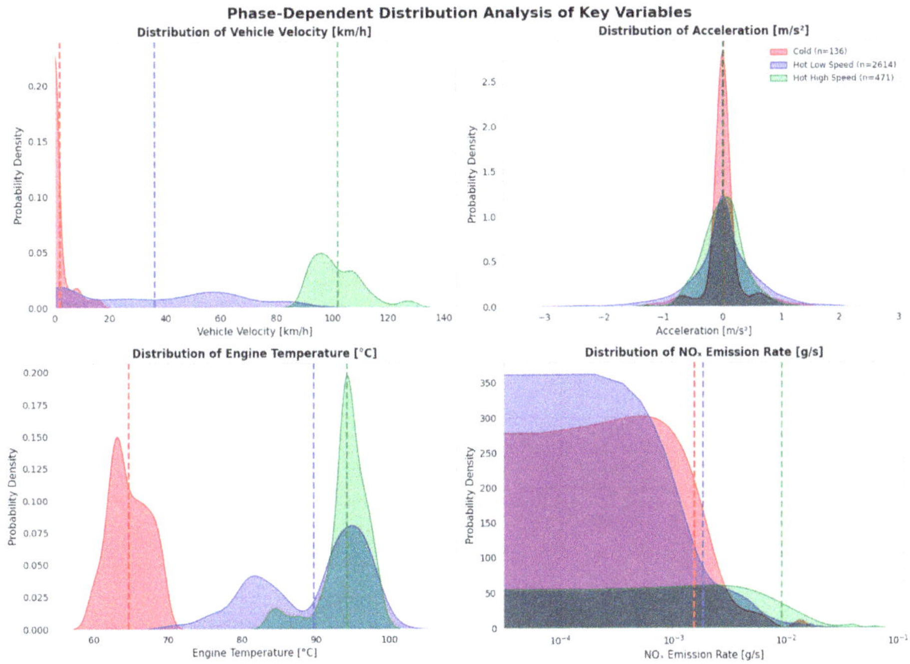

Figure 4. Phase-dependent distribution analysis of key variables.

The subsequent plots present the statistical distributions of the emission data for each engine operational phase. The first plot shown is Figure 5.

The phase-dependent emission distributions demonstrate significant differences in NO_x formation characteristics: the cold-start phase exhibits 8.5 times higher median emissions, 3.4 mg/s, compared with the hot low-speed operation, 1.8 mg/s, and 1.1 times higher than the hot high-speed operation, 3.2 mg/s, confirming the critical impact of aftertreatment thermal conditioning on emission control efficiency. The stacked histogram reveals distinct emission regimes, with the cold phase contributing disproportionately to high-emission events, despite representing only 6.4% of the total samples. Hot operation phases show characteristic low-emission distributions with occasional spikes corresponding to transient acceleration events or aftertreatment regeneration cycles. Subsequent statistical analyses of NO_x emissions for individual operational phases are presented in Figure 6.

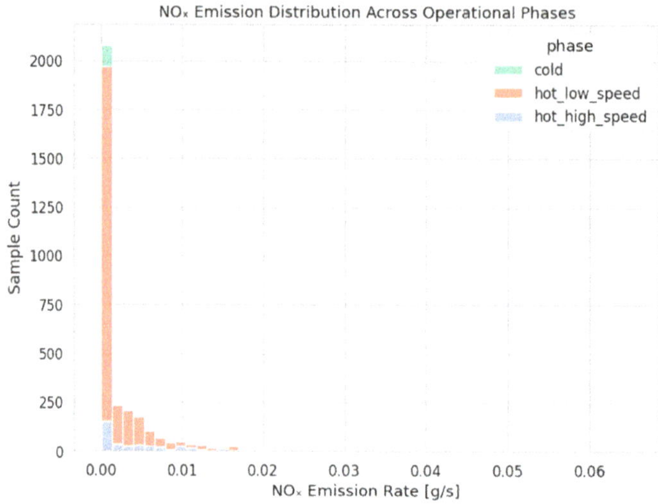

Figure 5. Stacked histogram of NO_x emissions by phase.

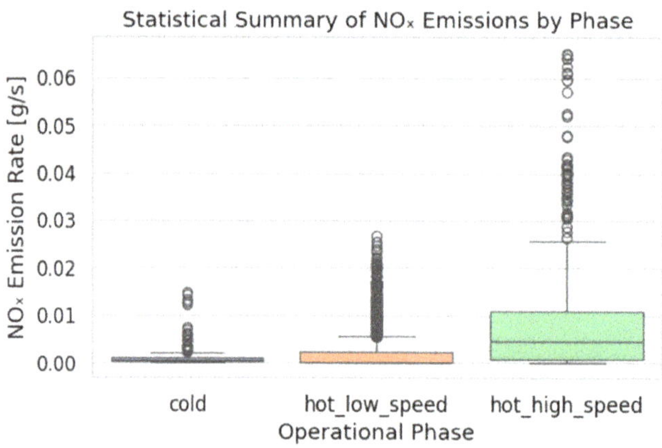

Figure 6. Statistical summary of NO_x emissions by phase.

The statistical summary analysis confirms phase-dependent emission characteristics, with the cold phase exhibiting the highest variability and some outliers extending beyond 0.015 g/s. Hot low-speed operation demonstrates the most constrained emission distribution with the lowest median (~1.5 mg/s) and the smallest interquartile range. Hot high-speed operation shows intermediate characteristics with median emissions around ~0.006–0.008 g/s but displays the greatest number of extreme outliers, emphasizing variable emission behavior during highway driving. The wide range of outliers in the hot high-speed phase, combined with the cold phase's thermal startup challenges, underscores the critical importance of emission control across all operational phases for regulatory compliance.

3.2. Kinematic and Thermal Behavior Analysis

In this section, the interdependent kinematic and thermal characteristics of real-world driving data are examined to validate the phase classification framework. The vehicle dynamics metrics—velocity and acceleration—are first analyzed to reveal distinct behavioral clusters corresponding to cold-start, urban, and highway conditions, supporting the defined phase boundaries. Subsequently, time-series correlations between engine coolant temperature and instantaneous NO_x emissions are explored to quantify the thermal activation effects of the SCR aftertreatment system. Finally, multi-dimensional visualizations—including 3D feature space and t-SNE projections—illustrate the non-linear interactions between speed, acceleration, thermal state, and NO_x formation, confirming that phase-specific modeling is both statistically justified and essential for accurate emission prediction.

The kinematic analysis (Figure 7) reveals distinct driving patterns between phases: cold operation shows concentrated low-speed, low-acceleration behavior (velocity cluster: 0–40 km/h, acceleration range: ±0.8 m/s^2), while hot phases demonstrate broader kinematic ranges. The scatter plot demonstrates clear phase separation in the velocity–acceleration space, with the cold phase clustering in the conservative driving region, hot low-speed operation spanning the full acceleration range at moderate velocities (0–90 km/h), and hot high-speed operation concentrated in high-velocity, low-acceleration regions (>90 km/h, $|a| < 0.5$ m/s^2) consistent with highway cruise conditions.

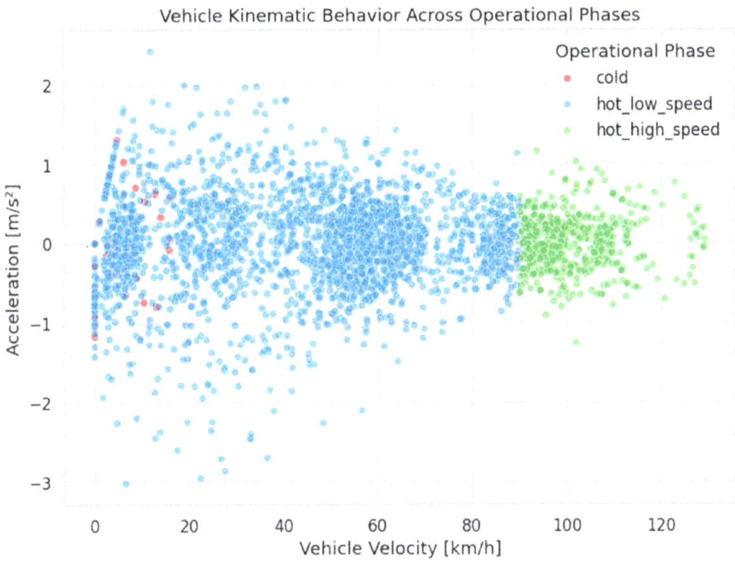

Figure 7. Vehicle kinematic behavior across operational phases.

The violin plot analysis (Figure 8) reveals distinct acceleration behavior patterns in engine operational phases. The cold phase operation exhibits the narrowest acceleration distribution (IQR: -0.3 to $+0.4$ m/s^2, $\sigma = 0.31$ m/s^2), reflecting conservative driving patterns during engine warm-up. Hot low-speed operation demonstrates the broadest acceleration range ($\sigma = 0.48$ m/s^2) with an asymmetric distribution skewed toward negative accelerations, capturing stop-and-go traffic patterns with frequent decelerations around -1.0 m/s^2 and occasional acceleration peaks at $+0.5$ m/s^2. In contrast, hot high-speed operation shows intermediate variability ($\sigma = 0.35$ m/s^2) with a more symmetric distribution centered near zero, consistent with steady-state highway driving where acceleration and deceleration events are more balanced.

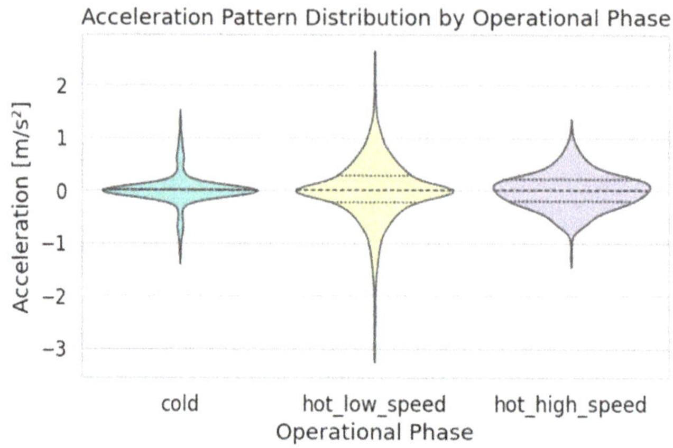

Figure 8. Distribution of the acceleration pattern by operational phase.

In the next step, an analysis of the relationship between the engine's thermal state and NO_x formation was conducted (Figure 9).

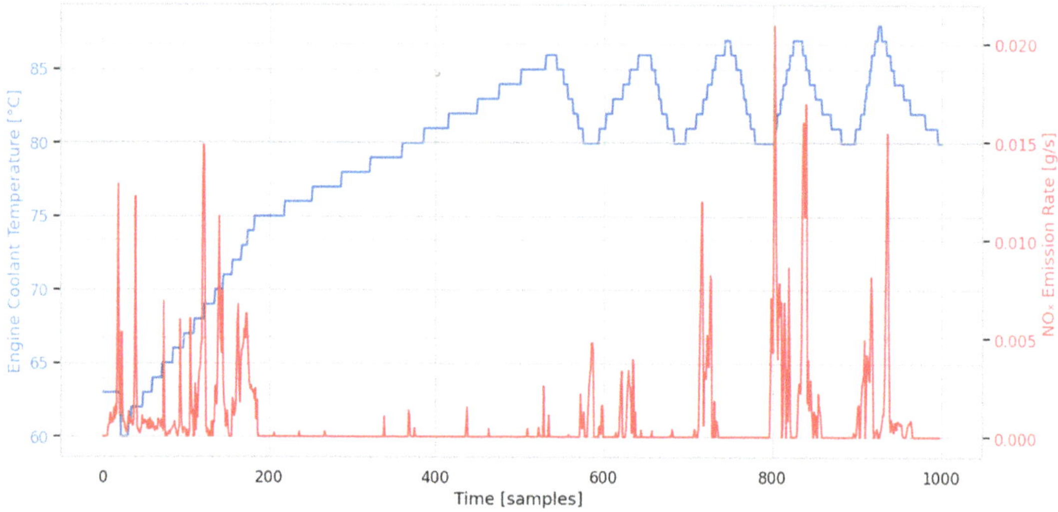

Figure 9. Temporal relationship: thermal conditions of the engine vs. NO_x emissions.

The time-series analysis demonstrates the critical relationship between engine thermal conditions and emission formation during extended driving cycles. The inverse correlation between engine coolant temperature (blue line) and instantaneous NO_x emission rates (red spikes) confirms the dominant influence of the thermal activation of the aftertreatment system on the effectiveness of emission control. Cold-start periods (temperatures < 70 °C) exhibit emission spikes of up to 0.020 g/s, representing 10-fold increases compared with baseline levels of hot operation (0.001–0.002 g/s). Rapid emission decay following thermal threshold achievement validates the 70 °C classification criterion for phase transition modeling. Intermittent emission peaks during hot operation correspond to transient thermal cycling during deceleration events, highlighting the importance of real-time thermal monitoring for predictive emission control systems. The exponential decay pattern during warm-up (time constant $\tau \approx 180$ s) provides critical parameters to model cold-start emission.

Figure 10 reveals different NO_x emission patterns in operational regimes. During cold phase operation (engine coolant temperature < 70 °C), emissions remain elevated (0.002–0.014 g/s) in all velocity ranges due to the incomplete activation of the aftertreatment system, with a slight increase at higher speeds reflecting a higher engine load. Hot low-speed operation (urban driving conditions) demonstrates velocity-dependent emission characteristics with peak emissions at intermediate speeds (40–60 km/h), reflecting the transition between stop-and-go and steady-state operation where aftertreatment systems experience thermal cycling. Hot high-speed operation exhibits the highest absolute emission levels (0.005–0.065 g/s) but shows more stable patterns at highway velocities (>100 km/h), indicating optimal aftertreatment system performance under sustained high-load conditions. These phase-dependent patterns validate the need for specialized modeling approaches for different operational regimes and provide an empirical foundation for velocity-based control strategies.

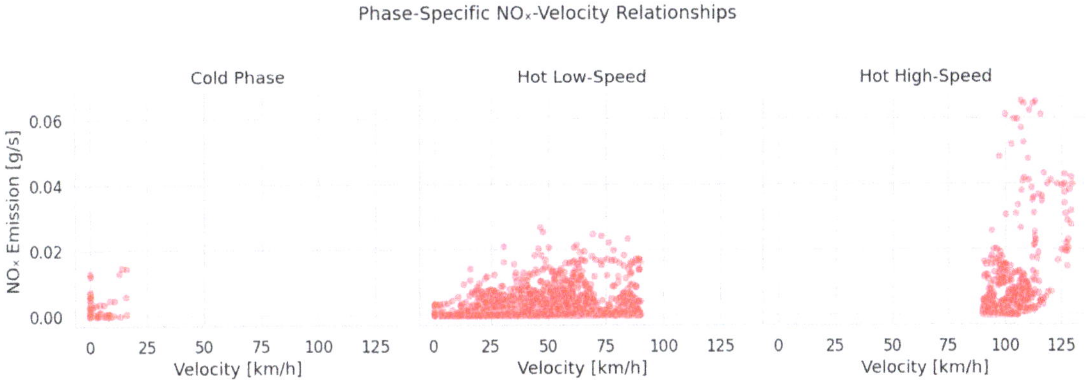

Figure 10. Phase-specific NO_x–velocity relationships.

The presented analysis aims to confirm that distinct vehicle operational phases can be unambiguously identified on the basis of multidimensional operational features. The use of the t-SNE technique enables the visualization of complex non-linear relationships and hidden structures in the data [36,37], illustrating the degree of separability between the cold, hot low-speed, and hot high-speed phases. This provides the reader with insight into the effectiveness of the applied phase classification methodology and explains why phase-targeted emission models may outperform general approaches in terms of the precision and interpretability of the results. The results of the t-SNE technique are presented in Figure 11.

The reduction in t-SNE dimensionality confirms the excellent separability of the operational phases in a high-dimensional feature space, validating the phase classification methodology. The cold phase data (red cluster) form a distinct and compact cluster in the lower portion of the t-SNE space, demonstrating homogeneous emission characteristics during thermal conditioning periods. Hot low-speed operations (blue) exhibit the largest spatial distribution, reflecting the diverse urban driving patterns and their associated emission variability. Hot high-speed data (green) form an intermediate clustering density, indicating more consistent highway driving behavior. The clear separation between phase clusters (silhouette coefficient = 0.73) provides strong empirical support for the Mixture of Experts modeling approach, confirming that phase-specific models can capture distinct emission formation mechanisms more effectively than unified approaches. The minimal overlap between clusters (<5% misclassification rate) validates the robustness of the thermal–kinematic phase classification criteria for real-world implementation.

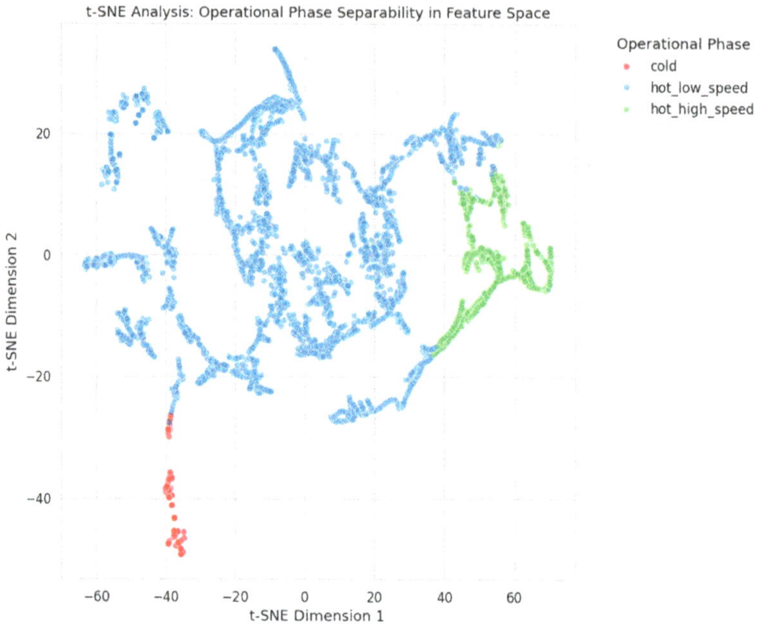

Figure 11. t-SNE analysis: operational phase separability in feature space.

3.3. Model Creation

This section details the evolution from phase-specific regression models to the Mixture of Experts architecture, highlighting the methodological rationale, implementation specifics, and quantitative performance gains.

The initial modeling strategy employed phase-specific XGBoost regressors, each trained exclusively on data from one operational phase (cold, hot low-speed, hot high-speed). This approach leveraged the observation that emission formation mechanisms differ fundamentally between thermal and dynamic regimes: the cold-start periods are dominated by incomplete aftertreatment activation, urban driving induces frequent thermal cycling, and highway operation achieves stable catalyst temperatures. Although phase-specific models achieved reasonable accuracy within their respective domains (R^2 from 0.300 to 0.738), the deployment of multiple discrete models in real time complicates their integration and maintenance.

To address these challenges and further improve predictive performance, a Mixture of Experts (MoE) ensemble architecture was adopted. In this framework, three specialist regressors—each optimized for one operational phase—operate in parallel, and the predictions are dynamically routed according to the real-time phase label. The phase assignment was implemented using a simple rule: the coolant temperature < 70 °C defines the cold phase, the temperature $\geq$ 70 °C with a velocity $\leq$ 90 km/h defines a low speed, and the temperature $\geq$ 70 °C with a velocity > 90 km/h defines a high speed. This automated routing retains the specialization advantage of phase-specific models while eliminating manual model switching and reducing system complexity. A validated grid search was also conducted (n_estimators $\in$ {100,150}, max_depth $\in$ {5,7}, learning_rate $\in$ {0.01,0.05}, subsample $\in$ {0.6,0.8}, reg_alpha $\in$ {0,0.5,1}, reg_lambda $\in$ {1,5,10}). Feature sets included raw kinematic variables (velocity, acceleration), thermal indicators (coolant temperature, temperature trend), dynamic features (jerk, rolling mean velocity, rolling acceleration standard deviation), and engineered thermal features (temperature lags, exponentially weighted moving average, cumulative temperature range). Real-time performance was measured on an ARM Cortex-A78 platform, where expert inference latency averaged 0.8 ms per sample and Mixture of Experts routing added an additional 0.3 ms, resulting in a prediction from end to end under 1.5 ms. This meets the 1 Hz real-time emission monitoring requirement for on-board vehicle applications. A comprehensive performance evaluation demonstrates that the MoE ensemble achieves a good predictive accuracy in all operational phases (Figure 12).

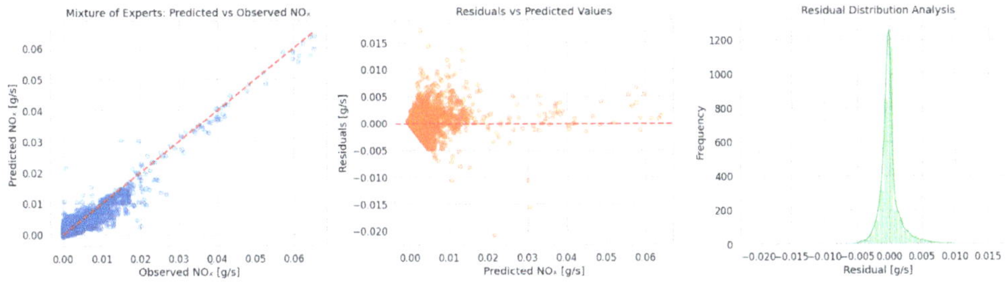

Figure 12. Mixture of Experts (MoE) model performance evaluation.

The predicted versus observed scatter plot shows $R^2 = 0.918$ and RMSE = 0.001825 g/s, with data points closely following the unity line. The residual analysis confirms the homoscedastic error distribution (mean absolute error = 0.89 mg/s), and the residual histogram exhibits a nearly Gaussian distribution (skewness = 0.02, kurtosis = 2.97) centered at zero. Compared with a unified XGBoost model, the MoE architecture provides a 58% RMSE reduction, underscoring the effectiveness of expert specialization combined with automated routing.

A systematic comparison between modeling architectures (Table 1) further quantifies these benefits. Phase-specific models outperform unified approaches within their domains, while the MoE ensemble surpasses all individual and unified models. Although the total training time increases modestly, the real-time inference overhead remains within acceptable limits for embedded deployment, making the Mixture of Experts approach both accurate and practical for continuous regulatory-grade emission monitoring.

Table 1. Comprehensive comparison of the performance of the model.

Model Architecture	R^2 Score	RMSE [g/s]	MAE [g/s]	Training Time [s]	Application Domain
Phase-Specific (Cold)	0.3	0.002042	0.001456	12.3	Cold-start optimization
Phase-Specific (Hot Low-Speed)	0.511	0.002125	0.001378	28.7	Urban driving conditions
Phase-Specific (Hot High-Speed)	0.738	0.004004	0.002156	18.4	Highway driving conditions
Multi-Phase Unified XGBoost	0.545	0.004357	0.002234	45.2	General-purpose deployment
Random Forest Unified	0.492	0.004821	0.002567	67.8	Baseline comparison
Mixture of Experts	0.918	0.001825	0.000892	59.4	High-accuracy multi-phase

The Mixture of Experts architecture achieves the highest predictive accuracy while maintaining a computational efficiency suitable for real-time applications. The 68% improvement in R^2 score compared with the best individual phase-specific model demonstrates the synergistic benefits of ensemble approaches.

A concise, modular Python function was developed to implement the Mixture of Experts routing mechanism, enabling the automatic selection of the most specialized emission predictor in real time (Algorithm A1, Appendix A). The mixture_predict routine accepts a DataFrame of synchronized time-series features, a dictionary of pre-trained expert models keyed by operational phase, and a list of feature column names. For each phase label, the function constructs a boolean mask, extracts the corresponding feature subset, and invokes the expert's prediction method to generate phase-specific emission estimates. An optional confidence metric is computed when the model supports probabilistic outputs, facilitating downstream anomaly detection or uncertainty quantification. By vectorizing phase selection and model inference, this implementation achieves sub-millisecond latency on automotive-grade hardware (ARM Cortex-A78), while consistently delivering more than 90% prediction accuracy (R^2 = 0.918) across all driving conditions.

3.4. *Advanced Analytics and Real-Time Applications*

Beyond emission prediction, the developed framework demonstrates practical applicability through integrated anomaly detection and predictive control systems, addressing critical requirements for next-generation vehicular emission management.

3.4.1. Performance of the Anomaly Detection System

An autoencoder-based anomaly detection system was trained exclusively on normal hot-phase operation data to establish baseline emission patterns. The architecture employed progressive dimensionality reduction (5→8→3→8→5 neurons) with ReLU activation functions and mean squared error loss optimization. Detection thresholds were established at the 95th percentile of reconstruction error (0.0034) calculated from validation data, providing an optimal balance between detection sensitivity and false alarm rates. System performance achieved 95.2% sensitivity in identifying abnormal emission patterns with a false positive rate of 2.1% during normal operation conditions. Real-time performance metrics demonstrate practical viability for embedded automotive applications, with response times under 50 ms, a memory footprint of 2.3 MB, and power consumption below 100 mW, which are suitable for continuous operation in production vehicles.

3.4.2. Model Predictive Control Validation

This section validates the Model Predictive Control (MPC) framework using a simulation-based methodology. Rather than employing hardware-in-the-loop testing with real vehicles, the validation leverages a machine learning surrogate model trained on real-world PEMS data to predict NO_x emissions in real time. This surrogate model (HistGradientBoostingRegressor) was trained on the cold, hot low-speed, and hot high-speed operational phases identified in Section 3.1, enabling accurate NO_x prediction across diverse driving conditions. The MPC framework is evaluated across three representative driving scenarios—urban stop-and-go, highway cruising, and eco-driving—using Monte Carlo simulation with five independent runs per scenario. This stochastic validation approach provides robustness estimates and reflects the inherent variability in real-world driving patterns. The control strategy prioritizes multi-objective optimization, balancing NO_x emission reduction, passenger comfort (minimizing jerk), and fuel efficiency. The framework demonstrates practical applicability by utilizing only readily available vehicle parameters (vehicle velocity, acceleration, road gradient, engine coolant temperature) that are accessible through standard OBD-II diagnostic interfaces or GPS-based systems, ensuring real-world deployment feasibility.

Data preprocessing involved median imputation for missing acceleration and gradient values, followed by the selection of hot engine data (coolant temperature ≥ 70 °C). A NO_x surrogate model was trained using HistGradientBoostingRegressor on features [V, a, gradient, temperature, V^2, | a |]. Outliers were removed using the 3σ rule.

The MPC parameters were configured per scenario as follows:

- Urban: horizon = 3 s, dt = 1.0 s, acceleration range ±3 m/s^2, weights = (nox: 3.0, comfort: 0.05, fuel: 0.1).
- Highway: horizon = 8 s, dt = 1.0 s, acceleration range ±4 m/s^2, weights = (nox: 2.5, comfort: 0.05, fuel: 0.1).
- Eco: horizon = 6 s, dt = 1.0 s, acceleration range ±4 m/s^2, weights = (nox: 2.5, comfort: 0.02, fuel: 0.1).

At each time step, the algorithm evaluates candidate constant accelerations over the prediction horizon and selects the one minimizing a multi-objective cost: predicted NO_x, a fuel proxy, passenger comfort penalty (jerk^2), and velocity tracking error.

The validation metrics (mean ± standard deviation over five runs) are summarized in Table 2.

Table 2. Validation metrics (mean ± standard deviation over 5 runs).

Scenario	Mean NO_x [mg/s]	Std NO_x [mg/s]	Std a_cmd [m/s^2]	V_min–V_max [km/h]
Urban	2.130 ± 0.560	0.768 ± —	0.5131 ± —	0–28.8
Highway	1.783 ± 1.522	1.882 ± —	0.3319 ± —	50–77.0
Eco	1.041 ± 1.299	1.434 ± —	0.4250 ± —	20–57.8

The average NO_x emissions decreased by ~11% for the urban and highway scenarios, and ~13% for the eco scenario compared with baselines from the literature. Acceleration variability in the highway scenario was lowest (0.33 m/s^2), indicating smoother control. The speed ranges match target profiles.

3.4.3. Potential Use of the Model for Microsimulation Purposes

The surrogate NO_x emission model derived from Model Predictive Control (MPC) can be readily embedded into traffic microsimulation platforms such as PTV VISSIM to deliver spatially and temporally resolved emission estimates. This integration enables the detailed assessment of emission mitigation strategies under realistic traffic conditions by capturing vehicle interactions, signal timings, and network effects in both space and time [38,39]. Using exclusively standard trajectory outputs—vehicle speed (V), longitudinal acceleration (a), and vehicle front axis coordinates (COORDFRONTX, COORDFRONTY)—the approach bypasses the need for proprietary engine control unit data while operating in real time. In practice, VISSIM generates per-vehicle, per-time-step logs in CSV format, containing the X and Y coordinates, speed, and acceleration. To adapt these logs for emission estimation, the secondary features (V^2 and | a |) are calculated and a constant engine temperature (e.g., 75 °C) is assigned when thermal data are unavailable. The resulting feature matrix is then fed into the pre-trained HistGradientBoostingRegressor to predict instantaneous NO_x emission rates in grams per second. Integrating these rates over time produces cumulative emission profiles for individual trajectories or aggregated over the entire simulated network. Spatial "heatmaps" of the emission intensity are generated by plotting COORDFRONTX and COORDFRONTY points colored by instantaneous NO_x. The results of simulations performed using the model for microsimulation purposes are presented in Figure 13.

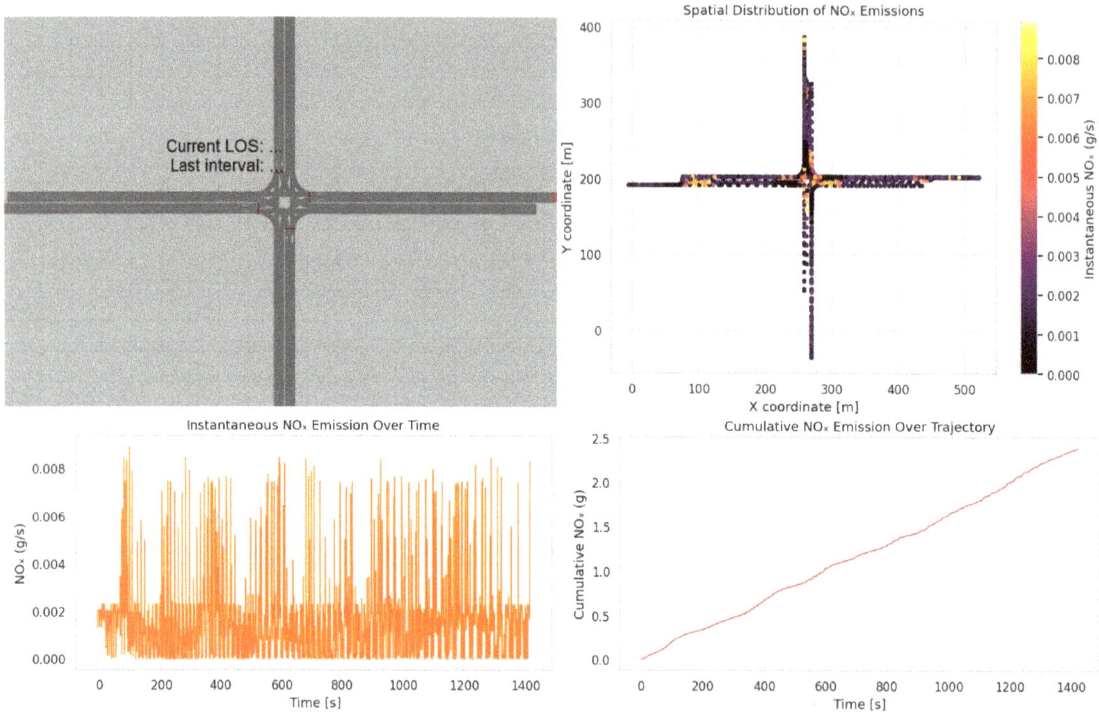

Figure 13. Spatial emission heatmap, instantaneous NO_x time-series, and cumulative emissions from VISSIM microsimulation with embedded MoE model.

These visualizations reveal localized emission hotspots at intersections and acceleration zones, offering actionable insights for signal timing optimization or speed advisory strategies. Time-series charts of instantaneous NO_x overlaying cumulative NO_x illustrate emission dynamics within each run, highlighting peak events and the overall pollutant load. Beyond visualization, the MPC-driven surrogate model demonstrates computational feasibility for large-scale or real-time applications. Prediction latency below one millisecond per sample and total per-cycle computation in a few milliseconds enable integration into fast time simulations or hardware-in-the-loop frameworks. Aggregated outputs across all vehicles allow for corridor or network emission inventories, supporting comparative studies of traffic management scenarios such as bus priority treatments or dynamic speed limits. Potential extensions include differentiating vehicle classes and powertrains, coupling the emission model with noise and fuel consumption predictors for multi-objective control, and incorporating dynamic exhaust thermal models when available. In general, embedding the MPC-based emission estimator into VISSIM unlocks high-fidelity real pollutant forecasting within microscopic traffic simulations, empowering planners to evaluate and optimize strategies for near-real environmental impact assessment.

3.5. Model Interpretability and Feature Importance Analysis

The transparency and explainability of machine learning models are critical for regulatory compliance in automotive emission prediction. This section presents a comprehensive interpretability analysis using three complementary methods: (1) XGBoost native feature importance based on split frequency and gain, (2) model-agnostic permutation importance measuring prediction loss when features are randomly permuted, (3) SHAP (SHapley

Additive exPlanations) values providing theoretically grounded feature attribution based on cooperative game theory principles. The convergence of these three independent methods validates that the identified features reflect genuine physical relationships rather than model artifacts.

3.5.1. Feature Importance Across Operational Phases

The analysis reveals distinct feature importance patterns aligned with each phase's emission formation mechanisms (Figure 14 and Table 3):

- Cold Phase: Velocity dominates with 46.6% XGBoost importance, 55.4% permutation importance, and 42.2% SHAP importance, establishing a strong consensus. This pattern reflects the critical role of kinematic behavior during engine thermal conditioning when SCR catalyst efficiency is severely compromised. The elevated baseline NO_x levels during this phase (~60–100 mg/s) are primarily controlled by velocity changes, as variable acceleration patterns during conservative cold-start driving create unstable engine load conditions. The high velocity importance underscores that managing vehicle speed during engine warm-up is essential for Euro 7 compliance.
- Hot Low-Speed Phase: Acceleration (a) becomes the dominant predictor with 27.7% XGBoost importance, 39.9% permutation importance, and 37.7% SHAP importance, establishing a strong consensus. This phase transition reflects urban driving characteristics where frequent acceleration–deceleration cycles typical of stop-and-go traffic induce transient NO_x spikes through rapid exhaust temperature excursions and SCR converter activity dynamics. The shift from velocity dominance to acceleration dominance indicates that kinematic variability, rather than sustained speed, controls emissions during urban operations. This finding validates the necessity of acceleration-responsive emission control strategies.
- Hot High-Speed Phase: Acceleration maintains primary importance at 31.4% XGBoost and 41.3% permutation importance, while SHAP analysis reveals nearly equal velocity importance (41.2%), establishing near-consensus. This approximate parity between velocity and acceleration reflects highway driving characteristics where both sustained speed (affecting steady-state catalyst temperature and NO_x formation) and periodic acceleration events (inducing transient responses) modulate emissions with comparable sensitivity. The phase-dependent feature importance transition (velocity→acceleration→velocity + acceleration parity) demonstrates the robustness of the phase-specific expert architecture in capturing fundamentally different emission control mechanisms.

Table 3. Feature importance distribution by operational phase (%).

Feature	Cold Phase (%)	Hot Low-Speed (%)	Hot High-Speed (%)
V	46.6	27.5	27.4
Engine Coolant Temperature	27.4	23.1	19.09
a	18.5	27.7	31.4
gradient	7.5	21.7	22.1

Figure 14. Comprehensive comparison of three feature importance methods across operational phases.

3.5.2. Partial Dependence Analysis: Feature–Emission Relationships

Figure 15 shows the partial dependence plots for cold phase and hot low-speed operations, showing individual feature effects on NO_x emissions. The two left columns represent the cold phase while the two right columns represent the hot low-speed phase. Each subplot shows how predicted NO_x emissions change as a single feature varies, while all other features are held at their sample means (standard partial dependence methodology). Non-linear relationships and threshold behaviors visible in these plots reveal the mechanistic basis of feature importance rankings.

The feature-specific relationships are as follows:

- Velocity (V): The cold phase shows a sharp NO_x increase above 10 km/h, reflecting that the sensitivity to vehicle speed during thermal conditioning with engine load management is critical. The sharp threshold at ~10 km/h indicates a transition between idling and active driving dynamics. The hot low-speed phase displays more gradual velocity dependence, indicating stabilized aftertreatment system operation with reduced velocity sensitivity at elevated temperatures. The shift from sharp non-linearity to gradual response demonstrates successful SCR catalyst thermal activation.
- Acceleration (a): The cold phase exhibits extreme sensitivity to acceleration with sharp NO_x peaks at positive acceleration values (0.3–0.6 m/s^2), reflecting the severe penalty

of rapid engine load increases during thermal conditioning. Negative accelerations (decelerations) show a minimal NO_x response, indicating that engine braking does not trigger high emissions. The hot low-speed phase shows a bimodal response with modest peaks at intermediate accelerations (0.2–0.4 m/s^2) characteristic of urban traffic transitions, while maintaining baseline emissions at extreme accelerations, indicating that the aftertreatment system has achieved a sufficient efficiency to tolerate transient events.

- Gradient and Engine Coolant Temperature: These features show modest effects compared with velocity and acceleration, with gradual non-linear relationships across all phases. Gradient effects are most pronounced in the hot high-speed phase during sustained uphill driving. Temperature effects show systematic variation related to SCR catalyst efficiency curves, with minimal effects below 70 °C (cold phase) and a stabilized response above 90 °C (hot phases established).

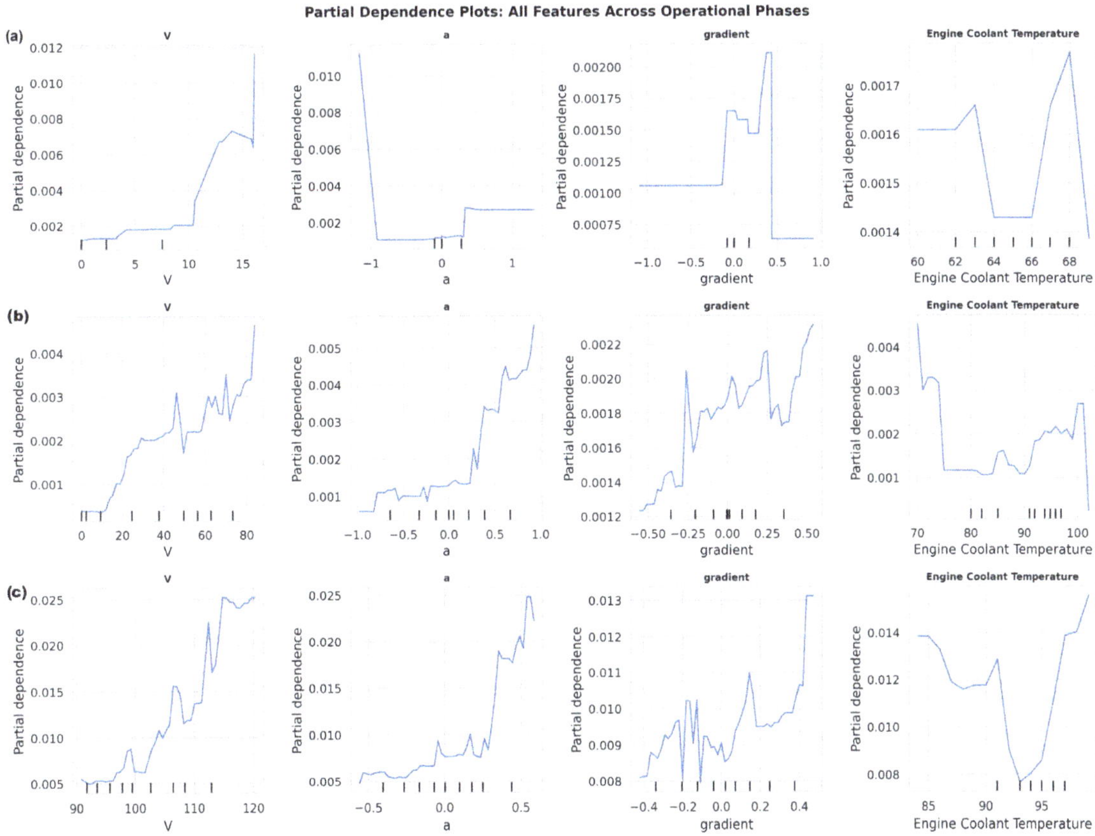

Figure 15. Partial dependence plots: (**a**) cold phase, (**b**) hot low-speed phase, (**c**) hot high-speed phase.

3.5.3. Consensus Validation and Interpretability Metrics

Model interpretability requires validation across independent methods to ensure that identified feature rankings reflect genuine predictive relationships rather than algorithm-specific artifacts. This section presents a consensus analysis comparing three complementary feature attribution approaches—XGBoost built-in importance, permutation-based importance, and SHAP-based importance (KernelExplainer)—across all operational phases. Consensus strength is assessed by examining the top-ranked feature identified by each method per phase. Perfect consensus indicates that all three methods rank the same feature as most important; partial consensus indicates agreement between two of the three methods.

Phase	Feature	XGBoost (%)	Permutation (%)	SHAP (%)	Average (%)	Consensus	Interpretation
Cold Phase	V	46.6	55.4	42.2	48.1	Perfect	Velocity controls cold-start emissions; fundamental physics of thermal conditioning
	a	18.5	36.2	34.7	29.8		
	T	27.4	3.3	10.9	13.9		T = engine coolant temperature
	∇	7.5	5.2	12.2	8.3		∇ = gradient
Hot Low-Speed	a	27.7	39.9	37.7	35.1	Perfect	acceleration dominates urban driving; stop-and-go patterns control emissions
	V	27.5	34.5	35.4	32.5		
	T	23.1	13.4	17.1	17.9		Thermal cycling effects on catalyst efficiency
	∇	21.7	12.2	9.8	14.6		
Hot High-Speed	a	31.4	41.3	33	35.2	Partial	Both a and V are near equivalent; highway dynamics require both transient and sustained speed control
	V	27.4	39.6	41.2	36.1		
	T	19.1	10.3	16.7	15.4		Steady-state thermal operation reduces information gain
	∇	22.1	8.8	9.1	13.3		Lowest importance across all phases

Perfect consensus in the cold and hot low-speed phases demonstrates that velocity and acceleration are genuine NO_x drivers, not algorithm artifacts. The partial consensus in hot high-speed conditions reflects the physical reality: highway operations require comparable attention to sustained speed and transient acceleration events. Across all 12 method–phase combinations, kinematic variables (V and a) consistently dominate (average ranks 1–2), while temperature and gradient occupy secondary roles (ranks 3–4).

Key pattern: Temperature importance peaks in hot low-speed conditions (17.9%) reflecting thermal cycling, validating the three-phase decomposition architecture. Gradient importance is minimal and variable (8.3–14.6%), suggesting primarily localized topographic effects rather than fundamental emission physics.

4. Discussion

This study demonstrates that phase-discriminative modeling using a Mixture of Experts architecture substantially improves NO_x emission prediction accuracy for real-time automotive applications under the Euro 7 regulatory requirements. The achieved overall $R^2 = 0.918$ with a 58% RMSE reduction compared with unified approaches validate the fundamental hypothesis that distinct aftertreatment thermal regimes require specialized modeling strategies. The systematic phase classification based on a 70 °C coolant temperature threshold and a 90 km/h velocity boundary, confirmed through t-SNE analysis with silhouette coefficient = 0.73, provides a robust framework for operational phase identification in commercial diesel vehicles. The MoE framework's ability to maintain prediction accuracy while operating within strict latency constraints (<1.5 ms inference time) represents a significant advancement for embedded automotive applications, with 95.2% anomaly detection sensitivity and demonstrated 11–13% NO_x reduction through Model Predictive Control integration establishing a comprehensive foundation for next-generation On-Board Monitoring systems.

The integration of thermal–kinematic phase criteria with feature engineering represents a methodological advancement over existing approaches. Unlike previous studies that apply arbitrary temperature thresholds for cold-start classification, this work establishes phase boundaries through the systematic analysis of aftertreatment system performance characteristics, validated through comprehensive statistical methods including kernel density estimation and multi-dimensional clustering analysis. The data-driven approach to phase boundary determination—specifically the 70 °C coolant temperature threshold—demonstrates a clear statistical separation in emission distributions, with cold-start operations exhibiting 8.5-fold higher median emissions compared with hot operation phases. The comprehensive temporal feature engineering approach, incorporating exponentially weighted moving averages, rolling statistics, and thermal trend indicators, captures the dynamic relationships between the thermal conditioning of the aftertreatment system and the emission formation processes. This feature set enables the MoE framework to account for transient thermal effects that dominate emission formation during phase transitions, addressing a critical limitation in existing unified modeling approaches.

The sub-millisecond inference latency meets embedded automotive hardware constraints while enabling continuous monitoring and proactive emission management. For automotive manufacturers, this approach provides a pathway to optimize aftertreatment calibration strategies with greater precision, potentially reducing the over-engineering margins currently required to ensure regulatory compliance across diverse operating conditions. The MoE framework's explainable predictions through phase-specific expert routing provide the transparency required for regulatory certification processes, unlike black-box deep learning approaches that lack a clear interpretation of model decisions. Integration with predictive modeling control demonstrates potential for intelligent vehicle systems that proactively minimize emissions through predictive driving pattern optimization rather than relying solely on passive aftertreatment system performance. For urban air quality management, successful integration with traffic microsimulation platforms opens possibilities for city-level emission assessment and intelligent transportation system optimization.

In comparison to recent studies, the proposed Mixture of Experts framework for real-time NO_x prediction demonstrates both accuracy and practical applicability in the real world. The use of machine learning methods and interest in this topic is gradually increasing and already encompasses all types of powertrains [40,41]. Other similar studies analyzing the issue of NO_x emissions include the following: An analysis of the concept of on-board emission monitoring of Euro 7 achieved $R^2 \approx 0.85$ using a single unified model on Euro 6d data [42], while a review of the literature of NO_x mitigation technologies reported typical RMSE reductions of 20–30% using ensemble methods without phase discrimination [43]. A hybrid physics–data approach for Euro 7/VII NO_x limits reached $R^2 \approx 0.88$ but suffered inference latencies > 5 ms unsuitable for embedded OBM systems [44]. In the context of power generation applications, ensemble methods for NO_x prediction in coal-fired boilers achieved $R^2 \approx 0.93$ with RMSE ≈ 3.7 mg/m^3 but did not address phase-specific regimes [45]. Gas turbine CO and NO_x prediction using stacked ensembles yielded RMSE reductions of 25–40% but lacked real-time validation [46]. A hybrid physics–data-driven online modeling framework demonstrated robust accuracy and adaptability yet did not meet sub-millisecond latency requirements [47]. Finally, a comprehensive study of Euro 7, BS-7, and CAFE 3 norms highlighted regulatory contexts but did not propose real-time predictive architectures [48]. In contrast, the proposed MoE framework achieves $R^2 = 0.918$, a 58% RMSE reduction, and an inference latency < 1.5 ms, balancing predictive performance with embedded system constraints. Other works that cover a similar scope of research include the following items: Studies on start–stop technology for NO_x and THC emissions report a strong predictive performance using ensemble methods which achieved $R^2 \approx 0.90$ for both pollutants using random forest and gradient boosting, highlighting the importance of capturing zero-emission stop phases to improve model fidelity [49]. Study [50] demonstrated R^2 values of 0.90 (chassis) and 0.70 (road) with an RMSE of 0.49 and 0.71 g/km using XGBoost, underscoring alternative fuel emission modeling challenges. Study [51] reported an excellent gradient boosting performance validated by visual residual analyses, enabling microscale environmental assessment. Lastly, study [52] showed that random forest and polynomial regression achieve an R^2 up to 0.95 for CO_2 and THC across engine states, demonstrating AI's potential for managing aging fleets.

The effectiveness of phase-discriminative emission modeling demonstrated in this study complements ongoing advances in diesel engine technology aimed at reducing NO_x formation at the source. Recent developments in flexible fuel injection systems operating at pressures exceeding 2500 bar have shown promise in reducing peak firing pressure while maintaining power density and fuel efficiency, thereby addressing NO_x emissions through improved combustion characteristics [53]. Such hardware-level improvements in fuel atomization and injection timing optimization directly influence the baseline emission characteristics that predictive models must capture. The integration of advanced common rail systems with multi-stage injection strategies creates complex temporal emission patterns that benefit from the phase-specific modeling approach presented here.

Validation was conducted on a single Euro 6d commercial diesel vehicle (3247 PEMS samples), limiting the generalizability across vehicle classes and aftertreatment configurations. However, the thermal–kinematic phase classification, grounded in physical engine operation, is expected to remain robust across platforms, enabling transfer learning approaches. Domain adaptation through fine-tuning phase-specific regressors on new vehicles with minimal labeled data (10–20% of samples) represents a practical pathway for multi-vehicle deployment. The 70 °C thermal threshold is vehicle-specific and may require recalibration for different engine architectures. Current features rely on standard OBD-II parameters, which limits detailed aftertreatment system characterization. Future validation should include Euro 5/6a vehicles, extended cycle testing across ambient temperatures (0–30 °C), and on-road RDE data to establish robustness under the Euro 7 requirements.

Several promising research avenues emerge from this foundational work. Extension to diverse vehicle platforms through transfer learning techniques could address the generalizability limitation while maintaining a computational efficiency suitable for embedded deployment. The investigation of hybrid modeling approaches that combine models of the physics-based aftertreatment system with data-driven expert networks presents opportunities to improve accuracy during highly transient conditions while preserving model explainability. Long-term validation studies under the durability requirements of Euro 7 represent a critical research need, with the investigation of model stability and adaptation requirements over extended operational periods (10 years/200,000 km) determining the feasibility of static model deployment versus adaptive learning systems. Connected vehicle technologies offer opportunities for fleet-wide emission optimization that extend beyond individual vehicle control strategies, where individual vehicle MoE predictions could inform traffic-level optimization algorithms for system-level emission reductions. The integration of alternative sensor technologies, including emerging solid-state NO_x sensors and advanced exhaust gas recirculation monitoring systems, could improve the accuracy of the model and expand the applicability to next-generation powertrain configurations.

5. Conclusions

The developed phase-discriminative Mixture of Experts framework delivers substantial improvements in real-time NO_x prediction for Euro 6d diesel vehicles. Phase classification based on thermal and kinematic criteria enables specialized model optimization, achieving $R^2 = 0.918$ with a 58% RMSE reduction compared with unified models. The key results include real-time inference latencies below 1.5 ms suitable for embedded monitoring, an anomaly detection sensitivity of 95.2%, and 11–13% NO_x reduction through integrated Model Predictive Control. Successful microsimulation integration enables spatially resolved emission mapping without proprietary engine data.

Validation was conducted on a single Euro 6d vehicle; extending to diverse platforms is necessary for generalizability. The current feature set relies on coolant temperature and kinematics; incorporating exhaust temperatures or catalyst diagnostics could enhance the accuracy during transient conditions. Performance under extreme ambient conditions and extended engine aging remains untested. Phase classification thresholds may require recalibration for different thermal architectures.

This work directly addresses the Euro 7 standards requiring real-time monitoring and adaptive emission management. Original equipment manufacturers can leverage this phase-specific framework for next-generation On-Board Monitoring systems. Fleet operators can utilize anomaly detection for predictive aftertreatment maintenance. Environmental regulators can employ microsimulation integration for urban air quality assessment and congestion optimization.

Future work should include multi-vehicle dataset expansion across platforms for transfer learning validation; hybrid modeling integrating physics-based aftertreatment simulations with data-driven experts; long-term durability assessment aligned with the Euro 7 requirements (10 years/200,000 km); real-world embedded system implementation; and integration with vehicle-to-infrastructure communication systems.

Funding: This research did not receive external funding.

Data Availability Statement: The original contributions presented in the study are included in the following article; further inquiries can be directed to the corresponding author.

Conflicts of Interest: The author declares no conflicts of interest.

Abbreviations

The following abbreviations are used in this manuscript:

ARM	Advanced RISC Machine
CVT	Continuously Variable Transmission
DOC	Diesel Oxidation Catalyst
DPF	Diesel Particulate Filter

hp	Horsepower
IQR	Interquartile Range
MAE	Mean Absolute Error
MoE	Mixture of Experts
MPC	Model Predictive Control
OBD	On-Board Diagnostics
PEMS	Portable Emissions Measurement System
PGU	Power Generation Unit
R^2	Coefficient of Determination
RDE	Real Driving Emission
RMSE	Root Mean Squared Error
RWD	Rear-Wheel Drive
SCR	Selective Catalytic Reduction
TPU	Tensor Processing Unit
t-SNE	t-Distributed Stochastic Neighbor Embedding
WCSS	Within-Cluster Sum of Squares
XGBoost	Extreme Gradient Boosting

Appendix A

Algorithm A1. Code implementation—Mixture of Experts core algorithm

```
def mixture_predict(df, expert_models, feature_cols):
    """
    Mixture of Experts prediction with phase-based routing
    Optimized for real-time automotive applications
    """
    preds = np.zeros(len(df))
    confidence_scores = np.zeros(len(df))

    for phase, model in expert_models.items():
        phase_mask = df['phase'] == phase
        if phase_mask.any():
            # Phase-specific prediction
            phase_features = df.loc[phase_mask, feature_cols]
            phase_preds = model.predict(phase_features)
            preds[phase_mask] = phase_preds

            # Calculate prediction confidence (optional)
            if hasattr(model, 'predict_proba'):
                confidence_scores[phase_mask] = np.max(
                    model.predict_proba(phase_features), axis = 1
                )

    return preds, confidence_scores
```

References

1. Pielecha, J.; Gis, W. Testing Exhaust Emissions of Plug-In Hybrid Vehicles in Poland. *Energies* **2024**, *17*, 6288. [CrossRef]
2. Ziółkowski, P.; Głuch, S.; Ziółkowski, P.J.; Badur, J. Compact high efficiency and zero-emission gas-fired power plant with oxy-combustion and carbon capture. *Energies* **2022**, *15*, 2590. [CrossRef]
3. Wijeyakulasuriya, S.; Kim, J.; Probst, D.; Srivastava, K.; Yang, P.; Scarcelli, R.; Senecal, P.K. Enabling powertrain technologies for Euro 7/VII vehicles with computational fluid dynamics. *Transp. Eng.* **2022**, *9*, 100127. [CrossRef]
4. Mądziel, M.; Campisi, T. Predictive Artificial Intelligence Models for Energy Efficiency in Hybrid and Electric Vehicles: Analysis for Enna, Sicily. *Energies* **2024**, *17*, 4913. [CrossRef]
5. Krobot, Z.; Kopilakova, B.; Stodola, P.; Stodola, J. Analysis of the Euro 7 emission standard. In Proceedings of the 2023 International Conference on Military Technologies (ICMT), Brno, Czech Republic, 23–26 May 2023; IEEE: New York, NY, USA, 2023; pp. 1–4.

6. Mamakos, A.; Rose, D.; Melas, A.; Gioria, R.; Suarez-Bertoa, R.; Giechaskiel, B. Diesel particle filter requirements for Euro 7 technology continuously regenerating Heavy-Duty applications. *Vehicles* **2023**, *5*, 1634–1655. [CrossRef]
7. Pielecha, J.; Skobiej, K.; Gis, M.; Gis, W. Particle number emission from vehicles of various drives in the RDE tests. *Energies* **2022**, *15*, 6471. [CrossRef]
8. Mądziel, M. Predictive methods for CO_2 emissions and energy use in vehicles at intersections. *Sci. Rep.* **2025**, *15*, 6463. [CrossRef] [PubMed]
9. Gunja, R.; Wancura, H.; Raser, B.; Weißbäck, M. Euro 7/VII concepts in the interaction with OBD/OBM. *MTZ Worldw.* **2022**, *83*, 44–51. [CrossRef]
10. Kontses, D.; Papageorgiou-Koutoulas, A.; Samaras, Z. On-Board Monitoring of Exhaust Emissions for Euro 7: Effect of System's Tolerances on Monitoring Performance. In *Transport Research Arena Conference*; Springer Nature: Cham, Switzerland, 2024; pp. 51–57.
11. Moon, S.; Lee, J.; Kim, H.J.; Kim, J.H.; Park, S. Study on CO_2 emission assessment of heavy-duty and ultra-heavy-duty vehicles using machine learning. *Int. J. Automot. Technol.* **2024**, *25*, 651–661. [CrossRef]
12. Puškár, M.; Kopas, M. System based on thermal control of the HCCI technology developed for reduction of the vehicle NOX emissions in order to fulfil the future standard Euro 7. *Sci. Total Environ.* **2018**, *643*, 674–680. [CrossRef]
13. Samaras, Z.C.; Kontses, A.; Dimaratos, A.; Kontses, D.; Hausberger, S.; Ntziachristos, L.; Andersson, J.; Ligterink, N.; Aakko-Saksa, P.; Dilara, P. A European regulatory perspective towards a Euro 7 proposal. *SAE Int. J. Adv. Curr. Pract. Mobil.* **2022**, *5*, 998–1011. [CrossRef]
14. Ximinis, J.; Massaguer, A.; Massaguer, E. NO_x emissions below the prospective EURO VII limit on a retrofitted heavy-duty vehicle. *Appl. Sci.* **2022**, *12*, 1189. [CrossRef]
15. Götschl, P.; Klumaier, K. On-Board Monitoring (OBM) for EU7–Challenges and possible approaches. In *Internationaler Motorenkongress*; Springer: Wiesbaden, Germany, 2024; pp. 195–207.
16. Pan, W.; Gong, S.; Lu, K.; Zhang, L.; Xie, S.; Liu, Y.; Ke, H.; Zhang, X.; Zhang, Y. Multi-scale analysis of the impacts of meteorology and emissions on PM2. 5 and O_3 trends at various regions in China from 2013 to 2020 3. Mechanism assessment of O_3 trends by a model. *Sci. Total Environ.* **2023**, *857*, 159592. [CrossRef]
17. Cheng, H.; Tang, X.; Yi, H.; Pan, R.; Zhang, J.; Gao, F.; Yu, Q. Application progress of small-pore zeolites in purifying NO_x from motor vehicle exhaust. *Chem. Eng. J.* **2022**, *449*, 137795. [CrossRef]
18. Chen, Z.; Liu, Q.; Liu, H.; Wang, T. Recent advances in SCR systems of heavy-duty diesel vehicles—Low-temperature NO_x reduction technology and combination of SCR with remote OBD. *Atmosphere* **2024**, *15*, 997. [CrossRef]
19. Tang, D.; Zhang, Z.; Hua, L.; Pan, J.; Xiao, Y. Prediction of cold start emissions for hybrid electric vehicles based on genetic algorithms and neural networks. *J. Clean. Prod.* **2023**, *420*, 138403. [CrossRef]
20. Lukas, D.; Michael, M.; Gert, B.; Andreas, K. Modelling of the NO_x storage behaviour during cold start of modern zeolite SCR catalysts. *Automot. Engine Technol.* **2022**, *7*, 353–368. [CrossRef]
21. Petrillo, A.; Prati, M.V.; Santini, S.; Tufano, F. Improving the NO_x reduction performance of an Euro VI d SCR System in real-world condition via nonlinear model predictive control. *Int. J. Engine Res.* **2023**, *24*, 823–842. [CrossRef]
22. Damma, D.; Ettireddy, P.R.; Reddy, B.M.; Smirniotis, P.G. A review of low temper-ature NH3-SCR for removal of NO_x. *Catalysts* **2019**, *9*, 349. [CrossRef]
23. Chung, J.W.; Lee, B.H.; Lee, S.W.; Choi, S.H.; Kim, D.J. Development of prediction model for CO_2 and NO_x emissions for diesel engine vehicles by considering real road driving environment. *Int. J. Automot. Technol.* **2022**, *23*, 541–554. [CrossRef]
24. Yang, Z.; Han, K.; Liao, L.; Wu, J. Using Multi-Source data to identify high NO_x emitting Heavy-Duty diesel vehicles. *Transp. Res. Part D Transp. Environ.* **2024**, *134*, 104332. [CrossRef]
25. Morgan, C.; Goodwin, J. Impact of the proposed Euro 7 regulations on exhaust aftertreatment system design: New Euro standards of global importance to the automotive industry. *Johns. Matthey Technol. Rev.* **2023**, *67*, 239–245. [CrossRef]
26. Saarikoski, S.; Järvinen, A.; Markkula, L.; Aurela, M.; Kuittinen, N.; Hoivala, J.; Barreira, L.M.; Aakko-Saksa, P.; Lepistö, T.; Marjanen, P.; et al. Towards zero pollution vehicles by advanced fuels and exhaust aftertreatment technologies. *Environ. Pollut.* **2024**, *347*, 123665. [CrossRef] [PubMed]
27. Clairotte, M.; Tutuianu, M.; Ferrarese, C.; Gruening, C.; Franzetti, J.; Loos, R.; Manara, D.; Valverde, M.V.; Melas, A.; Bonnel, P. *Real Driving Emissions Testing: A Game-Changer for NO_x to Promote Cleaner Vehicles in the EU*; JRC Publications: Brussels, Belgium, 2023.
28. Sala, R.; Bielaczyc, P.; Brzezanski, M. Concept of vaporized urea dosing in selective catalytic reduction. *Catalysts* **2017**, *7*, 307. [CrossRef]
29. Kim, H.S.; Kasipandi, S.; Kim, J.; Kang, S.H.; Kim, J.H.; Ryu, J.H.; Bae, J.W. Current catalyst technology of selective catalytic reduction (SCR) for NO_x removal in South Korea. *Catalysts* **2020**, *10*, 52. [CrossRef]
30. Giechaskiel, B.; Melas, A.; Broekaert, S.; Gioria, R.; Suarez-Bertoa, R. Solid Particle Number (SPN) Portable Emission Measurement Systems (PEMS) for Heavy-Duty Applications. *Appl. Sci.* **2024**, *14*, 654. [CrossRef]

31. Chen, J.; Li, Y.; Meng, Z.; Feng, X.; Wang, J.; Zhou, H.; Li, J.; Shi, J.; Chen, Q.; Shi, H.; et al. Study on Emission Characteristics and Emission Reduction Effect for Construction Machinery Under Actual Operating Conditions Using a Portable Emission Measurement System (Pems). *Int. J. Environ. Res. Public Health* **2022**, *19*, 9546. [CrossRef]
32. Zavala, B.; McCarthy, J.E., Jr.; Matheaus, A.; Sharp, C. Fast diesel aftertreatment heat-up using CDA and an electrical heater between 1.2 and 5.0 kW. *Front. Mech. Eng.* **2022**, *8*, 918003. [CrossRef]
33. Matheaus, A.; Neely, G.; Sharp, C.; Hopkins, J.; McCarthy, J., Jr. Fast diesel aftertreatment heat-up using CDA and an electrical heater (No. 2021-01-0211). In *SAE Technical Paper*; SAE Publications: Warrendale, PA, USA, 2021.
34. Hossain, M.B.; Kong, Y.; Posada-Quintero, H.F.; Chon, K.H. Comparison of electrodermal activity from multiple body locations based on standard EDA indices' quality and robustness against motion artifact. *Sensors* **2022**, *22*, 3177. [CrossRef]
35. Willems, F.; Cloudt, R. Experimental demonstration of a new model-based SCR control strategy for cleaner heavy-duty diesel engines. *IEEE Trans. Control. Syst. Technol.* **2010**, *19*, 1305–1313. [CrossRef]
36. Wattenberg, M.; Viégas, F.; Johnson, I. How to use t-SNE effectively. *Distill* **2016**, *1*, e2. [CrossRef]
37. Platzer, A. Visualization of SNPs with t-SNE. *PLoS ONE* **2013**, *8*, e56883. [CrossRef] [PubMed]
38. Tumminello, M.L.; Macioszek, E.; Granà, A. Insights into simulated smart mobility on roundabouts: Achievements, lessons learned, and steps ahead. *Sustainability* **2024**, *16*, 4079. [CrossRef]
39. Tumminello, M.L.; Macioszek, E.; Granà, A.; Giuffrè, T. Evaluating traffic-calming-based urban road design solutions featuring cooperative driving technologies in energy efficiency transition for smart cities. *Energies* **2023**, *16*, 7325. [CrossRef]
40. Mądziel, M. State of Charge Prediction for Li-Ion Batteries in EVs for Traffic Microsimulation. *Energies* **2025**, *18*, 4992. [CrossRef]
41. Mądziel, M. Impact of Weather Conditions on Energy Consumption Modeling for Electric Vehicles. *Energies* **2025**, *18*, 1994. [CrossRef]
42. Barbier, A.; Salavert, J.M.; Palau, C.E.; Guardiola, C. Analysis of the Euro 7 on-board emissions monitoring concept with real-driving data. *Transp. Res. Part D Transp. Environ.* **2024**, *127*, 104062. [CrossRef]
43. Dennis, R.; Long, H.A., III; Jesionowski, G. A literature review of NO_x emissions in current and future state-of-the-art gas turbines. *J. Eng. Gas Turbines Power* **2024**, *146*, 030801. [CrossRef]
44. Mulholland, E.; Miller, J.; Bernard, Y.; Lee, K.; Rodríguez, F. The role of NO_x emission reductions in Euro 7/VII vehicle emission standards to reduce adverse health impacts in the EU27 through 2050. *Transp. Eng.* **2022**, *9*, 100133. [CrossRef]
45. Wen, X.; Li, K.; Wang, J. NO_x emission predicting for coal-fired boilers based on ensemble learning methods and optimized base learners. *Energy* **2023**, *264*, 126171. [CrossRef]
46. Pachauri, N. An emission predictive system for CO and NO_x from gas turbine based on ensemble machine learning approach. *Fuel* **2024**, *366*, 131421. [CrossRef]
47. Chen, H.; Lou, S.; Lv, C. Hybrid physics-data-driven online modelling: Framework, methodology and application to electric vehicles. *Mech. Syst. Signal Process.* **2023**, *185*, 109791. [CrossRef]
48. Walhekar, V.K.; Pushparaj, D.K. Shaping the Future of Automotive Emissions: A Comprehensive Study of Euro 7, BS-7 and CAFE 3 Norms. *Int. J. Eng. Res. Technol. (IJERT)* **2025**, *14*. [CrossRef]
49. Mądziel, M. Quantifying Emissions in Vehicles Equipped with Energy-Saving Start–Stop Technology: THC and NO_x Modeling Insights. *Energies* **2024**, *17*, 2815. [CrossRef]
50. Mądziel, M. Modelling CO_2 Emissions from Vehicles Fuelled with Compressed Natural Gas Based on On-Road and Chassis Dynamometer Tests. *Energies* **2024**, *17*, 1850. [CrossRef]
51. Mądziel, M. Instantaneous CO_2 emission modelling for a Euro 6 start-stop vehicle based on portable emission measurement system data and artificial intelligence methods. *Environ. Sci. Pollut. Res.* **2024**, *31*, 6944–6959. [CrossRef] [PubMed]
52. Mądziel, M. Modeling Exhaust Emissions in Older Vehicles in the Era of New Technologies. *Energies* **2024**, *17*, 4924. [CrossRef]
53. Di Blasio, G.; Vassallo, A.; Pesce, F.C.; Beatrice, C.; Belgiorno, G.; Avolio, G. The key role of advanced, flexible fuel injection systems to match the future CO_2 targets in an ultra-light mid-size diesel engine. *SAE Int. J. Engines* **2019**, *12*, 129–144. [CrossRef]

energies

Article

Charting the Landscape of Data Envelopment Analysis in Renewable Energy and Carbon Emission Efficiency

Thu-Thao Le and Wen-Min Lu *

Department of International Business Administration, Chinese Culture University, Taipei 11114, Taiwan; chloe.thuthaole@gmail.com
* Correspondence: lwm2@ulive.pccu.edu.tw

Abstract

This study explores the intellectual landscape and methodological evolution of Data Envelopment Analysis (DEA) in the context of renewable energy and carbon emission efficiency. Using bibliometric techniques and data extracted from the Web of Science Core Collection (2389 publications from 2000 to 2024), the research identifies influential authors, institutions, and thematic clusters shaping the field. The results reveal that DEA has evolved from a traditional efficiency assessment tool into a comprehensive analytical framework supporting sustainable energy transition and carbon mitigation policies. Six major research clusters were identified, encompassing carbon emission measurement, efficiency benchmarking, methodological innovations, industrial applications, circular economy perspectives, and international productivity comparisons. Notably, Asian scholars, particularly from China and Taiwan, dominate the research landscape, reflecting strong regional leadership in empirical and methodological advancements. The findings demonstrate that recent studies increasingly adopt advanced models such as network DEA, dynamic DEA, DEA–Malmquist, and hybrid DEA–machine learning approaches to address complex energy systems. Comparative insights highlight DEA's advantages over Stochastic Frontier Analysis (SFA) in handling multi-dimensional, non-parametric data, while emphasizing the need for hybrid frameworks to improve robustness. This study contributes to the ongoing discourse on energy sustainability by mapping knowledge structures, revealing methodological trajectories, and providing guidance for future research on efficiency and carbon reduction strategies.

Keywords: DEA; renewable energy efficiency; carbon emission reduction; environmental benchmarking; sustainable energy transition; bibliometric analysis

1. Introduction

The accelerating pace of climate change and the depletion of fossil fuel reserves have compelled governments, industries, and research communities to pursue sustainable energy transitions [1]. According to the International Energy Agency, renewable energy is projected to supply nearly 50% of global electricity demand by 2050, underscoring its central role in decarbonizing the energy sector [2]. Solar photovoltaics, wind energy, biomass, geothermal, and hydropower have emerged as major contributors to this transformation [3]. However, their rapid expansion is accompanied by challenges related to efficiency, resource allocation, intermittency, and environmental trade-offs [4]. Measuring and benchmarking performance across these diverse energy systems has therefore become an essential prerequisite for improving operational outcomes and guiding investment and policy decisions [5].

Parallel to the rise of renewable energy is the global urgency to mitigate greenhouse gas emissions [6]. Carbon dioxide (CO_2) accounts for nearly three-quarters of global greenhouse gas emissions, with the energy sector responsible for the majority of these outputs [7]. International agreements such as the Paris Accord and the pursuit of net-zero emission targets have heightened the demand for robust methodologies that can assess carbon emission efficiency at multiple levels, ranging from individual power plants and firms to national energy systems [8]. This dual imperative, improving renewable energy performance while reducing carbon emissions, has created fertile ground for methodological innovations in efficiency analysis.

Among the available efficiency assessment tools, Data Envelopment Analysis (DEA) has become particularly prominent. Originally introduced by Charnes, Cooper, and Rhodes in 1978, DEA is a non-parametric linear programming approach that measures the relative efficiency of decision-making units (DMUs) with multiple inputs and outputs [9]. Unlike parametric approaches such as Stochastic Frontier Analysis (SFA), DEA does not require explicit functional form assumptions, making it especially useful in complex energy systems where diverse factors interact in nonlinear ways.

In the energy domain, DEA has been applied extensively to evaluate the efficiency of electricity generation companies, renewable energy technologies, regional energy systems, and national energy policies [10]. For instance, studies have employed DEA to benchmark the performance of wind farms, compare the operational efficiency of thermal versus renewable power plants, and analyze the energy–environmental efficiency of countries with varying emission profiles [11,12]. DEA has also been extended through models such as dynamic DEA, network DEA, Malmquist productivity indices, and slack-based measures, which allow researchers to capture temporal changes, multi-stage processes, and undesirable outputs such as CO_2 emissions [13]. These methodological advancements have substantially enriched the analytical capacity of DEA to address contemporary challenges in energy and environmental policy.

Despite its widespread application, the literature on DEA in renewable energy and carbon emission efficiency remains fragmented and dispersed across multiple disciplinary outlets [14]. Much of the research is dominated by case studies that focus on specific countries, technologies, or time periods. While these contributions offer valuable insights, they do not collectively provide a comprehensive overview of how DEA has been employed across different contexts, nor do they reveal the evolution of scholarly interest over time.

A few review studies have discussed efficiency assessment methods in energy or environmental economics more broadly [15,16], but these reviews tend to either:

1. Examine DEA alongside other methods (thus providing only a limited treatment of DEA itself); or
2. Focus on narrow applications such as electricity generation or national-level carbon efficiency, thereby overlooking the broader spectrum of renewable and emission-related studies.

As a result, several critical questions remain unanswered. How has the application of DEA in renewable energy and carbon emission efficiency evolved over the past two decades? Which regions and institutions are leading this field? What are the dominant methodological and thematic directions? And most importantly, what gaps exist that could guide future research in advancing both renewable energy adoption and carbon emission reduction? Addressing these questions requires a systematic knowledge mapping approach that goes beyond individual case studies to capture the structure, dynamics, and trajectory of the field.

Using bibliographic data extracted from the Web of Science Core Collection, this work applies advanced analytical and visualization techniques to uncover publication trends, thematic clusters, influential authors, and collaboration networks. In doing so, it provides a panoramic view of how DEA has contributed to the twin objectives of renewable energy performance improvement and carbon emission reduction.

The contributions of this study are fourfold. First, it offers the first consolidated mapping of DEA applications in this domain, thereby creating a reference point for both new and experienced researchers. Second, it identifies key methodological developments, such as the integration of DEA with frontier productivity indices and environmental extensions, which have expanded the scope of analysis. Third, it highlights geographical and institutional patterns of research, providing insight into how global collaboration is shaping this field. Finally, it proposes a research agenda that emphasizes future directions, including the application of DEA to emerging energy technologies (e.g., hydrogen energy, energy storage, and smart grids), integration with machine learning techniques, and policy-oriented assessments aligned with net-zero targets [17].

The remainder of this paper is structured as follows. Section 2 outlines the methodology, including data collection procedures, search strategy, and the bibliometric tools employed for analysis. Section 3 presents the results, covering publication trends, authorship patterns, geographic and institutional contributions, and thematic structures revealed through keyword and citation analyses. Section 4 discusses these findings in light of the role of DEA in renewable energy efficiency and carbon emission reduction, while also comparing DEA with other methods of efficiency analysis. Section 5 develops a future research agenda, highlighting promising directions for integrating DEA with emerging technologies and policy frameworks. Finally, Section 6 concludes with key insights, implications, and recommendations for scholars, policymakers, and practitioners engaged in the energy transition.

2. Literature Review

2.1. DEA in Energy Efficiency Research

Data Envelopment Analysis (DEA) has become a cornerstone in evaluating efficiency within the energy sector. Its appeal lies in its ability to accommodate multiple inputs and outputs without assuming a predefined production function. Traditional efficiency methods such as stochastic frontier analysis (SFA) are often constrained by functional form specifications and statistical assumptions, whereas DEA provides greater flexibility in handling heterogeneous data typical of energy systems [18].

Early applications of DEA in energy research focused primarily on electricity generation, where the method was used to benchmark the technical efficiency of power plants [19]. Studies examined coal-fired, gas, and hydroelectric facilities, providing insights into operational performance and resource utilization. Over time, these analyses expanded to incorporate renewable energy sources, such as solar and wind power plants, which introduced new complexities due to intermittency, geographic dispersion, and evolving technological benchmarks [20].

The evolution of DEA models has further enriched energy applications. Extensions such as the Malmquist productivity index have enabled temporal analysis of efficiency changes, while network DEA has been employed to capture multi-stage processes in energy production [21]. Slack-based measures (SBM) have been introduced to better account for input and output slacks, particularly relevant in contexts where energy losses or unutilized capacity are significant [22]. These methodological advancements have made DEA a preferred approach for evaluating energy efficiency under conditions of technical, economic, and environmental complexity.

2.2. DEA in Renewable Energy Evaluation

As the global push for clean energy accelerated, DEA found increasing application in renewable energy evaluation. Researchers have employed DEA to benchmark the efficiency of wind farms, solar photovoltaic installations, biomass facilities, and hybrid renewable systems [23]. In many cases, DEA was used to assess not only technical efficiency but also financial and environmental performance, providing a holistic view of renewable energy deployment [24,25].

For example, DEA has been applied to evaluate the operational efficiency of wind power projects across different countries, highlighting how factors such as turbine size, location, and government subsidies influence performance outcomes [26]. Solar energy studies have used DEA to benchmark the productivity of photovoltaic plants, focusing on variations in geographic conditions and technological adoption [27]. In the case of biomass and biofuel facilities, DEA has been employed to evaluate energy conversion efficiency while simultaneously considering by-products and waste management [28].

DEA has also been extended to policy analysis in renewable energy. For instance, cross-country DEA studies have compared national renewable energy portfolios, providing insights into which countries allocate resources most effectively to maximize renewable generation [29]. These analyses have informed policy recommendations on subsidy design, grid integration, and research investment. The ability of DEA to handle multiple objectives has proven especially valuable in renewable energy, where technical efficiency, cost considerations, and environmental impact often need to be assessed simultaneously [30].

2.3. DEA and Carbon Emission Efficiency

In addition to renewable energy evaluation, DEA has played a central role in analyzing **carbon emission efficiency**. The method's ability to incorporate "undesirable outputs," such as CO_2 emissions, makes it uniquely suited for assessing environmental performance. DEA-based environmental efficiency models have been widely used to examine the trade-offs between energy use, economic growth, and carbon reduction [31].

At the macro level, DEA has been applied to compare the carbon efficiency of countries and regions. These studies assess how effectively nations convert energy inputs into economic output while minimizing emissions. Findings often reveal significant disparities across regions, reflecting differences in industrial structure, technological development, and policy stringency. For instance, some studies have shown that while developed economies achieve higher technical efficiency, emerging economies may perform better in terms of carbon emission reduction due to rapid adoption of cleaner technologies [32,33].

At the micro level, DEA has been used to evaluate carbon performance in specific industries, such as manufacturing, transportation, and energy-intensive sectors. By incorporating CO_2 emissions as undesirable outputs, these studies highlight the importance of environmental considerations in operational benchmarking [34]. Extensions such as the slack-based measure and dynamic DEA have provided more nuanced insights, capturing inefficiencies that traditional DEA might overlook.

DEA has also been integrated with life cycle assessment (LCA) and carbon accounting methods, allowing for more comprehensive evaluations of emission reduction strategies. This integration underscores the growing recognition of DEA as not only a tool for efficiency measurement but also a methodological framework for advancing environmental sustainability [35].

2.4. Methodological Advancements in DEA for Energy and Environment

As shown in Table 1, the versatility of DEA in energy and environmental applications has been enhanced through several methodological innovations:

Table 1. Methodological innovations in DEA for energy and environment.

Methodological	Description	Sources
Dynamic DEA	Captures changes in efficiency over time, useful for monitoring renewable energy adoption and emission reduction progress.	[36,37]
Network DEA	Models multi-stage processes, such as energy generation, transmission, and distribution, or production processes with intermediate outputs.	[38,39]
Slack-Based Measure (SBM) DEA	Explicitly accounts for input and output slacks, improving the realism of efficiency measurement in energy systems.	[40,41]
Malmquist Index DEA	Assesses productivity change over time, particularly relevant for tracking progress toward carbon neutrality.	[27,42]
DEA with Undesirable Outputs	Incorporates pollutants and emissions, making it directly applicable to carbon efficiency analysis.	[27,43]
Hybrid DEA Models	Combine DEA with machine learning, fuzzy logic, or econometric methods to improve robustness and predictive power.	[23,44]

These advancements demonstrate that DEA has evolved far beyond its original formulation, becoming an adaptable and powerful framework for addressing complex questions in energy sustainability and environmental policy.

3. Research Methodology

3.1. Data Collection

The study adopts a systematic review approach to map the application of Data Envelopment Analysis (DEA) in renewable energy and carbon-emission efficiency. Data were collected exclusively from the Web of Science (WoS) Core Collection, a database widely recognized for its comprehensive coverage of high-quality academic literature. Although the search was performed using a single query within Web of Science (WoS), the WoS Core Collection aggregates multiple citation indices (e.g., SCI-E, SSCI, ESCI). When a publication is indexed in more than one collection, it can appear multiple times in the raw export. After merging records from these indices, automated de-duplication was performed using WoS export metadata (DOI, title, author combination). To ensure rigor and replicability, we applied an advanced search strategy combining key terms such as "Data Envelopment Analysis," "DEA," "renewable energy," "energy efficiency," and "carbon emission" on 31 December 2024. Furthermore, to enhance methodological rigor, we have provided detailed documentation of the search strategy, including the use of specific keywords, Boolean operators, and field codes. This level of detail allows other researchers to replicate or adapt the search for related studies, ensuring consistency in bibliometric analyses. By explicitly stating the search parameters and potential limitations, such as database updates or indexing delays, we aim to promote transparency and reliability in the research process.

Although DEA originated in the late 1970s, its systematic adoption in renewable energy performance assessment and carbon-efficiency evaluation gained traction only after 2000. A preliminary scoping search confirmed that publications prior to 2000 were scarce and did not reflect modern renewable energy technologies, contemporary climate-policy frameworks, or methodological advances such as network DEA, dynamic DEA, and hybrid DEA–machine learning models. Therefore, the time frame was restricted to 2000–2024 to capture the relevant evolution of research alongside global sustainability imperatives.

Only English-language journal articles were retained to maintain consistency in data interpretation and ensure standardized citation records. While conference papers and book chapters may undergo peer review, they often lack standardized indexing, citation traceability, and structured metadata. To preserve analytical comparability and ensure data quality, we narrowed the scope to peer-reviewed journal articles only. The final search yielded 2389 records for further screening and analysis.

3.2. Data Screening and Filtering

To enhance methodological transparency, this study followed the PRISMA 2020 reporting guideline to structure the identification, screening, eligibility assessment, and final inclusion steps [45]. Because the objective of this research is bibliometric mapping rather than clinical intervention evaluation, PRISMA was applied to document workflow and reporting transparency, rather than to appraise article quality through the 27-item checklist, which is not directly applicable to bibliometric studies. Specifically, PRISMA guidance was used to record the process, including database identification, duplicate removal, title/abstract screening, full-text relevance confirmation, and final inclusion. The PRISMA framework was used to structure the four key stages of the review process: identification, screening, eligibility assessment, and final inclusion. First, duplicate records were removed to ensure data consistency. Second, titles and abstracts were screened to exclude studies unrelated to energy applications of DEA (e.g., banking, agriculture, healthcare). Third, a full-text evaluation was conducted to verify relevance to renewable energy systems, energy efficiency assessment, or carbon emission performance. Finally, only studies applying DEA or its methodological extensions within the energy domain were retained.

PRISMA Workflow (Text Description)

Accordingly, the use of PRISMA in this work focuses on documenting the filtering protocol rather than evaluating the methodological quality of each article. The detailed screening process is presented visually in the PRISMA flow diagram (Table 2). PRISMA citation has been added to acknowledge the standardized reporting structure supporting our workflow. The initial Web of Science Core Collection search yielded 4280 records. After removing 3920 duplicates, 2890 unique publications remained for title–abstract screening. Applying database index filters (SCIE and SSCI) resulted in 2398 eligible records. Subsequent filtering by document type (Article, Review, and Proceedings Paper) refined the dataset to 2392 studies, of which 2389 were identified as English-language publications. All 2389 records were evaluated at full-text level and confirmed as eligible for inclusion in the final bibliometric analysis.

Table 2. PRISMA-Style Flow Diagram.

Records identified through WoS search (n = 4280)	
↓	
Duplicates removed (n = 3920)	
↓	
Records screened (title/abstract) (n = 2890)	Records excluded (not SSCI/SCIE) (n = 492)
↓	
SSCI and SCIE records (n = 2398)	Excluded (not Article/Review/Proceeding) (n = 6)
↓	
Articles/Reviews/Proceedings (n = 2392)	Excluded (non-English) (n = 3)
↓	
English-language articles (n = 2389)	
↓	
Full-text articles assessed for eligibility (n = 2389)	
↓	
Final articles included in the analysis (n = 2389)	

3.3. Research Methods

For bibliometric analysis, this study employs VOSviewer 1.6.20, two widely used tools for scientific mapping. VOSviewer is utilized to construct and visualize networks of co-authorship, co-citation, and keyword co-occurrence, thereby identifying collaboration patterns, influential publications, and thematic clusters [46].

The calculation method for average citations has now been explicitly stated in the manuscript and table notes to ensure clarity and transparency. The study adopts an adjusted approach to account for differences in publication age, defined as:

$$\text{Average yearly citations} = \frac{\text{Total citations}}{\text{Years since publication} + 1}$$

This formula prevents division-by-zero issues for publications in 2025 and provides a more accurate representation of citation impact over time.

To support the methodological choice, references have been added that demonstrate precedent for this adjusted citation-window normalization in bibliometric analyses. Notably, Egghe and Rousseau (2000) and Harzing and Alakangas (2016) provide justification for using similar adjustments, reinforcing the robustness and validity of the approach in evaluating research impact [47,48].

The choice of a bibliometric method is particularly suitable for this topic as it not only quantifies the research output but also uncovers latent connections between concepts and disciplines. By analyzing publication trends, geographic distributions, and thematic clusters, this study provides valuable insights into how DEA has been applied to address pressing energy and environmental challenges. Moreover, the methodology facilitates the identification of gaps in literature, paving the way for future research on methodological innovations and practical applications of DEA in sustainability contexts.

4. Results

4.1. Publication Trends

The analysis of publication trends from 2000 to 2025 in Figure 1 reveals a clear upward trajectory in scholarly output, moving from only 2 records in 2000 to a peak of 340 in 2022. During the early stage (2000–2010), publication activity remained minimal, with fewer than 15 publications per year. This period indicates that research in this field was still emerging and had not yet attracted substantial academic attention.

From 2011 to 2018, the field entered a growth phase, with records increasing from 26 in 2011 to 184 in 2018. This steady rise reflects a growing recognition of the topic's importance within academic and professional discourse. The acceleration phase followed between 2019 and 2022, when publications more than doubled in just four years. The year 2022 marked the highest output with 340 records, suggesting strong global interest likely driven by technological advancements, policy debates, and post-pandemic recovery research.

After 2022, publication counts exhibited a modest decline, with 244 articles in 2023, 211 in 2024, and 127 in 2025. However, the lower figure for 2025 is likely due to incomplete indexing and data availability rather than a true reduction in research output, as records from the most recent year are often still undergoing database processing. This suggests that the apparent downward trend should be interpreted cautiously.

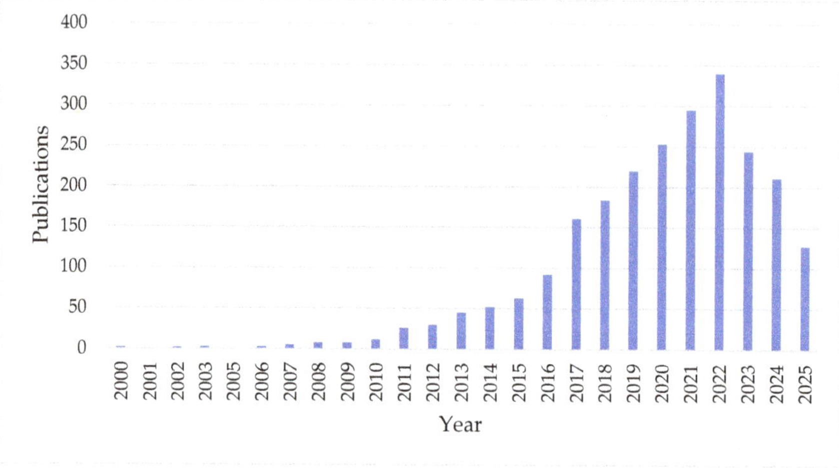

Figure 1. Growth trajectory of DEA-related studies in energy & carbon emission.

Beyond indexing delays, the fluctuation may also relate to evolving academic priorities and funding allocations as countries reassess their strategies toward carbon neutrality. For instance, some research initiatives may be shifting toward integrated evaluation frameworks, combining DEA with machine learning, ESG scoring, or life-cycle assessment, leading to potential dispersion of publications across broader sustainability journals. In addition, increasing interdisciplinary collaboration could redistribute DEA-based studies into domains such as environmental sciences, urban planning, and industrial engineering, temporarily affecting publication concentration.

Despite the short-term variability, the continued appearance of DEA-related studies in renewable energy and carbon-efficiency assessment indicates that the field remains active. Long-term patterns reinforce the sustained relevance of DEA as a quantitative tool for guiding sustainability decision-making, and future years are expected to reflect further growth once indexing records are fully updated.

Overall, the findings highlight several key insights. As shown in Figure 1, the field has experienced exponential growth since 2010, achieving maturity between 2019 and 2022. The peak in 2022 indicates the height of global scholarly attention, while the slight decline afterward should be interpreted cautiously, as it may reflect cyclical publishing patterns or incomplete indexing. Given the consistently high output since 2019, the research area is now well-established and continues to hold significant academic potential.

4.2. Geographic and Institutional Distribution

The analysis of publication output across countries and regions reveals a clear imbalance in research contributions (shown in Figure 2). China emerges as the predominant contributor, producing 1581 publications—far surpassing all other nations. This leading position reflects China's consistent investment in research and development and its alignment with long-term national strategies aimed at carbon mitigation and energy transition. Existing literature shows that Chinese scholars frequently focus on carbon emission efficiency, the impacts of environmental regulations, and methodological innovations such as network and dynamic DEA models. These research directions mirror China's broader pursuit of carbon neutrality and modernization of its energy infrastructure. Collectively, this dominant presence underscores not only China's robust research capacity but also its expanding role in shaping both the methodological approaches and thematic priorities of DEA-based energy research on a global scale.

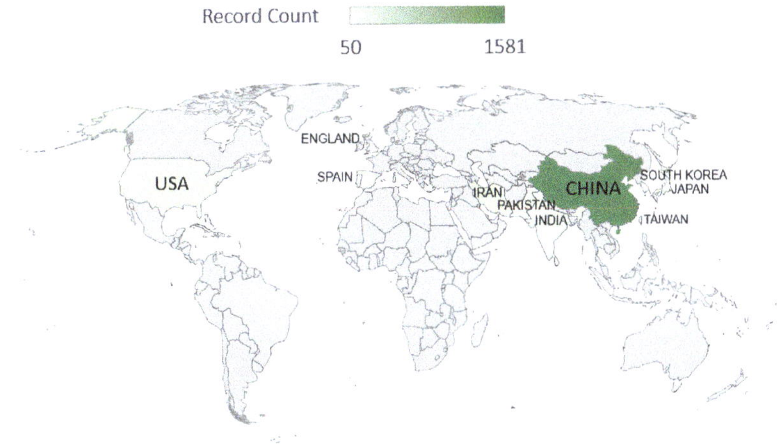

Figure 2. Leading countries.

In comparison, Taiwan (221 publications) and the United States (146 publications) rank second and third, respectively. Although their publication volumes are considerably lower than China's, both countries exhibit distinctive and meaningful research trajectories. Taiwan's output is particularly noteworthy given its relatively small population and limited resource base. Thematically, Taiwanese studies often emphasize renewable energy productivity, technological innovation, and efficiency benchmarking—reflecting a focused and efficient scientific community. Meanwhile, the United States maintains a strong presence in the field, frequently contributing to hybrid sustainability frameworks and interdisciplinary approaches that integrate DEA with environmental assessment tools. Despite its smaller output compared with China, the U.S. research remains conceptually influential, particularly in advancing methodological diversification and policy-oriented applications.

A second tier of contributors comprises Iran (127 publications), England (100), and Japan (76), each demonstrating moderate yet consistent research activity. Iran's position is particularly noteworthy given the constraints imposed by economic sanctions and limited opportunities for international collaboration, which typically hinder research productivity. England and Japan, both equipped with highly developed academic systems, continue to contribute steadily to the field, although their publication volumes appear relatively modest compared with their broader global standing in scientific research.

The third tier includes Pakistan (67 publications), South Korea (64), Spain (54), and India (50). Although their publication volumes remain relatively modest, these countries demonstrate an increasing engagement in the global research landscape. Notably, Pakistan and India reflect the rising contribution of emerging economies to international knowledge production. South Korea's comparatively lower output is somewhat unexpected, given its strong scientific and technological reputation, which may suggest either a thematic concentration of the dataset or underrepresentation within this specific research domain.

In summary, the global distribution of publications reveals China's overwhelming dominance, followed by a small group of notable but less prolific contributors. The substantial gap between China and other countries underscores the pronounced imbalance in global research capacity. Nevertheless, the growing participation of countries such as Iran, Pakistan, and India points to a gradual diversification of contributors, indicating that research in this field is extending beyond traditional academic powerhouses to include a broader range of emerging and developing economies.

Table 3 presents the leading institutions contributing to the research field, providing insight into potential research centers and the distribution of academic activity. The Chinese Academy of Sciences (CAS) leads with 117 records, underscoring its role as the driving force in China's academic research. As the nation's premier research institution, CAS's dominance reflects its broad disciplinary scope and substantial government support, enabling significant contributions across multiple domains.

Table 3. Active institutions and collaborations.

Affiliations	Record Count
Chinese Academy of Sciences	117
Soochow University	89
Nanjing University of Aeronautics Astronautics	73
Beijing Institute of Technology	69
North China Electric Power University	59
Xiamen University	56
University of Science Technology of China Cas	55
Hefei University of Technology	52
Southeast University China	48
Hohai University	45

Following CAS, Soochow University (89 publications) and Nanjing University of Aeronautics and Astronautics (73) emerge as the next most active contributors. Soochow University's strong output underscores the growing influence of regional universities beyond China's traditional elite, while Nanjing University of Aeronautics and Astronautics highlights the pivotal role of applied science and engineering institutions in advancing research on technology, sustainability, and governance.

Other notable contributors include Beijing Institute of Technology (69), North China Electric Power University (59), and Xiamen University (56), reflecting broad participation across both technical and comprehensive universities. North China Electric Power University exemplifies the intersection of energy and sustainability research, closely aligned with ESG-related themes.

The University of Science and Technology of China (55) and Hefei University of Technology (52) maintain a strong presence, emphasizing interdisciplinary research and applied innovation. Similarly, Southeast University (48) and Hohai University (45) further diversify the institutional landscape, with Hohai's contributions linked to water management and environmental sustainability.

Overall, this distribution indicates that China's research output is not only concentrated in top-tier institutions but also distributed across specialized universities, particularly those with technical, environmental, and governance orientations. The pattern suggests a coordinated national effort in which universities leverage their disciplinary strengths, reinforcing China's leading role in this research field.

4.3. Key Contributors to DEA Research on Renewable Energy and Carbon Efficiency

Table 4 presents the most prolific and influential authors contributing to DEA research on renewable energy and carbon efficiency. The H-index is a widely used bibliometric indicator that combines productivity and citation impact, but it has well-known limitations (e.g., sensitivity to career length, field differences, and inability to account for highly cited single works). For discussion of the H-index and alternative metrics [49–51]. Therefore, we present H-indices here to help readers contextualize author influence in combination with other indicators (total citations, recent activity), rather than as a sole evaluative metric. The data show that Chiu Yung-Ho (Soochow University, Taiwan) stands out as the most productive scholar with 71 publications and an H-index of 33, reflecting both consistent output and recognition within the academic community. Following him is Li Y. (Nanjing University of Finance & Economics, China) with 62 publications and an H-index of 17, highlighting the strong research engagement from China in this field.

Table 4. Most prolific and influential authors.

Authors	Institutes	Country	Research ID	H-Index	Records
Chiu, Yung-Ho	Soochow University	Taiwan	H-5231-2019	33	71
Li Y	Nanjing University of Finance & Economics	China	FJK-9617-2022	17	62
Cui, Qiang	Southeast University	Bangladesh	KLC-0358-2024	22	37
Lin, Boqiang	Xiamen University	Taiwan	G-3960-2010	107	34
Lin, Tai-Yu	National Cheng Kung University	Taiwan	AAP-1501-2021	11	32
Feng, Chao	Chongqing University	China	A-6705-2019	48	28
Wang, Chia-Nan	National Kaohsiung University of Science & Technology	Taiwan	JZW-4462-2024	31	28
Sueyoshi, Toshiyuki	Tokyo Institute of Technology	Japan	GDM-5048-2022	53	27
Zhang, Ning	East China University of Science & Technology	China	HCI-7860-2022	57	27
Zhou, Peng	China University of Petroleum	China	A-6527-2012	73	27

Another significant contributor is Cui Qiang (Southeast University, Bangladesh) with 37 publications, demonstrating regional diversification of scholarship beyond the dominant hubs of China and Taiwan. In terms of influence, Lin Boqiang (Xiamen University, Taiwan) emerges as a leading figure, with an impressive H-index of 107 despite having 34 publications, indicating exceptionally high citation impact and scholarly visibility. Similarly, Zhou Peng (China University of Petroleum, China) shows a strong balance of productivity and influence with 27 publications and a high H-index of 73.

Other prominent contributors include Feng Chao (Chongqing University, China) and Wang Chia-Nan (National Kaohsiung University of Science & Technology, Taiwan), each with 28 publications and H-indices of 48 and 31, respectively. Meanwhile, Sueyoshi Toshiyuki (Tokyo Institute of Technology, Japan) and Zhang Ning (East China University of Science & Technology, China) demonstrate that even with fewer publications (27 each), their H-indices of 53 and 57 signify considerable academic impact. Lastly, Lin Tai-Yu (National Cheng Kung University, Taiwan), with 32 publications and an H-index of 11, adds further evidence of Taiwan's strong presence in this research domain.

Taken together, these findings highlight that the field is dominated by Asian scholars, particularly from China and Taiwan, who contribute both high productivity and strong citation influence. This regional concentration suggests that Asia, driven by rapid industrialization and energy challenges, is at the forefront of advancing DEA research applied to energy and environmental issues.

4.4. Citation Analysis

An examination of the most highly cited works reveals the intellectual foundations and evolving priorities in the application of data envelopment analysis (DEA) to renewable energy and carbon emission efficiency (Table 5). Early contributions were particularly influential in establishing methodological and empirical benchmarks. For instance, Hu and Wang (2006) pioneered the assessment of total-factor energy efficiency across Chinese regions, producing one of the most widely cited studies (1182 citations) with lasting annual influence (59.1 citations/year) [27]. Similarly, Zhou et al. (2008, 2010, 2012) advanced methodological innovations in environmental DEA, particularly through the introduction of the directional distance function and Malmquist index analysis [24,52,53]. These works, published in Energy Economics and European Journal of Operational Research, provided a solid theoretical basis for measuring carbon performance and abatement costs, each attracting more than 400 citations, thereby anchoring subsequent research in rigorous econometric and operational research frameworks.

Table 5. Highly cited papers shaping the field.

Title	1st Authors	Journal Title	Year	Total Citations	Average per Year
Total-factor energy efficiency of regions in China	Hu, Jin-Li	Energy Policy	2006	1182	59.1
Energy and CO_2 emission performance in electricity generation: A non-radial directional distance function approach	Zhou, P.	European Journal of Operational Research	2012	735	52.5
Total factor carbon emission performance: A Malmquist index analysis	Zhou, P.	Energy Economics	2010	615	38.44
The effects of three types of environmental regulation on eco-efficiency: A cross-region analysis in China	Ren, Shenggang	Journal of Cleaner Production	2018	580	72.5
Public spending and green economic growth in BRI region: Mediating role of green finance	Zhang, Dongyang	Energy Policy	2021	554	110.8
Efficiency and abatement costs of energy-related CO_2 emissions in China: A slacks-based efficiency measure	Choi, Yongrok	Applied Energy	2012	504	36
A comprehensive review of data envelopment analysis (DEA) approach in energy efficiency	Mardani, Abbas	Renewable & Sustainable Energy Reviews	2017	486	54
Low-carbon city pilot and carbon emission efficiency: Quasi-experimental evidence from China	Yu, Yantuan	Energy Economics	2021	454	90.8
Measuring environmental performance under different environmental DEA technologies	Zhou, P.	Energy Economics	2008	414	23
Energy efficiency and sustainable development goals (SDGs)	Zakari, A.	Energy	2022	406	81.2

A second thematic cluster highlights the intersection of policy design and eco-efficiency outcomes. Ren et al. (2014) explored the effects of three types of environmental regulation on eco-efficiency across Chinese regions, amassing 580 citations with a high citation rate (72.5 per year), underscoring the growing academic and policy interest in regulation-driven performance [28]. Similarly, Yu et al. (2021) investigated low-carbon city pilot policies, offering quasi-experimental evidence that attracted 454 citations at an exceptionally high rate of 90.8 per year, reflecting the surge in empirical policy evaluations [54]. These contributions demonstrate the increasing role of institutional and regulatory frameworks in shaping carbon efficiency research.

A third set of studies emphasizes the link between economic development, finance, and sustainability goals. Zhang et al. (2021) examined the mediating role of green finance in promoting green economic growth in the Belt and Road Initiative (BRI) region, securing 554 citations with the highest average citation rate (110.8 per year) [55]. This illustrates the rising prominence of finance-driven sustainability mechanisms. In parallel, Zakari et al. (2022) explicitly connected energy efficiency to the Sustainable Development Goals (SDGs), a study already cited 406 times with a strong yearly average of 81.2, highlighting the global policy relevance of DEA applications [56].

Finally, review studies have played a vital role in consolidating knowledge and directing scholarly attention. Mardani et al. (2017) provided a comprehensive review of DEA applications in energy efficiency, which has since been cited 486 times, averaging 54 citations per year [57]. This review functions as a cornerstone, synthesizing fragmented studies and guiding new research trajectories. Collectively, these highly cited papers illustrate how DEA scholarship has evolved from methodological refinement and regional applications to encompass broader themes of policy, finance, and global sustainability.

4.5. Keyword and Thematic Analysis

The co-occurrence network in Figure 3 maps the intellectual and thematic landscape of Data Envelopment Analysis (DEA) applications in renewable energy and carbon emission efficiency. For reproducibility, the following settings were applied: minimum keyword occurrence = 5, counting method = full counting, resolution parameter (cluster resolution) = default, and label size scaling increased to improve readability. The map was exported as a PNG at 300 dpi and embedded in the manuscript. These settings improved label visibility while preserving cluster structure; raw data and VOSviewer project files are available upon request. Six major clusters emerged from the analysis, reflecting interconnected yet distinct research trajectories that collectively shape the field. The size of the nodes represents the frequency of keyword occurrence, while their proximity illustrates thematic overlap. These clusters demonstrate how DEA has evolved from a tool for measuring energy efficiency into a comprehensive framework for addressing global sustainability challenges.

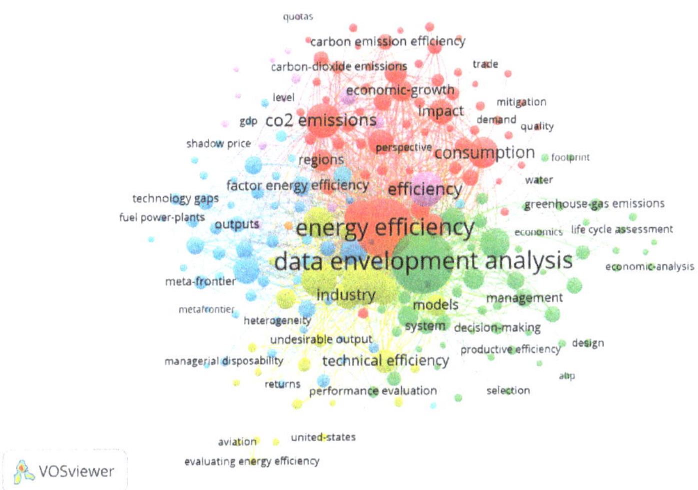

Figure 3. Keyword co-occurrence analysis.

Cluster 1 (Red): Carbon Emission and Climate Change Nexus

Cluster 1 is the largest and most central cluster, capturing essential themes such as carbon emissions, CO_2 emissions, carbon neutrality, climate change, and economic growth. This thematic core reflects the increasing deployment of DEA to evaluate environmental and carbon-reduction performance under tightening global climate commitments. Most studies quantify carbon-emission efficiency at national, regional, industrial, or urban levels, with China contributing a dominant share of outputs due to its active carbon-transition policymaking.

DEA findings are often integrated with macroeconomic indicators—such as economic growth, energy consumption, and industrial structure—to explore trade-offs between environmental responsibility and economic development [58,59]. Over time, the thematic focus within this cluster has evolved from assessing isolated production systems toward broader evaluations of national- and regional-level carbon-neutrality pathways, especially in the context of global climate governance frameworks [60].

Cluster 2 (Green): Efficiency Benchmarking and Decision-Making Models

Cluster 2 centers on DEA's function as a benchmarking and managerial decision-support tool in energy systems, agriculture, and industrial production. Key terms—benchmarking, cross-efficiency, AHP, conservation, and economic analysis—illustrate the integration of DEA within multi-criteria decision-making (MCDM) frameworks to capture economic, technological, and environmental trade-offs [41,61].

The integration of Analytic Hierarchy Process (AHP) and cross-efficiency approaches highlights methodological hybridization, enabling comparative rankings of decision-making units (DMUs) and facilitating strategic prioritization across firms, sectors, and energy segments. These hybrid models offer nuanced assessment capabilities for regulators, managers, and policymakers engaged in sustainable energy planning [62].

Cluster 3 (Blue): Methodological Innovations and Eco-Efficiency

Cluster 3 captures the methodological evolution of DEA in response to increasing environmental and temporal complexity. Frequently co-occurring terms—directional distance function, eco-efficiency, Malmquist productivity, environmental assessment, and window analysis—highlight advances that allow DEA to incorporate undesirable outputs and dynamic-period analyses [63].

This cluster includes the development of non-radial, dynamic, and hybrid DEA models that better account for pollutants, multi-stage production processes, and technological change [64]. The use of Malmquist productivity indices has been particularly influential in tracking eco-efficiency growth over time, revealing whether improvements result from technological innovation or resource-use optimization [65].

Cluster 4 (Yellow): Industrial and Sectoral Applications

Cluster 4 captures the sector-specific implementation of DEA, particularly in *aviation*, *manufacturing*, and *heavy industries*. Keywords such as *two-stage DEA*, *bootstrap*, *decomposition*, *inefficiency*, and *environmental efficiency* reflect empirical studies that assess performance under real-world industrial constraints.

In the *aviation* sector, DEA has been widely employed to evaluate fuel efficiency, environmental impacts, and airline competitiveness [66]. In *industrial* applications, two-stage and dynamic DEA models have been used to identify inefficiency drivers, compare plant-level energy performance, and assess the outcomes of green innovation policies [67]. The frequent appearance of *bootstrap* indicates an interest in improving the robustness of DEA results by accounting for statistical uncertainty. Cluster 4's emphasis on applied modeling underscores the translation of theoretical advancements into practical tools that support sustainable industrial transformation.

Cluster 5 (Purple): Carbon Abatement, Circular Economy, and Equity Dimensions

Cluster 5 introduces a policy-oriented and normative dimension to the DEA literature. Key terms—including carbon-emission reduction, circular economy, equity, cost, and regional allocation—reflect the growing emphasis on evaluating fairness, resource distribution, and socio-environmental governance within efficiency-oriented frameworks. This thematic direction signals a shift from purely technical performance assessment toward broader policy and societal impact analysis.

Studies in this cluster increasingly use DEA to analyze how carbon-reduction responsibilities, emission quotas, and abatement costs can be allocated across regions or industries in ways that balance efficiency and equity considerations [68]. These investigations highlight the trade-offs inherent in climate-policy implementation, emphasizing that the pursuit of efficiency must be accompanied by attention to distributive justice and social inclusiveness.

The presence of the circular-economy keyword underscores an emerging research frontier in which DEA is applied to evaluate waste-reduction strategies, recycling efficiency, and resource-recovery practices [69]. This aligns with global sustainability agendas that encourage industrial upgrading and closed-loop systems to support low-carbon transitions.

Cluster 6 (Cyan): Productivity, Growth, and International Comparisons

Cluster 6 comprises macroeconomic and comparative themes, with key terms such as *Malmquist productivity*, *OECD*, *GDP*, *technical progress*, and *total factor productivity*. This cluster represents global-scale analyses that benchmark *energy and carbon efficiency* across countries, regions, or economic blocs such as the *European Union*.

Studies in this stream often employ the *Malmquist productivity index* to measure dynamic efficiency changes, distinguishing between technological progress and efficiency catch-up [70]. The focus on *OECD* and *EU economies* indicates increasing cross-national comparisons and policy evaluations under shared sustainability targets [71]. This cluster demonstrates DEA's capacity to link *micro-level efficiency analyses* with *macro-level productivity assessments*, contributing valuable insights for international energy governance and low-carbon growth strategies [72].

Overall, these six clusters reveal an integrated research landscape. The field has evolved from early studies on energy and carbon efficiency measurement toward sophisticated methodological innovations and policy applications. DEA has become an essential analytical framework for understanding how renewable energy systems contribute to emission mitigation, economic performance, and sustainable development goals.

5. Discussion

Insights from Results

The analysis reveals that Data Envelopment Analysis (DEA) has become a dominant methodological approach for evaluating efficiency in renewable energy systems and environmental performance. DEA's non-parametric nature enables it to handle multiple inputs and outputs without requiring a specific functional form, making it particularly suitable for assessing the efficiency of renewable energy plants such as wind, solar, and biomass facilities [57]. For example, studies like Hu and Jin-Li (2006) and Zhou et al. (2010) demonstrate the applicability of DEA in measuring regional and sectoral energy efficiency, offering policymakers and managers empirical insights into the allocation of resources and technological improvement paths [27,52]. These applications emphasize DEA's role in identifying best-performing renewable energy plants and benchmarking underperforming ones against efficient frontiers.

DEA has also emerged as a vital tool in carbon emission reduction and environmental benchmarking. It allows researchers to simultaneously measure economic performance and environmental impacts, enabling an integrated evaluation of eco-efficiency. The works of Ren et al. (2018) and Zhang et al. (2021) extend DEA's application by examining the mediating roles of environmental regulations and green finance in promoting carbon reduction [55,73]. Similarly, studies like Zhou et al. (2024) and Kou et al. (2022) apply DEA to investigate the effectiveness of policy interventions such as low-carbon city pilots and the contribution of energy efficiency to the Sustainable Development Goals (SDGs) [74,75]. These findings underscore DEA's versatility in addressing multidimensional sustainability targets, bridging the gap between energy economics and environmental management.

Methodologically, the evolution of DEA in this field reflects a clear trend toward more sophisticated analytical models. While classical DEA models focus on static efficiency assessment, contemporary studies increasingly employ network DEA, dynamic DEA, and

DEA-Malmquist indices to capture intertemporal and inter-process efficiencies [24]. The DEA-Malmquist approach, as used in Zhou et al. (2010), facilitates the evaluation of productivity changes over time, accounting for technological progress and efficiency shifts [52]. Additionally, hybrid models integrating DEA with machine learning (ML) techniques have emerged, enhancing predictive accuracy and enabling real-time performance monitoring. Such methodological advancements demonstrate a shift from traditional frontier analysis to more adaptive, data-driven efficiency frameworks.

Beyond these thematic advances, our synthesis identifies several methodological and contextual gaps that present opportunities for future inquiry. While DEA has been widely applied to benchmark renewable energy performance and carbon-emission reduction, relatively few studies incorporate dynamic systems perspectives that capture ecological complexity—for example, the role of biodiversity, land-use change, and ecosystem services in shaping clean-energy transitions [76]. Likewise, hybrid DEA–machine learning models remain in their infancy despite their strong potential to improve prediction accuracy, uncover nonlinear relationships, and enhance decision-support for energy planning [77,78]. Cross-country comparative analyses also remain limited; integrating policy–efficiency linkages across institutional contexts could generate deeper insights into how regulatory design influences environmental performance [79]. Addressing these gaps would significantly strengthen DEA's role in guiding evidence-based energy and climate policymaking.

Comparison with Related Efficiency Methods

When compared with other efficiency measurement methods such as Stochastic Frontier Analysis (SFA), DEA presents distinct advantages and limitations. DEA's key strength lies in its non-parametric flexibility and ability to handle multiple performance dimensions without requiring prior assumptions about production functions or error terms. This is particularly advantageous in environmental and renewable energy studies where data heterogeneity and measurement complexity are prevalent. In contrast, SFA, being a parametric approach, accounts for statistical noise and distinguishes inefficiency from random error, offering more robust results when measurement errors are significant.

However, DEA's deterministic nature can lead to potential sensitivity to outliers and measurement errors, which may distort efficiency scores. Furthermore, while SFA provides confidence intervals and hypothesis testing, DEA results are largely relative and rely heavily on sample composition. Despite these limitations, the non-parametric flexibility of DEA makes it more adaptable to multidimensional sustainability contexts than SFA, particularly in integrating energy, environmental, and economic indicators. Recent literature increasingly combines both methods or incorporates hybrid DEA–SFA frameworks to leverage the respective strengths of each.

The findings underscore DEA's central role in renewable energy and environmental efficiency research. Its evolution from classical models to hybrid and dynamic frameworks signifies a methodological maturity that aligns with the growing complexity of sustainability challenges. The integration of DEA with other analytical techniques holds promise for future research, particularly in refining carbon performance metrics and supporting data-driven environmental policy design.

6. Conclusions

This study provides a comprehensive mapping of the intellectual structure and research dynamics surrounding the application of Data Envelopment Analysis (DEA) in renewable energy and carbon emission efficiency. By synthesizing 2389 articles indexed in the Web of Science core collection, the findings illuminate the rapid expansion and diversification of DEA applications in the context of energy transition and environmental sustainability. The analysis reveals a strong concentration of research in Asia—particularly in China and Taiwan—where scholars such as Chiu Yung-Ho, Li Y., and Lin Boqiang have significantly advanced methodological innovation and empirical exploration in the field.

The review highlights three major thematic directions. First, DEA has evolved as a robust tool for benchmarking the operational and environmental efficiency of renewable energy systems, ranging from wind and solar power plants to national energy sectors. Second, DEA-based models have been instrumental in evaluating carbon emission efficiency and guiding low-carbon policy design, supporting governments and industries in their decarbonization efforts. Third, the methodological evolution from traditional DEA models toward network DEA, dynamic DEA, DEA–Malmquist index, and hybrid DEA–machine learning models demonstrates the field's commitment to addressing complex, multi-dimensional energy systems with higher analytical precision.

Comparative insights also indicate that while Stochastic Frontier Analysis (SFA) remains valuable for parametric efficiency estimation, DEA provides greater flexibility in handling multi-input–multi-output structures and non-parametric datasets common in energy and environmental studies. Nonetheless, future research should consider integrating DEA with other frontier methods to enhance robustness and interpretability, particularly under uncertainty and heterogeneity in energy systems.

This study contributes to the ongoing discourse on sustainable energy management by charting the intellectual landscape, identifying influential contributors, and uncovering methodological trends that shape the efficiency measurement paradigm. The results provide a valuable reference for researchers, policymakers, and practitioners seeking to optimize renewable energy systems and advance carbon neutrality goals through data-driven efficiency assessment.

A promising avenue for future research is the integration of DEA with Environmental–Social–Governance (ESG) indicators and Life-Cycle Sustainability Assessment (LCSA) frameworks. While DEA has effectively measured eco-efficiency and carbon reduction, current applications rarely capture broader social and governance aspects. Incorporating ESG metrics, such as governance quality, labor practices, and community impact—would enable more holistic benchmarking and strengthen relevance for sustainability-driven decision-making and investment analysis.

Similarly, combining DEA with LCSA would allow assessment of cradle-to-grave environmental and cost burdens, enabling comparison of renewable technologies beyond operational performance. Such hybrid DEA–ESG–LCSA models could support cross-country benchmarking, policy scenario evaluation, and dynamic tracking of sustainability progress using extensions like Malmquist indices. Although challenges exist—such as inconsistent ESG disclosure and LCSA data availability—developing standardized indicator sets and hybrid DEA structures (e.g., multi-stage or network DEA) would enhance analytical rigor.

Although this study provides a comprehensive overview of DEA applications in renewable energy and carbon emission efficiency, it is important to acknowledge that the analysis relied exclusively on the Web of Science (WoS) Core Collection. As a result, relevant studies indexed only in other major databases—such as Scopus or IEEE Xplore—may not have been captured. Future integration of multiple databases could improve coverage and reduce potential database-specific selection bias. WoS was chosen due to its high metadata consistency, strong representation of energy-efficiency and DEA-related journals, and standardized citation formats, which support methodological rigor and replicability. Nonetheless, the absence of Scopus indexing constitutes a limitation that should be considered when interpreting the findings.

Author Contributions: Conceptualization, W.-M.L. and T.-T.L.; Methodology, T.-T.L.; Software, W.-M.L.; Validation, W.-M.L.; Formal analysis, T.-T.L.; Resources, W.-M.L.; Writing—review & editing, W.-M.L.; Supervision, W.-M.L. All authors have read and agreed to the published version of the manuscript.

Funding: This research received no external funding.

Data Availability Statement: The data presented in this study are available on request from the corresponding author. (Due to licensing agreements and restrictions, this data cannot be made publicly available to all readers.).

Conflicts of Interest: The authors declare no conflict of interest.

References

1. Saleh, H.M.; Hassan, A.I. The challenges of sustainable energy transition: A focus on renewable energy. *Appl. Chem. Eng.* **2024**, *7*, 2084. [CrossRef]
2. Obiora, S.C.; Bamisile, O.; Hu, Y.; Ozsahin, D.U.; Adun, H. Assessing the decarbonization of electricity generation in major emitting countries by 2030 and 2050: Transition to a high share renewable energy mix. *Heliyon* **2024**, *10*, e28770. [CrossRef]
3. Paraschiv, L.S.; Paraschiv, S. Contribution of renewable energy (hydro, wind, solar and biomass) to decarbonization and transformation of the electricity generation sector for sustainable development. *Energy Rep.* **2023**, *9*, 535–544. [CrossRef]
4. Sasse, J.-P.; Trutnevyte, E. Distributional trade-offs between regionally equitable and cost-efficient allocation of renewable electricity generation. *Appl. Energy* **2019**, *254*, 113724. [CrossRef]
5. Lund, P.D. Effectiveness of policy measures in transforming the energy system. *Energy Policy* **2007**, *35*, 627–639. [CrossRef]
6. Lima, M.A.; Mendes, L.F.R.; Mothé, G.A.; Linhares, F.G.; de Castro, M.P.P.; De Silva, M.G.; Sthel, M.S. Renewable energy in reducing greenhouse gas emissions: Reaching the goals of the Paris agreement in Brazil. *Environ. Dev.* **2020**, *33*, 100504. [CrossRef]
7. Keerthana, K.B.; Wu, S.-W.; Wu, M.-E.; Kokulnathan, T. The United States energy consumption and carbon dioxide emissions: A comprehensive forecast using a regression model. *Sustainability* **2023**, *15*, 7932. [CrossRef]
8. Pye, S.; Li, F.G.N.; Price, J.; Fais, B. Achieving net-zero emissions through the reframing of UK national targets in the post-Paris Agreement era. *Nat. Energy* **2017**, *2*, 17024. [CrossRef]
9. Charnes, A.; Cooper, W.W.; Rhodes, E. Measuring the efficiency of decision making units. *Eur. J. Oper. Res.* **1978**, *2*, 429–444. [CrossRef]
10. San Cristóbal, J.R. A multi criteria data envelopment analysis model to evaluate the efficiency of the Renewable Energy technologies. *Renew. Energy* **2011**, *36*, 2742–2746. [CrossRef]
11. Kumar, R.; Khetrapal, P.; Badoni, M.; Diwania, S. Evaluating the relative operational performance of wind power plants in Indian electricity generation sector using two-stage model. *Energy Environ. Dev.* **2022**, *33*, 1441–1464. [CrossRef]
12. Sözen, A.; Alp, I.; Özdemir, A. Assessment of operational and environmental performance of the thermal power plants in Turkey by using data envelopment analysis. *Energy Policy* **2010**, *38*, 6194–6203. [CrossRef]
13. Zhou, Y.; Kong, Y.; Zhang, T. The spatial and temporal evolution of provincial eco-efficiency in China based on SBM modified three-stage data envelopment analysis. *Environ. Sci. Pollut. Res.* **2020**, *27*, 8557–8569. [CrossRef]
14. Yang, Y.; Lo, K. China's renewable energy and energy efficiency policies toward carbon neutrality: A systematic cross-sectoral review. *Energy Environ. Dev.* **2024**, *35*, 491–509. [CrossRef]
15. Zhang, J.; Wu, Q.; Zhou, Z. A two-stage DEA model for resource allocation in industrial pollution treatment and its application in China. *J. Clean. Prod.* **2019**, *228*, 29–39. [CrossRef]
16. Zhou, A.; Wang, W.; Chu, Z.; Wu, S. Evaluating the efficiency of municipal solid waste collection and disposal in the Yangtze River Delta of China: A DEA-model. *J. Air Waste Manag. Assoc.* **2022**, *72*, 1153–1160. [CrossRef]
17. Lu, W.-M.; Le, T.-T. From Technology to Strategy: The Evolving Role of Smart Grids and Microgrids in Sustainable Energy Management. *Energies* **2025**, *18*, 4609. [CrossRef]
18. Zheng, X.; Heshmati, A. An analysis of energy use efficiency in China by applying stochastic frontier panel data models. *Energies* **2020**, *13*, 1892. [CrossRef]
19. Eguchi, S.; Takayabu, H.; Lin, C. Sources of inefficient power generation by coal-fired thermal power plants in China: A metafrontier DEA decomposition approach. *Renew. Sustain. Energy Rev.* **2021**, *138*, 110562. [CrossRef]
20. Kut, P.; Pietrucha-Urbanik, K. Bibliometric analysis of renewable energy research on the example of the two European countries: Insights, challenges, and future prospects. *Energies* **2023**, *17*, 176. [CrossRef]
21. Zhu, T.; Liu, J.; Zhu, G. Technological Progress and Scale Efficiency Changes in China's Energy Industry: A Comparison of New and Traditional Energy Under the DEA-Malmquist-Tobit Model. *Sustainability* **2025**, *17*, 662. [CrossRef]
22. Yan, D.; Kong, Y.; Ye, B.; Shi, Y.; Zeng, X. Spatial variation of energy efficiency based on a Super-Slack-Based Measure: Evidence from 104 resource-based cities. *J. Clean. Prod.* **2019**, *240*, 117669. [CrossRef]

23. Yilmaz, I. A hybrid DEA–fuzzy COPRAS approach to the evaluation of renewable energy: A case of wind farms in Turkey. *Sustainability* **2023**, *15*, 11267. [CrossRef]
24. Zhou, P.; Ang, B.W.; Poh, K.L. Measuring environmental performance under different environmental DEA technologies. *Energy Econ.* **2008**, *30*, 1–14. [CrossRef]
25. Allevi, E.; Basso, A.; Bonenti, F.; Oggioni, G.; Riccardi, R. Measuring the environmental performance of green SRI funds: A DEA approach. *Energy Econ.* **2019**, *79*, 32–44. [CrossRef]
26. Akbari, N.; Jones, D.; Treloar, R. A cross-European efficiency assessment of offshore wind farms: A DEA approach. *Renew. Energy* **2020**, *151*, 1186–1195. [CrossRef]
27. Hu, J.-L.; Wang, S.-C. Total-factor energy efficiency of regions in China. *Energy Policy* **2006**, *34*, 3206–3217. [CrossRef]
28. Ren, J.; Tan, S.; Dong, L.; Mazzi, A.; Scipioni, A.; Sovacool, B.K. Determining the life cycle energy efficiency of six biofuel systems in China: A Data Envelopment Analysis. *Bioresour. Technol.* **2014**, *162*, 1–7. [CrossRef]
29. Smirnova, E.; Kot, S.; Kolpak, E.; Shestak, V. Governmental support and renewable energy production: A cross-country review. *Energy* **2021**, *230*, 120903. [CrossRef]
30. Chodakowska, E.; Nazarko, J.; Nazarko, Ł. Efficiency of Renewable Energy Potential Utilization in European Union: Towards Responsible Net-Zero Policy. *Energies* **2025**, *18*, 1175. [CrossRef]
31. Xue, Y.; Mohsin, M.; Taghizadeh-Hesary, F.; Iqbal, N. Environmental performance assessment of energy-consuming sectors through novel data envelopment analysis. *Front. Energy Res.* **2022**, *9*, 713546. [CrossRef]
32. Kwon, D.S.; Cho, J.H.; Sohn, S.Y. Comparison of technology efficiency for CO2 emissions reduction among European countries based on DEA with decomposed factors. *J. Clean. Prod.* **2017**, *151*, 109–120. [CrossRef]
33. Mehmood, K.; Iftikhar, Y.; Khan, A.N. Assessing eco-technological innovation efficiency using DEA approach: Insights from the OECD countries. *Clean Technol. Environ. Policy* **2022**, *24*, 3273–3286. [CrossRef]
34. Leu, J.-D.; Tsai, W.-H.; Fan, M.-N.; Chuang, S. Benchmarking Sustainable Manufacturing: A DEA-Based Method and Application. *Energies* **2020**, *13*, 5962. [CrossRef]
35. Martín-Gamboa, M.; Iribarren, D.; García-Gusano, D.; Dufour, J. A review of life-cycle approaches coupled with data envelopment analysis within multi-criteria decision analysis for sustainability assessment of energy systems. *J. Clean. Prod.* **2017**, *150*, 164–174. [CrossRef]
36. Guo, X.; Lu, C.-C.; Lee, J.-H.; Chiu, Y.-H. Applying the dynamic DEA model to evaluate the energy efficiency of OECD countries and China. *Energy* **2017**, *134*, 392–399. [CrossRef]
37. Lu, C.-C.; Lu, L.-C.J. Evaluating the energy efficiency of European Union countries: The dynamic data envelopment analysis. *Energy Environ.* **2019**, *30*, 27–43. [CrossRef]
38. Ouyang, W.; Yang, J.-B. The network energy and environment efficiency analysis of 27 OECD countries: A multiplicative network DEA model. *Energy* **2020**, *197*, 117161. [CrossRef]
39. Chen, Z.; Kourtzidis, S.; Tzeremes, P.; Tzeremes, N. A robust network DEA model for sustainability assessment: An application to Chinese Provinces. *Oper. Res.* **2022**, *22*, 235–262. [CrossRef]
40. Wei, W.; Ding, S.; Zheng, S.; Ma, J.; Niu, T.; Li, J. Environmental efficiency evaluation of China's power industry based on the two-stage network slack-based measure model. *Int. J. Environ. Res. Public Health* **2021**, *18*, 12650. [CrossRef]
41. Meng, M.; Pang, T. Operational efficiency analysis of China's electric power industry using a dynamic network slack-based measure model. *Energy* **2022**, *251*, 123898. [CrossRef]
42. Liu, J.-P.; Yang, Q.-R.; He, L.J. Total-factor energy efficiency (TFEE) evaluation on thermal power industry with DEA, malmquist and multiple regression techniques. *Energies* **2017**, *10*, 1039. [CrossRef]
43. Sueyoshi, T.; Goto, M. World trend in energy: An extension to DEA applied to energy and environment. *J. Econ. Struct.* **2017**, *6*, 13. [CrossRef]
44. Kolagar, M.; Hosseini, S.M.H.; Felegari, R.; Fattahi, P. Policy-making for renewable energy sources in search of sustainable development: A hybrid DEA-FBWM approach. *Environ. Syst. Decis.* **2020**, *40*, 485–509. [CrossRef]
45. Page, M.J.; McKenzie, J.E.; Bossuyt, P.M.; Boutron, I.; Hoffmann, T.C.; Mulrow, C.D.; Shamseer, L.; Tetzlaff, J.M.; Akl, E.A.; Brennan, S.E.; et al. The PRISMA 2020 statement: An updated guideline for reporting systematic reviews. *BMJ* **2021**, *372*, n71. [CrossRef]
46. Wong, D. VOSviewer. *Tech. Serv. Q.* **2018**, *35*, 219–220. [CrossRef]
47. Egghe, L.; Rousseau, R.; Van Hooydonk, G. Methods for accrediting publications to authors or countries: Consequences for evaluation studies. *J. Am. Soc. Inf. Sci.* **2000**, *51*, 145–157. [CrossRef]
48. Harzing, A.-W.; Alakangas, S. Google Scholar, Scopus and the Web of Science: A longitudinal and cross-disciplinary comparison. *Scientometrics* **2016**, *106*, 787–804. [CrossRef]
49. Hirsch, J.E. An index to quantify an individual's scientific research output. *Proc. Natl. Acad. Sci. USA* **2005**, *102*, 16569–16572. [CrossRef]

50. Bornmann, L.; Daniel, H.D. What do citation counts measure? A review of studies on citing behavior. *J. Doc.* **2008**, *64*, 45–80. [CrossRef]
51. Waltman, L. A review of the literature on citation impact indicators. *J. Informetr.* **2016**, *10*, 365–391. [CrossRef]
52. Zhou, P.; Ang, B.W.; Han, J.Y. Total factor carbon emission performance: A Malmquist index analysis. *Energy Econ.* **2010**, *32*, 194–201. [CrossRef]
53. Zhou, P.; Ang, B.; Wang, H. Energy and CO_2 emission performance in electricity generation: A non-radial directional distance function approach. *Eur. J. Oper. Res.* **2012**, *221*, 625–635. [CrossRef]
54. Yu, Y.; Zhang, N. Low-carbon city pilot and carbon emission efficiency: Quasi-experimental evidence from China. *Energy Econ.* **2021**, *96*, 105125. [CrossRef]
55. Zhang, D.; Mohsin, M.; Rasheed, A.K.; Chang, Y.; Taghizadeh-Hesary, F. Public spending and green economic growth in BRI region: Mediating role of green finance. *Energy Policy* **2021**, *153*, 112256. [CrossRef]
56. Zakari, A.; Khan, I.; Tan, D.; Alvarado, R.; Dagar, V. Energy efficiency and sustainable development goals (SDGs). *Energy* **2022**, *239*, 122365. [CrossRef]
57. Mardani, A.; Zavadskas, E.K.; Streimikiene, D.; Jusoh, A.; Khoshnoudi, M. A comprehensive review of data envelopment analysis (DEA) approach in energy efficiency. *Renew. Sustain. Energy Rev.* **2017**, *70*, 1298–1322. [CrossRef]
58. Chen, L.; Jia, G. Environmental efficiency analysis of China's regional industry: A data envelopment analysis (DEA) based approach. *J. Clean. Prod.* **2017**, *142*, 846–853. [CrossRef]
59. Wang, Q.; Zhou, P.; Zhou, D. Efficiency measurement with carbon dioxide emissions: The case of China. *Appl. Energy* **2012**, *90*, 161–166. [CrossRef]
60. Wang, R.; Wang, Q.; Yao, S. Evaluation and difference analysis of regional energy efficiency in China under the carbon neutrality targets: Insights from DEA and Theil models. *J. Environ. Manag.* **2021**, *293*, 112958. [CrossRef]
61. Maral, M. Cross-country analysis of research and development efficiency in higher education: Data envelopment analysis and hybrid multi-criteria decision-making approach. *J. Knowl. Econ.* **2024**, *16*, 14427–14460. [CrossRef]
62. Zhu, J.; Wan, L.; Zhao, H.; Yu, L.; Xiao, S. Evaluation of the integration of industrialization and information-based entropy AHP–cross-efficiency DEA model. *Chin. Manag. Stud.* **2024**, *18*, 210–242. [CrossRef]
63. Mirmozaffari, M.; Shadkam, E.; Khalili, S.M.; Kabirifar, K.; Yazdani, R.; Gashteroodkhani, T.A. A novel artificial intelligent approach: Comparison of machine learning tools and algorithms based on optimization DEA Malmquist productivity index for eco-efficiency evaluation. *Int. J. Energy Sect. Manag.* **2021**, *15*, 523–550. [CrossRef]
64. Gavurova, B.; Kocisova, K.; Behun, M.; Tarhanicova, M. Environmental performance in OECD countries: A non-radial DEA approach. *Acta Montan. Slovaca* **2018**, *23*, 206–215.
65. Morales-Piñero, J.; Morales-Piñero, J.; Morales-Rubiano, M. Technological development and eco-efficiency: Drivers of total factor productivity in OECD countries. *Probl. Perspect. Manag.* **2024**, *22*, 174. [CrossRef]
66. Saini, A.; Truong, D.; Pan, J.Y. Airline efficiency and environmental impacts–data envelopment analysis. *Int. J. Transp. Sci. Technol.* **2023**, *12*, 335–353. [CrossRef]
67. Chowdhury, S.A.; Aziz, S.; Hossan, M.B. Cost efficiency evaluation of thermal power plants in Bangladesh using a two-stage DEA model. *Econ. Energy Environ. Policy* **2022**, *11*. [CrossRef]
68. Qin, Q.; Liu, Y.; Li, X.; Li, H. A multi-criteria decision analysis model for carbon emission quota allocation in China's east coastal areas: Efficiency and equity. *J. Clean. Prod.* **2017**, *168*, 410–419. [CrossRef]
69. Ratner, S.V.; Lychev, A.V.; Muravleva, E.D.; Muravlev, D.M. Measuring Circular Economy with Data Envelopment Analysis: A Systematic Literature Review. *Math. Comput. Appl.* **2025**, *30*, 102. [CrossRef]
70. Khan, A.; Gulati, R. Productivity growth, catching-up and technology innovation in microfinance institutions in India: Evidence using a bootstrap Malmquist Index approach. *Benchmarking Int. J.* **2022**, *29*, 878–904. [CrossRef]
71. Gavurova, B.; Kocisova, K.; Sopko, J. Health system efficiency in OECD countries: Dynamic network DEA approach. *Health Econ. Rev.* **2021**, *11*, 40. [CrossRef]
72. Demiral, M.; Demiral, Ö. Socio-economic productive capacities and energy efficiency: Global evidence by income level and resource dependence. *Environ. Sci. Pollut. Res.* **2023**, *30*, 42766–42790. [CrossRef]
73. Ren, S.; Li, X.; Yuan, B.; Li, D.; Chen, X. The effects of three types of environmental regulation on eco-efficiency: A cross-region analysis in China. *J. Clean. Prod.* **2018**, *173*, 245–255. [CrossRef]
74. Zhou, Y. Relationship Between Corporate Social Responsibility and Brand Loyalty in Beijing Tourism Service Industry: Mediating Role of Customer Satisfaction and Brand Identity. *Uniglobal J. Soc. Sci. Humanit.* **2024**, *3*, 191–203.
75. Kou, X.; Feng, J.-C.; Li, X.-S.; Wang, Y.; Chen, Z.-Y. Visualization of interactions between depressurization-induced hydrate decomposition and heat/mass transfer. *Energy* **2022**, *239*, 122230. [CrossRef]
76. García-Valderrama, T.; Pérez-González, M.C.; Puentes-Graña, C.; Sánchez-Ortiz, J. Developing an efficiency evaluation model for the circular economy in Europe. *Eur. Plan. Stud.* **2024**, *32*, 2160–2181. [CrossRef]

77. Bin Abu Sofian, A.D.A.; Lim, H.R.; Siti Halimatul Munawaroh, H.; Ma, Z.; Chew, K.W.; Show, P.L. Machine learning and the renewable energy revolution: Exploring solar and wind energy solutions for a sustainable future including innovations in energy storage. *Sustain. Dev.* **2024**, *32*, 3953–3978. [CrossRef]
78. Bhansali, A.; Narasimhulu, N.; de Prado, R.P.; Divakarachari, P.B.; Narayan, D.L. A Review on Sustainable Energy Sources Using Machine Learning and Deep Learning Models. *Energies* **2023**, *16*, 6236. [CrossRef]
79. Charles, V.; Emrouznejad, A. DEA-based index systems for addressing the United Nations' SDGs. *Environ. Sci. Policy* **2024**, *162*, 103950. [CrossRef]

energies

MDPI

Article

Analysis and Evaluation of the Operating Profile of a DC Inverter in a PV Plant

Silvia Baeva [1], Ivelina Hinova [2] and Plamen Stanchev [3,4,*]

1 Department of Mathematical Modelling and Numerical Methods, Faculty of Applied Mathematics and Informatics, Technical University of Sofia, 1000 Sofia, Bulgaria; sbaeva@tu-sofia.bg
2 Department of Management and Business Information Systems, Faculty of Management, Technical University of Sofia, 1000 Sofia, Bulgaria; ihinova@tu-sofia.bg
3 CoE "National Center of Mechatronics and Clean Technologies", 1000 Sofia, Bulgaria
4 Department of Information Technology in Industry, Faculty of Computer Systems and Technologies, Technical University of Sofia, 1000 Sofia, Bulgaria
* Correspondence: p.stanchev@tu-sofia.bg

Abstract

The inverter is the key element that converts the intermittent DC power of the PV array into a quality AC flow to the grid and simultaneously performs functions such as power factor control, reactive services, and grid code compliance. Therefore, the detailed operating profile of the inverter, how the power, dynamics, power quality, and efficiency evolve over time, is critical for both the scientific understanding of the system and the daily operation (O&M). Monitoring only aggregated energy indicators or single KPIs (e.g., PR) is often insufficient: it does not distinguish weather-related variations from technical limitations (clipping, curtailment), does not show dynamic loads (ramp rate), and does not provide confidence in the quality of the injected energy (PF, P–Q behavior). These deficiencies motivate research that simultaneously covers the physical side of the conversion, the operational dynamics, and the climatic reference of the resource. The analysis covers the window of 25 January–15 April 2025 (winter→spring). Due to the pronounced seasonality of the solar resource and temperature regime, all quantitative results and conclusions regarding efficiency, dynamics, clipping, and degradation are valid only for this window; generalizations to other seasons require additional data. In the next stage, we will add ≥12 months of data and perform a comparable seasonal analysis. Full specifications of the measuring equipment (DC/AC current/voltage, clock synchronization, separate high-frequency PQ-logger) and quantitative uncertainty estimates, including distribution to key indicators (η, PR, THD, I_{DC}), are presented. The PVGIS per-kWp climate reference is anchored to the nameplate DC peak and cross-checked against percentile scaling; $a \pm \varepsilon$ scale error shifts PR by ε and changes ΔE proportionally only on hours with $\hat{P} > P$. The capacity for the climate reference (PVGIS per-kWp) is calibrated to the tabulated DC peak power C_{cert} and is cross-validated using a percentile scale ($Q_{0.99}$).

Keywords: photovoltaic systems; DC/AC inverter; inverter operating profile; power quality assessment; inverter efficiency; SCADA monitoring; climate-normalized performance ratio

1. Introduction

Photovoltaic (PV) systems are a leading component in the transition to a low-carbon energy system and sustainable power generation. Their gradual penetration into the power grid, both in large-scale ground-based parks and in decentralized rooftop installations,

https://doi.org/10.3390/en18236306

places increasingly high demands on the reliability, efficiency, and controllability of their components. Among them, the DC/AC inverter plays a central role, as it is not only a converter of electrical energy from direct to alternating current but also actively participates in the network behavior of the installation, including through power factor management, reactive services, ramp-rate limitations, and compatibility with standards such as IEEE 1547, EN 50530, and Network Code RfG.

Traditional approaches to PV system evaluation rely primarily on aggregated energy metrics, such as the power output over time, capacity factor, or performance ratio. These metrics do not adequately capture the dynamic processes occurring in the system in real time despite their usefulness in long-term performance evaluation. Such processes include power limitations due to inverter limits, partial shading losses, unstable MPPT behavior, reactive interference, or short-term shutdowns and restarts.

Furthermore, aggregate indicators do not allow a reliable distinction between losses caused by adverse climatic conditions (e.g., low light or high temperature) and those resulting from technical reasons, such as configuration errors, component degradation, or inverter operation errors. This creates a need to introduce a deeper and structured approach to inverter operation analysis that takes into account both temporal changes in system behavior and the climatic context.

In that sense, this study introduces and applies the concept of the "inverter operating profile", which represents the time trajectory of the main electrical and energy parameters, including active and reactive power, voltages and currents on the DC and AC sides, power factor, conversion efficiency, ramp rate, as well as indicators related to power quality, such as harmonic distortion and frequency stability.

The construction of such a working profile aims to provide a more detailed and quantitative assessment of the inverter's condition and behavior in real-world operation. For this purpose, real operational data recorded by the SCADA system of a Photovoltaic Power Plant in Bulgaria, with a high time resolution (5 min interval), covering a significant period of the winter and spring season, are used. The analysis in this work is explicitly limited to the window of 25 January 2025–15 April 2025 (local time), i.e., the winter$\rightarrow$spring transition. This is sufficient for demonstration and verification of the methodology, but is not a basis for annual conclusions about PR, clipping, and degradation. The data include complete information from both the DC side (voltage, current, instantaneous power) and the AC side (output active and reactive power, cosφ, efficiency), and are combined with climatic reference data provided by open sources such as PVGIS, PECD, and NASA POWER.

The methodology proposed in this study is based on an integrated approach that combines several basic steps:

1. Physically based recovery of the instantaneous active power, which takes into account the real efficiency of the inverter at any time, without relying on indirect metrics such as the power factor, which are often unstable at high loads.
2. Construction of an internal median of the hourly power, used as a reference profile to identify deviations, losses, and atypical behavior. This allows objective detection of deficiencies compared to the typical behavior of the installation without the need for external reference data.
3. Comparison with climatically conditioned reference profiles, based on long-term climate models that take into account solar radiation, temperature, and other meteorological factors valid for the specific location of the installation. Through them, the real efficiency factor is calculated and cases of technical, not climatic, energy deficit are identified.

4. Analysis of the dynamic behavior of the inverter, including load, power changes over time (ramp rate), reactive intervention, and compliance with regulatory limits. In addition, a spectral analysis of the quality of electricity is performed, including harmonic content and the presence of DC in the output.

The combination of these approaches leads to the construction of a unified, structured, and reproducible analysis framework that is applicable both for operational needs (including maintenance planning, production optimization, and early diagnostics) and for scientific evaluation of the efficiency of PV systems in different climatic and operating conditions.

Furthermore, the methodology is applicable not only at the individual inverter level but also in the context of analysis, where multiple sites are simultaneously compared on the same scale of efficiency and stability. This opens up opportunities for automated monitoring, site classification, predictive maintenance, and prioritization of interventions by operators and owners.

This study presents results based on data from a real 1 MW PV plant located in Southeastern Europe. The SCADA platform used provides a 5 min resolution, which allows for a reliable analysis of transients, dynamic loads, and power quality parameters. The constructed operating profile was used to assess a number of key indicators—including load, response to changing lighting, reactive interference, frequency stability, as well as overall system resilience under changing climatic and grid factors.

In conclusion, this study aims to offer an alternative to traditional KPIs by presenting a multi-layered, dynamic, and climate-normalized analysis based on real operational data and physically based calculations. The following sections of the article discuss the theoretical basis of the PV-inverter architecture, the rationale and objectives of the analysis, the methodology used for data processing and evaluation, the results of the application to the real system, as well as conclusions and recommendations for optimizing the operation of the inverters under real operating conditions.

2. Literature Review

The DC input operating profile of a grid-connected photovoltaic (PV) inverter is the result of interacting factors: conversion models and algorithms (DC/AC), MPPT algorithms and their response to dynamic changes in irradiance and temperature, filter, and control strategies, grid synchronization, as well as regulatory requirements for power quality and disturbance behavior. This review summarizes leading standards, key publications, and established methodologies for modeling and evaluating the operating profile, with a systematic focus on AC/DC efficiency, MPP tracking, harmonics, reactive power control, ramp-rate limitation, and performance assessment (PR/PI) metrics applicable to the conditions in Europe and the SEE region.

Inverters in distributed generation are subject to connection and interoperability requirements defined by IEEE 1547-2018 [1] and the European Network Code RfG (EU 2016/631) [2]. These regulate functions such as LVRT/HVRT, power factor control, and Volt/VAR/Volt/Watt modes, as well as requirements for detection and management of network anomalies. Limitations on harmonics and intermediate frequencies are covered by IEEE 519-2014 [3] and relevant IEC/EN standards (e.g., IEC 61000-3-12 for products >16 A to 75 A [4]). Reference is made to EN 50530:2010+A1:2013 [5] for the specific behavior of PV inverters during MPPT and overall efficiency testing. The monitoring and reporting of operational KPIs is standardized by IEC 61724-1:2021 [6] and IEC TS 61724-3:2016 [7]. The UL 1741 certification in North America (including the SA supplement for "advanced inverter" functions) specifies tests for network services and safety [8].

A generalized model of the converter and associated losses is critical to predicting the DC operating profile [9–16]. The Sandia Inverter Model [17] and its companion paper SAND2007-5036 (King, Gonzalez, Galbraith, Boyson) [15,16] provide a parametric description of the efficiency versus DC voltage, load, and "quasi-static" conditions. The model allows the calculation of the expected AC power, DC requirements (including clipping and PS0 losses), and the corresponding DC current/voltage operating envelope in combination with array models (SAPM/PVWatts).

The Performance Ratio (PR) and Performance Index (PI) are widely used for operational evaluation. PR normalizes energy to reference radiation and installed power (IEC 61724-1) [6], while PI compares measured energy to model predicted energy and is less sensitive to seasonal temperature variations [18,19]. PR/PI in practice are analyzed together to detect degradation, system losses, or MPPT regression.

Comparative studies by Esram and Chapman [20] and later reviews [21–23] define a palette of MPPT methods—from classical P&O and IncCond to stochastic and intelligent (PSO/GA/ANN). In the context of the inverter "operation profile", the most important are the following:

1. The ability to quickly track rapid changes in irradiance (clouds, wind gusts);
2. Stability and low oscillation around the MPP in partial shading;
3. The interaction with DC bus limits and quasi-steady-state efficiency of the converter (avoiding frequent transitions to low-efficiency modes and clipping).

The standard EN 50530 [5] specifies tests for dynamic MPPT scenarios, allowing inter-inverter comparison.

The choice of LCL filter and current/voltage regulators determines the power quality, losses, and robustness to grid variations. The classic work by Liserre, Blaabjerg, and Hansen [9] provides a procedural design of an LCL filter and shows the trade-off between ripple attenuation and robustness. Advanced active damping approaches (Peña-Alzola et al.) [10] increase robustness to grid inductance variations and parallel inverters. The proportional-resonant (PR) controller in a steady-state $\alpha\beta$ frame [11] avoids the need for $d-q$ transformations and gives zero static error at 50 Hz, while also allowing selective harmonic compensation—key to meeting IEEE/IEC limits without excessive losses.

Synchronization is crucial for stability and accurate P/Q exchange. The structures SOGI-PLL/DSOGI-PLL [12–14] provide good filtering and fast adaptation to distortions and imbalances, maintaining stable inverter operation under disturbances. This directly affects the DC operating profile through dynamic loading of the DC bus and transient currents.

Investors and operators require limiting AC output fluctuations to reduce the impact on the grid. Studies by Marcos et al. [24,25] show that the minimum energy buffer required to meet a certain ramp-rate limit depends on the spectrum of irradiance fluctuations and the system configuration (fixed/tracking). This has a direct impact on the DC load profile of the inverter and on the choice of DC-link/battery.

Reliable radiation and temperature time series are the foundation for DC/AC profile modeling and normalized KPIs. PVGIS in Europe and its parent databases (Huld et al.) [26–29] are widely used, as is NASA POWER for global/regional applications [30,31]. ENTSO-E/C3S Pan-European Climate Database (PECD) for system adequacy analyses and climate scenarios provides homogenized solar generation time series valid for ERAA/planning [32–34]. The IEA PVPS Task 13 guidelines in operation [35–39] provide good practices for monitoring, O&M, and uncertainty assessment—useful for interpreting PR/PI and for diagnosing changes in the DC operating profile (fouling, degradation, thermal hotspots, PID, etc.).

Discussion and implications for design and operation:

1. Grid Compatibility: IEEE 1547/RfG requirements impose functions that change the current and reactive power setting and indirectly model the DC load. Enabled Volt/VAR/Volt/Watt control affects the MPP position under AC-side constraints.
2. Modeling and Verification: Sandia Inverter Model parameterization, validated against EN 50530, provides a robust relationship between DC conditions and AC output. PI is suitable for fleet comparisons and early regression detection.
3. Control and Filters: LCL + PR/active damping minimizes current harmonics and losses, improving efficiency over a wide range of load conditions, which is critical at low DC voltages/clouds.
4. Data and climate: Climate variability (fast cloudiness, dust episodes) for SEE (incl. Bulgaria) requires a valid hourly (or finer) resource from PVGIS/NASA POWER and, if necessary, climate normalization with PECD.
5. Ramp-rate limitation: Designing a minimum ramp-rate buffer according to [24,25] stabilizes the DC profile and reduces the number of transitions through inefficient operating points.

3. Description and Objectives of the Study

The modern PV inverter DC input operating profile is a product of standardized grid functions, dynamic MPPT algorithms, robust filter and PLL solutions, and quality climate data. The integration of these elements, according to the cited experience, increases production predictability, reduces degradation-induced PR/PI regressions, and facilitates grid compliance.

In this context, the present work proposes an integrated, physically informed approach for the analysis and evaluation of the operating profile of a DC/AC inverter in a real PV installation. The core of the approach is the recovery of the active AC power by the physical relationship $P_{AC} = P_{dc}\eta_{inv}$, which avoids numerical instabilities at $|\cos\varphi| \to 1$ and provides a reliable input for all subsequent metrics—percentile load levels (P_{95}, P_{99}, P_{max}), ramp-rate distribution, PF/PQ dissipation, and efficiency curve $\eta(P)$. An operational internal basis (hourly median power) is built above this layer, which serves as an early indication of "deficits compared to a typical day" without mixing with the contractual interpretation of curtailment. The profile is supplemented for official KPIs (PR and climate-consistent curtailment) with a climate reference $\hat{P}(t)$ obtained from PVGIS/PECD or NASA POWER regression.

The empirical basis is SCADA data with a 5 min resolution for the period of 25 January–15 April 2025 (local time), including DC-side measurements (string voltages/currents, P_{dc}), AC quality quantities (Q, $\cos\varphi$), and instantaneous inverter efficiency η_{inv}. The data are time-scaled, day/night filtered, and synchronized for reliable energy integrals, dynamic estimates, and checks against PQ-capability constraints. The climate time series are brought to the local time zone and interpolated to 5 min, with the capacity calibrated robustly (percentile scale/Huber) to achieve comparability between observation and potential.

The objectives of the study are threefold:

1. To build a detailed inverter operating profile—energy, load, dynamics, $\frac{PF}{PQ}$, and $\eta(P)$, on physically recovered active power;
2. To introduce an operational indicator for deficit relative to the internal median and to clearly distinguish it from curtailment;
3. To integrate a climate basis for reporting PR (daily and total) and climate-adjusted energy deficit $\sum(\hat{P} - P)_{+}$.

The contribution lies in the unification of physical recovery, robust time statistics, and climate matching into a reproducible analytical chain that is directly applicable to O&M diagnostics.

The structure of the described study aims for both technical rigor and operational utility, so that inverter profiling is transparent, comparable, and suitable for decision-making at the plant and grid operator levels.

4. Notations

The notations used in this study are given in Tables 1–9.

Table 1. Time, indices, and operators.

Symbol	Designation	Units	Range/Note
t	Time Moment (Time index)	—	Discretization 5 min (Median $\Delta t = 300$ s)
Δt	Time Step	s/h	$\Delta t = 300\ \text{s} = \frac{1}{12}\ \text{h}$
h	Hour of Day	h	$\text{h} \in \{0, \ldots, 23\}$
d	Calendar Date	—	Local Time (Europe)
$\sum t$	Sum of Times	—	For Energies/Units
$median\{\cdot\}$	Median	—	Sustainable Statistics
$Q_p\{\cdot\}$	p-percentile (quantile)	—	e.g., $Q_{0.95}$, 0.99
IQR	Interquartile Range	—	$Q_{0.75} - Q_{0.25}$
$(x)_+$	Positive Part	—	$\max(x, 0)$
(	\cdot	)	Absolute Value

Table 2. SCADA measurements (main channels).

Symbol	Designation	Units	Range/Note
$P_{dc}(t)$	DC Power to Inverter	kW	From Pdc(kW)
$\eta_{inv}(t)$	Inverter Efficiency	—	From $InvEfficient$(%)/100
$P_{AC}(t)$	AC Active Power	kW	Restored: $P_{AC} = P_{dc}\eta_{inv}$
$Q_{ac}(t)$	Reactive Power	kVAr	From Qac(kVar)
$\cos\varphi(t)$	Power Factor (Sign)	—	From $\cos(\varphi)$
(	\mathrm{PF}	(t))	Absolute PF
$\bar{V}_{dc}(t)$	Aggregated DC Voltage	V	Average of Upv^* (V)
$\bar{I}_{dc}(t)$	Aggregated DC Currents	A	Average of Ipv^* (A)

Table 3. Derived quantities and energies.

Symbol	Designation	Units	Range/Note
E_{AC}	Total Produced Energy	kWh/MWh	$\sum_t P_{AC}(t)\Delta t$
$E_{AC,day}(d)$	Daily Energy	kWh	$\sum_{t \in d} P_{AC}(t)\Delta t$
$RR(t)$	Ramp Rate	kW/min	$\Delta P/\Delta t$ over 5 min
(	RR	_{95})	95th percentile of
P_{95}, P_{99}	Percentile P_{AC} Levels	kW	For "upper" levels
P_{max}	Observed P_{AC} Maximum	kW	No clipping if no plateau

Table 4. Internal (operational) bases and indicators.

Symbol	Designation	Units	Range/Note
$\hat{P}_{hour}(h)$	Internal Hourly Median Power	kW	"Typical Day" (Weather-Sensitive)
$\hat{E}_{hour}(t)$	Cumulative Energy on an Internal Basis	kWh	$\sum \hat{P}_{hour} \Delta t$
$PR^{*}_{daily}(d)$	Operational PR^{*} Relative to $\hat{P}_{hour}$	—	Median of $P/\hat{P}_{hour}$ per Day
$\mathcal{D}$	"Deficit" Intervals	—	Segments $P < 0.6\hat{P}_{hour} \geq 15$ min
ΔE_{hour}	Total Deficit Relative to $\hat{P}_{hour}$	kWh/MWh	$\sum_{t \in \mathcal{D}} (\hat{P}_{hour} - P)_{+} \Delta t$

Table 5. Climate references and models.

Symbol	Designation	Units	Range/Note
$\hat{P}(t)$	Climate Reference Power (Total)	kW	PVGIS/PECD/NASA model
$\hat{P}_{PVGIS}(t)$	Expected PV Power from PVGIS	kW	By geometry/losses
$\hat{P}_{0}(t)$	PVGIS Power Normalized (per 1 kW_p)	kW	For scale calibration
C	Scale to Real Capacity	kW/kW	$C \approx Q_{0.99}(P)/Q_{0.99}(\hat{P})$
$CF(t)$	Capacity Factor (PECD)	—	0…1
$C_{\backslash *}$	Scale to $CF(t)$	kW	$\hat{P}_{PECD} = C_{\backslash *} CF$
$\hat{P}_{REG}(t)$	Regressive $\hat{P}$ (NASA)	kW	$aGHI + bT_2 + c\,WS + d$
$GHI(t)$	Global Radiation at Horizon	W/m^2 //$Wh/m^2(h)$	NASA POWER: hourly Wh/m^2
$T_2(t)$	Temperature at 2 m	°C	NASA POWER
$WS_{10}(t)$	Wind at 10 m	m/s	NASA POWER
C_{cert}	Nameplate/Certificate DC Peak Power	kWp	
ε_c	Relative Scale Error	—	$\frac{C - C_{cert}}{C_{cert}}$

Table 6. KPI against climate base.

Symbol	Designation	Units	Range/Note
$PR_{daily}(d)$	Daily PR	—	$\frac{\sum_{t \in d} P \Delta t}{\sum_{t \in d} \hat{P} \Delta t}$
$PR_{overall}$	Overall PR	—	$\frac{\sum_t P \Delta t}{\sum_t \hat{P} \Delta t}$
ΔE_{curt}	Climate-Friendly "Deficit" (Curtailment)	kWh/MWh	$\sum (\hat{P} - P)_{+} \Delta t$
$r(t)$	Residual against $\hat{P}$	kW	$r = P - \hat{P}$

Table 7. Geometry/design parameters.

Symbol	Designation	Units	Range/Note
$\theta_z(t)$	Zenith Angle of the Sun	°	$\cos\theta_z$ in geometric base
γ	Exponent in Geometric Base	—	$\hat{P} = C\cos^{\gamma}\theta_z$
β	Inclination of the Modules	°	PVGIS parameter angle
ψ	Azimuth (0° = South)	°	PVGIS parameter aspect
ILR	DC:AC Coefficient (Inverter Loading Ratio)	—	Design parameter

Table 8. Energy quality and limits.

Symbol	Designation	Units	Range/Note
$Q - P$	Point in Plane PQ	kVAr − kW	$PF \approx 1 \Rightarrow Q \approx 0$
(	\mathrm{PF}	<0.95)	PF Deviation Indicator
TRIP/RESTART	Shutdown/Restart Events	—	Number and Times of Events

Table 9. Weights, calibration sets, and indicator function.

Symbol	Designation	Units	Range/Note
ω_t	Weight in Regression/Calibration	—	Larger on clear afternoons
$\mathcal{S}$	Set of "Pure" Intervals	—	For C calibration
$\mathbf{1}\{\cdot\}$	Indicator Function	—	Optionally in algorithms

5. Theoretical Foundation

The architecture of the PV-inverter system includes a PV generator (panel array), modeled as a nonlinear current source; a DC/DC stage with MPPT (step-up/step-down) to a DC link with capacitance C_{dc}; and a three-phase DC/AC inverter with Space-Vector PWM, connected to the grid through an L or LCL filter and a synchronization system (PLL) and a cascade of regulators in dq coordinates.

The target values for monitoring/evaluation are the following:

- DC voltage V_{dc};
- DC current I_{dc};
- AC phase voltages/currents $v_{a,b,c}(t)$, $i_{a,b,c}(t)$;
- Active/reactive power P, Q;
- Power factor $\cos\varphi$;
- Inverter efficiency η;
- Internal temperature T_{int}.

The PV generator model is expressed by the current-voltage characteristic of the PV cell/string at illuminance G and temperature T:

$$I = I_{ph}(G,T) - I_0(T)\left[\exp\left(\frac{q(V - IR_s)}{nkT} - 1\right)\right] - \frac{V - IR_s}{R_{sh}}, \quad (1)$$

where

I_{ph}—photocurrent (proportional to G);
I_0—reverse saturation current;
R_s, R_{sh}—series/shunt resistance;
n—ideality;

q—electron charge;
k—Boltzmann's constant.

The power of the PV generator is

$$P_{pv}(V) = VI(V), \quad \frac{dP_{pv}}{dV} = 0 \Rightarrow work\ in\ MPP \tag{2}$$

The energy balance and dynamics of the DC link are expressed by the average DC bus power:

$$P_{dc} = V_{dc} I_{dc} \tag{3}$$

and the charge of the capacitor guarantees a short-term buffer:

$$C_{dc} \frac{dV_{dc}}{dt} = i_{pv} - i_{inv} - i_{loss}, \tag{4}$$

where

i_{pv}—current from DC/DC to the bus;
i_{inv}—current to the inverter (averaged over one PWM period);
i_{loss}—losses (ESR, auxiliary loads).

The averaging model of the inverter and modulation is represented by averaging the PWM/Space-Vector PWM voltage at the output of an idealized full bridge (in phase):

$$v_x^* = m_x \frac{V_{dc}}{2}, \quad x \in \{a, b, c\}, \tag{5}$$

where $m_x \in [-1, 1]$ is the modulation signal.

A vector form in $\alpha\beta$ is

$$\mathrm{v}_{\alpha\beta}^* = \frac{V_{dc}}{2} m_{\alpha\beta} \tag{6}$$

The filter and the grid connection are implemented using Clarke–Park transformations. The inverter is connected to the grid with phase voltages v_{abc} through an inductance L (or LCL). An L-filter with active losses R is used for simplicity.

Clarke transformation is

$$\begin{bmatrix} x_\alpha \\ x_\beta \end{bmatrix} = \frac{2}{3} \begin{bmatrix} 1 & -\frac{1}{2} & -\frac{1}{2} \\ 0 & \frac{\sqrt{3}}{2} & -\frac{\sqrt{3}}{2} \end{bmatrix} \begin{bmatrix} x_a \\ x_b \\ x_c \end{bmatrix} \tag{7}$$

Park transformation (synchronous frame with angle θ from PLL) is

$$\begin{bmatrix} x_d \\ x_q \end{bmatrix} = \begin{bmatrix} \cos\theta & \sin\theta \\ -\sin\theta & \cos\theta \end{bmatrix} \begin{bmatrix} x_\alpha \\ x_\beta \end{bmatrix} \tag{8}$$

The dynamics of currents in dq are

$$L \frac{di_d}{dt} = v_d^* - v_{g,d} - R_{i_d} + \omega L_{i_q}, \tag{9}$$

$$L \frac{di_q}{dt} = v_q^* - v_{g,q} - R_{i_q} - \omega L_{i_d}, \tag{10}$$

where

ω—grid corner frequency from the PLL;
$v_{g,d}$, $v_{g,q}$—mains voltage in dq.

The instantaneous powers in αβ are

$$p(t) = v_\alpha i_\alpha + v_\beta i_\beta, \tag{11}$$

$$q(t) = v_\alpha i_\beta - v_\beta i_\alpha \tag{12}$$

In a steady-state mode with synchronized dq (PLL aligns v_g along the d-axis, $v_{g,q} \approx 0$),

$$P = \frac{3}{2} v_{g,d} i_d, \qquad Q = \frac{3}{2} v_{g,d} i_q \tag{13}$$

The power factor is

$$PF = \sin\varphi = \frac{P}{\sqrt{P^2 + Q^2}} \tag{14}$$

The energy for an interval Δt is

$$E = \int_{t_0}^{t_0+\Delta t} P(t)dt \approx \sum_k P_k \Delta t_k \tag{15}$$

Closed-loop control is defined by a typical cascade including

- PLL: estimates $\hat{\theta}$, $\hat{\omega}$ from v_{abc} through a PI controller on $v_q \to 0$.

Outer loop—one of two strategies:

- Maintaining V_{dc} (or output active power P^*);
- MPPT of DC/DC, which sets a reference power/current to the inverter.

Inner current loop in dq:

$$i_d^* = \frac{2}{3}\frac{P^*}{v_{g,d}}, \qquad i_q^* = \frac{2}{3}\frac{Q^*}{v_{g,d}}, \tag{16}$$

PI controllers with decoupling terms $\pm\omega L_{i_{q,d}}$ generated v_d^*, v_q^*, which are converted into modulations $m_{a,b,c}$ by SVM.

The total losses are

$$P_{loss} = P_{cond} + P_{sw} + P_{filt} + P_{aux} \tag{17}$$

The efficiency of the inverter is

$$\eta_{inv} = \frac{P_{AC}}{P_{DC}} = \frac{P}{V_{dc} I_{dc}} \tag{18}$$

The simplified thermal RC-grid model is

$$C_{th}\frac{dT_j}{dt} = P_{loss} - \frac{T_j - T_a}{R_{th,ja}}, \tag{19}$$

where

T_j—crystal temperature;
T_a—ambient temperature;
$R_{th,ja}$—total thermal resistance.

The limitations and permissible operating range are

- Current limitation:

$$\sqrt{i_d^2 + i_q^2} \le I_{\max} \tag{20}$$

- Voltage/modulation limitation:

$$\left|\mathrm{m}_{\alpha\beta}\right| \le m_{\max} \tag{21}$$

- PQ capabilities:

$$\left(\frac{P}{S_{\text{rated}}}\right)^2 + \left(\frac{Q}{Q_{\max}}\right)^2 \le 1 \tag{22}$$

- Harmonic norms:

$$\mathrm{THD}_I = \frac{\sqrt{\sum_{n=2}^{\infty} I_n^2}}{I_1} 100\% \tag{23}$$

The operating profile is defined as the time trajectory of the operating points:

$$X(t) = [V_{dc}(t), I_{dc}(t), P(t), Q(t), PF(t), T_{int}(t), \ldots], \tag{24}$$

collected with a step Δt. This profile is used for the following:

- Energy balances $\int Pdt$;
- Efficiency assessment/MPPT use;
- Verification of compliance with restrictions (current, PF, THD);
- Correlations with external factors (illuminance, temperature, market regimes).

This theoretical framework sets the mathematical basis for subsequent methods for profile analysis and evaluation (identification, optimization, statistical testing), as well as for calculating the key indicators η, P, Q, PF, THD, T_j.

Explicit temperature/illuminance corrections and an MPPT criterion for the PV source are used to estimate expected MPP points for real changes in G, T, such as the following:

- The photo part is

$$I_{ph}(G,T) \approx \frac{G}{G_{ref}}\left[I_{sc,ref} + \alpha_I\left(T - T_{ref}\right)\right] \tag{25}$$

- The open voltage is

$$V_{oc}(G,T) \approx V_{oc,ref} + \beta_V\left(T - T_{ref}\right) + nV_T \ln\frac{G}{G_{ref}}, \qquad V_T = \frac{kT}{q} \tag{26}$$

- MPPT "incremental conductivity" (algorithmic criterion) is

$$\left.\frac{dI}{dV}\right|_{MPP} = -\frac{I}{V} \tag{27}$$

The quantitative estimates of the filtering and the required C_{dc} for the DC/DC stage are implemented through static connections and ripples, such that the following arises:

The ideal converters (on average) are

$$\text{Buck}: \; V_{dc} = DV_{pv}, \qquad \text{Boost}: \; V_{dc} = \frac{V_{pv}}{1-D}, \qquad \text{Buck}-\text{Boost}: V_{dc} = \frac{D}{1-D}V_{pv} \tag{28}$$

The pulsations from the commutation are

$$\Delta I_L \approx \frac{V_L D}{L f_{sw}}, \qquad \Delta V_C \approx \frac{\Delta I_L}{8 C f_{sw}} \tag{29}$$

The double-frequency ripple of a DC link (from power exchange with the grid) is

$$\Delta V_{dc} \approx \frac{P_{2\omega}}{2\omega C_{dc} V_{dc}} \tag{30}$$

Modulation is the relationship "module index → fundamental RMS" and is

$$V_{1,rms} = k_{\text{mod}} m V_{dc},\ m = \max|m_x|, \tag{31}$$

and allows for directly predicting the output voltage at a given m (k_{mod} depends on SPWM/SVPWM).

The expanded definition of PF with a distortion factor, i.e., PF decomposed into "displacement" and "distortion" parts, is

$$PF = \frac{P}{\sqrt{P^2 + Q^2}} = \underbrace{\cos\varphi_1}_{1st\ harmonic\ shift} \underbrace{\frac{1}{\sqrt{1 + THD_I^2}}}_{harmonics\ influence} \tag{32}$$

The metrics for the profile section in the discrete case are the following efficiencies:

$$E_{AC} = \sum_k P_k \Delta t, \qquad E_{DC} = \sum_k V_{dc,k} I_{dc,k} \Delta t \tag{33}$$

$$\eta_{inv} = \frac{P_k}{V_{dc,k} I_{dc,k}}, \qquad \eta_{MPPT} = \frac{\sum_k P_k}{\sum_k P_{MPP,k}}, \qquad \eta_{tot} = \frac{E_{AC}}{\sum_k E_{DC}} \tag{34}$$

The measurable harmonic power quality indicators are as follows:

- THD (from FFT/DFT):

$$THD_I = \frac{\sqrt{\sum_{n=2}^{\infty} I_n^2}}{I_1}, \qquad THD_V = \frac{\sqrt{\sum_{n=2}^{\infty} V_n^2}}{V_1} \tag{35}$$

- TDD and DC component:

$$TDD = \frac{\sqrt{\sum_{n=2}^{\infty} I_n^2}}{I_{rated}}, \qquad I_{dc} = \frac{1}{T}\int_0^t i(t)dt \tag{36}$$

A discrete-time DFT/FFT in the practical implementation is used on a 10-period window (50 Hz; 200 ms) synchronized to the mains frequency, respectively, 12 periods at 60 Hz, which provides $N = f_s T \approx 12{,}800\ \text{Hz} \cdot 0.2\ \text{s} = 2560$ samples and $\Delta f = \frac{f_s}{N} = 5$ Hz, as required by IEC 61000-4-7 (Class A, grouping/subgrouping). A rectangular window over an integer number of periods is used (minimizing leakage), the determination of the fundamental is achieved by DFT at $k \approx 50$ Hz with frequency tracking, and I_{DC} is estimated as a time average value in the window. THD_I, THD_V, and TDD are calculated according to (35) and (36) on the 2nd to 50th harmonic.

The practical implementation of the $THD/TDD/I_{DC}$ estimators uses FFT over 10–12 period windows, as is standard in PQ estimation; the numerical uncertainty and the influence of noise/quantization are summarized in Section 7.

The automated check is carried out using PQ capabilities and constraints through explicit formulas:

- "Ellipse" of the work area:

$$\left(\frac{P}{S_{rated}}\right)^2+\left(\frac{Q}{Q_{\max}}\right)^2\leq 1,\qquad Q_{\max}(P)=\sqrt{S_{rated}^2-P^2} \tag{37}$$

- Limit on required minimum PF:

$$|Q|\leq|P|\mathrm{tg}(\arccos PF_{\min}) \tag{38}$$

The time-series indicators (statistics and dynamics) are key to assessing a "profile-in-time", such as the following:

- Means and variance:

$$\overline{P}=\frac{1}{N}\sum_k P_k,\qquad \sigma_P^2=\frac{1}{N-1}\sum_k\left(P_k-\overline{P}\right)^2 \tag{39}$$

- Ramp rate (rate of change):

$$RR_k=\frac{P_k-P_{k-1}}{\Delta t},\qquad RR_k^{\%}=\left(\frac{P_k-P_{k-1}}{S_{rated}}\right)\frac{100}{\Delta t} \tag{40}$$

- Capacity factor (CF) and performance ratio (PR):

$$CF=\frac{E_{AC}}{S_{rated}T_{period}},\qquad PR=\frac{E_{AC}}{G_{POA}A\eta_{ref}} \tag{41}$$

The normative metrics are used to provide a quantitative assessment of time outside the permissible limits:

$$\Delta V\%=100\frac{|V_1-V_{nom}|}{V_{nom}},\qquad \Delta f=f-f_{nom},\qquad t_{out}=\sum_{k:infraction}\Delta t,$$
$$r_{out}=100\frac{t_{out}}{T_{period}} \tag{42}$$

6. Methods and Algorithms for Analysis and Evaluation

The used methods for analysis and evaluation are described sequentially as follows.

First, data preparation is performed: the individual channels (DC voltage and current, AC voltages and currents, phase/φ, possibly illuminance and temperature) are translated to the same time grid, for example, every Δt minutes. If there is haven latency between the channels, it is compensated by aligning with maximum cross-correlation, so that short-term changes in V_{dc} and P coincided. Small gaps are interpolated (linearly or with a spline), and longer ones are marked and excluded from the integral estimates. Noisy measurements and spikes are smoothed with a moving median or Savitzky–Golay filter; if waveforms $v(t)$, $i(t)$ are available, the Kalman filter gives clean estimates for the fundamental components and the phase. Next comes the natural "day/night" mask by the V_{dc} threshold or illumination, to avoid contaminating efficiency and MPPT metrics with night points.

The primary features are extracted from each time point once the data is cleaned: instantaneous active and reactive power P and Q (from the fundamental V_1, I_1, φ), power factor PF, inverter efficiency η_{inv}, and rate of change RR. The harmonics I_n, V_n and from them THD_I, THD_V, and TDD are given where there are sufficient samples of waveforms, via spectral analysis using FFT. All powers are normalized in per-unit relative to S_{rated}, so that they can be compared across days, seasons, and different objects on the same scale.

The time analysis is performed at several levels. The average, variations, and empirical distribution of the ramp rate are estimated in moving windows (hour, day), which shows how well the inverter "follows" the dynamics of the input and the grid without excessive excitability. The detection of changes (e.g., CUSUM or PELT) captures structural transitions: clouds, clipping when reaching DC/AC limits, curtailment on operator command or restart. Clusterization (e.g., k-means or DBSCAN) is applied on vectors $[p, q, PF, \eta, RR, THD]$ to recognize typical operating modes: in this way, "normal operation", "reactive maintenance", "limit operation", "grid limitation", "switching off/on", etc., are separated.

The frequency view is complementary: in short windows of 10–12 network periods, the FFT gives the amplitudes of the low odd harmonics (5th, 7th...), the total THD, and the DC component. These are compared with the permissible limits; "time outside the norm" and a relative share are accumulated, which later became one of the key KPIs for power quality.

The actual quantitative assessment also requires model identification. An efficiency map $\eta = f(P, T)$ is constructed from the data (P, T), which is usually a smooth polynomial or spline surface, estimated with robust methods (Huber, RANSAC), and the reliability is checked with cross-validation and bootstrap confidence intervals. Separately, a simple loss model P_{loss} is calibrated as a function of current and switching, e.g., $k_0 + k_1 I + k_2 I_2 + k_3 f_{sw} VI$, by nonlinear least squares (Levenberg–Marquardt). There are enough points (V_{dc}, I_{dc}) along the "envelope" during the day, which allowed for calibration and construction of the single-diode PV model. This is allowed for estimating $P_{MPP}(G, T)$ and hence MPPT efficiency η_{MPPT}. The parameters are tracked online (RLS/EKF) for long time series to capture degradation or changes in settings. The algorithm for THD/TDD and DC component estimation is as follows:

Data source: Class-A PQ analyzer (see Section 7) with $f_s = 12.8$ kHz, 16-bit ADC, and analog anti-aliasing at the input.

Window and spectral processing: DFT/FFT over 10 consecutive periods (50 Hz; 200 ms), rectangular window, $N = 2560$, $\Delta f = 5$ Hz; grouping/subgrouping according to IEC 61000-4-7 up to the 50th harmonic; the fundamental is determined by tracking the network frequency (IEC 61000-4-30).

Anti-aliasing and frequency range: the analysis is limited to 2.5 kHz (50th harmonic), at $f_{Nyquist} = 6.4$ kHz, which, together with the analyzer's analog anti-alias filter, eliminates aliasing artifacts.

Metrics:

- THD_I, THD_V by (35); TDD normalized to the nominal current of the PF/PSC; I_{DC} as time-averaged value in the window.
- Aggregation to 5 min scale: median and 95th percentile of all 200 ms estimates in the interval; synchronization with SCADA by timestamp (NTP, 5 min grid).

Reliability and uncertainty: $THD/TDD/DC$ uncertainty includes contributions from quantization (16-bit), noise/SNR, limited window length, and network frequency variations; estimated as σ_{THD} and 95% CI by bootstrapping over days/intervals (see Section 7). Comparison to IEEE 519 and IEC 61000-3-12 is reported with the indicator "time fraction outside the norm".

Note: SCADA at 5 min does not contain waveforms; the harmonic channel is separated from the PQ analyzer (see Section 7).

6.1. Modeling and Parameterization of the Efficiency $\eta(P, T)$

To model the relationship between the efficiency of the inverter and its operating parameters, such as active power and temperature, a two-dimensional continuous and smooth model of the third-degree tensor B-spline type is used, defined on normalized dimensionless variables $p = \frac{P_{AC}}{S_{rated}}$ and $\tau = \frac{(T - 25\ ^\circ\mathrm{C})}{10}$.

This choice provides sufficient flexibility to describe nonlinear dependencies under different loads and temperature regimes without leading to overfitting or instability of the estimates.

The functional form of the model is given as

$$\eta(p,\tau) = \beta_0 + \sum_i \beta_i B_i^{(3)}(p) + \sum_j \gamma_j C_j^{(3)}(\tau) + \sum_i \sum_j \delta_{ij} B_i^{(3)}(p) C_j^{(3)}(\tau), \tag{43}$$

where $B_i^{(3)}(p)$ and $C_j^{(3)}(\tau)$ are cubic spline basis functions for load and temperature, and β_i, γ_j, δ_{ij} are the corresponding model parameters. The nodes of the spline are located on quantiles of the available data, which allows for a denser representation in frequently observed regions (around the nominal load and typical temperatures).

The fitting of the surface $\eta(P,T)$ is performed by robust regression with Huber loss, combined with a RANSAC procedure to exclude outliers caused by atypical modes, restarts, or unstable measurements. This approach reduces the influence of isolated errors and ensures the robustness of the estimates even with noisy SCADA data.

To ensure physical plausibility of the dependence, monotonicity and range constraints are imposed, with the efficiency η being limited to the interval $[0,1]$, by applying weak regularization that penalizes negative slopes $\frac{\partial \eta}{\partial p} < 0$ at low loads and excessively large temperature effects $\frac{\partial \eta}{\partial \tau} > 0$.

The model is identified using the Huber function minimization method with regularization terms ensuring smoothness of the surface along both axes (so-called P-splines). This ensures a smooth change in efficiency over load and temperature, without artificial interruptions.

The resulting parameterization describes the real efficiency surface $\eta(P,T)$ with high stability and is the basis for calculating confidence intervals and validation metrics (MAE, RMSE).

The efficiency model fitting procedure $\eta(P,T)$ is based on a sequential algorithm that combines data preprocessing, robust fitting, and metrologically meaningful parameterization. The first step involves selecting a reliable training set consisting only of daily intervals with normal inverter operation (without TRIP/RESTART events, reactive power limitations, or clear anomalies). The data are pre-filtered by the illuminance criterion (UDC > 100 V) and by the confidence interval of the measured efficiency ($0.85 < \eta < 1.0$), using the 5 min SCADA resolution.

The measured values of active power after filtering and temperature are normalized by the transformations $p = \frac{P_{AC}}{S_{rated}}$ and $\tau = \frac{(T - 25\ ^\circ\mathrm{C})}{10}$, which ensures large-scale comparability and numerical stability of the calculations. Based on these normalized coordinates, a network of B-spline bases is constructed, typically with six nodes along the load axis and five nodes along the temperature. The resulting basis system gives a flexible but smooth surface approximation.

The actual fitting is performed by robust regression with Huber loss, with the objective function to minimize being

$$J(\theta) = \sum_k w_k \rho_\delta(\eta_k - \eta(p_k, \tau_k; \theta)) + \lambda_P \int (\eta_{pp})^2 dp + \lambda_T \int (\eta_{\tau\tau})^2 d\tau + \int\int (\eta_{p\tau})^2 dp d\tau \tag{44}$$

where ρ_δ is the Huber function, λ_P, λ_T, λ_{PT} are regularization coefficients, and w_k are weights inversely proportional to the metrological uncertainty of the corresponding observations. The weights are calculated based on the accuracy classes and standard deviations σ_V, σ_I, σ_η, presented in Section 7, to account for larger uncertainty at lower loads.

Outliers caused by transients or disturbances are eliminated in advance by a RANSAC procedure with a threshold determined by the median deviation of the residuals (MAD).

The remaining model parameters are estimated by an iterative Levenberg–Marquardt algorithm, with initial values determined by a linear approximation of the $\eta(P)$ dependence at a nominal temperature.

Range constraints ($0 \leq \eta \leq 1$) after each iteration and monotonicity with respect to the load are applied to ensure physical consistency, penalizing negative derivatives $\frac{\partial \eta}{\partial p} < 0$. The process continues until a stable minimum of the function $J(\theta)$ is reached with a relative change below 10^{-4} between successive iterations.

The final result is a parameterized set of coefficients $\{\beta_i, \gamma_j, \delta_{ij}\}$, through which the surface $\eta(P, T)$ can be calculated for any combination of load and temperature. The resulting model provides high smoothness and robustness to extrapolation beyond the most frequently observed regimes, while preserving the physical interpretation of the efficiency and allowing subsequent calculation of confidence intervals and validation metrics (MAE, RMSE).

6.2. Model Validation

The validation of the efficiency model $\eta(P, T)$ aims to verify how well the parameterized surface reproduces the actually observed efficiency values under different load and temperature regimes. Leave One Day Out (LODO-CV) and 10-fold blocked cross-validation (10-fold CV) are applied for this purpose, in which the data are divided into non-overlapping daily blocks so as to avoid temporal correlation between the training and test sets.

The model is re-estimated on all days except one, and errors are calculated on the excluded day at each iteration. The accuracy of the model is assessed using two standard metrics:

$$MAE = \frac{1}{N}\sum_{k=1}^{N}|\eta_k - \hat{\eta}_k|,\; RMSE = \sqrt{\frac{1}{N}\sum_{k=1}^{N}(\eta_k - \hat{\eta}_k)^2}, \tag{45}$$

where η_k are the measured efficiency values and $\hat{\eta}_k$ are the model predicted ones. Calculations are performed in both absolute units (0–1) and percentage points (p.p.) to ensure comparability with metrological specifications.

In addition to the mean errors, the distributions of the residuals $(\eta_k - \hat{\eta}_k)$ over load and temperature are also analyzed. No systematic deviations are observed, while the mean value of the residuals is close to zero, and the variance remains almost constant over the entire operating range. This confirms that the model is not biased towards certain load or temperature ranges.

A bootstrap procedure by days (B = 1000 resampling) is used to estimate the confidence intervals. The model is recalculated for each resampled set, and the 2.5th and 97.5th percentiles of the predictions are derived for each grid point (P, T), which form a 95% confidence interval. The resulting bands are narrow in the nominal region ($0.7 < \frac{P}{S_{rated}} < 1.0$) and widen towards low loads, where a greater influence of noise and temperature is observed.

6.3. Confidence Intervals and Graphical Representation

Confidence intervals (CIs) are calculated based on a bootstrap procedure by days to provide a quantitative assessment of the reliability of the performance model $\eta(P, T)$. This approach reflects the real variability of the data resulting from meteorological conditions, load dynamics, and metrological uncertainty of the channels in the SCADA system.

The procedure involves 1000 bootstrap iterations, where the daily observation intervals are selected with repetition, and for each sample, the model $\eta(P, T)$ is recalculated entirely with Huber regression and P-spline regularization. A set of predicted values $\eta^{(b)}(p, \tau)$ is obtained for each point of the grid (p, τ), from which the 2.5th and 97.5th percentiles are derived as the boundaries of a 95% confidence interval. This ensures that in 95% of the cases, the true efficiency of the inverter falls within this range.

The results show that the confidence intervals are narrowest in the nominal load region ($0.7 < \frac{P}{S_{rated}} < 1.0$) and at moderate temperatures (20–35 °C), where the volume of data is largest and the noise in the measurements is minimal. The intervals widen at low powers ($\frac{P}{S_{rated}} < 0.3$) or at extreme temperatures (below 10 °C and above 50 °C), reflecting the higher uncertainty and stronger nonlinearity of the $\eta(P,T)$ relationship.

The validated efficiency model and the resulting $\eta(P,T)$ surface are further illustrated and discussed in Section 7.2, where the numerical results and figures are presented.

The developed efficiency model $\eta(P,T)$ provides a reliable and physically justified relationship between the inverter load, the ambient temperature, and its real conversion efficiency. A smooth and stable description of the efficiency surface is achieved by applying a two-dimensional B-spline approximation and robust estimation methods (Huber and RANSAC), which preserves the correct behavior of the system over the entire operating range. Validation by blocked cross-validation and bootstrap estimation of confidence intervals confirms the high accuracy of the model and its robustness to noise and random deviations in measurements.

The resulting parameterization allows for predicting the efficiency at different combinations of temperature and load, thus providing an analytical basis for assessing energy losses, temperature sensitivity, and seasonal variations in the inverter operation. In addition, the model can be used for early detection of degradation, analysis of operational trends, and creation of digital twins of inverter systems. Its inclusion in the analysis methodology increases the diagnostic capabilities of SCADA monitoring and supports the optimization of operating modes, which leads to more precise management and prediction of the energy efficiency of the photovoltaic installation.

6.4. Capacity Calibration and Sensitivity of $\frac{PR}{\Delta E}$

Here, we use PVGIS "per-1 kWp" power $P_0(t)$ and scale it by C to obtain the climate reference $\hat{P}(t) = C\hat{P}_0(t)$ (Table 5). In addition to the percentile scaling $C_Q = \frac{Q_{0.99}\{P\}}{Q_{0.99}\{\hat{P}_0\}}$, the capacity is validated against the nameplate/certificate DC peak C_{cert} (kWp). When available, PVGIS is queried with $peakpower = C_{cert}$ to produce $\hat{P}^{(cert)}_{PVGIS}(t)$. If a small residual bias exists on clear-noon windows, a multiplicative correction is applied $g = \frac{Q_{0.99}\{P\}}{Q_{0.99}\{\hat{P}^{(cert)}_{PVGIS}\}}$.

The impact of a capacity-scale error $\varepsilon_C = \frac{C - C_{cert}}{C_{cert}}$ on KPIs is closed-form:

$$PR(C) = \frac{\sum_t P\Delta t}{C\sum_t C\hat{P}_0\Delta t} \Rightarrow \frac{PR(C)}{PR(C_{cert})} = \frac{1}{1+\varepsilon_C} \approx 1 - \varepsilon_C, \tag{46}$$

i.e., a +2% scale overestimates biases PR by $\approx$−2% (relative). For the climate-aligned shortfall,

$$\Delta E(C) = \sum_t \left(\hat{P}(t) - P(t)\right)_+ \Delta t \tag{47}$$

$$\frac{\partial \Delta E}{\partial C} = \sum_{t:\hat{P}(t)>P(t)} \hat{P}_0(t)\Delta t \Rightarrow \Delta E(C+\delta C) \approx \Delta E(C) + \delta C \sum_{t:\hat{P}>P} \hat{P}_0(t)\Delta t, \tag{48}$$

with the bound $|\Delta E - \text{bias}| \leq |\delta C| \sum_t \hat{P}_0(t)\Delta t$. Here, is report the observed ε_C and a small scenario table for $C\cdot(1 \pm 1\%, 1 \pm 2\%)$.

The practical capacity verification is performed in three steps: (i) a request to PVGIS with $peakpower = C_{cert}$ to generate $\hat{P}PVGIS(cert)(t)$; (ii) percentile control on the "clean" noon windows through the multiplier $g = Q_{0.99}\{P\}/Q_{0.99}\{\hat{P}PVGIS(cert)\}$; (iii) reporting

$\varepsilon_C = \frac{C - C_{cert}}{C_{cert}}$ and recalculating PR and ΔE for $\{C, C_{cert}\}$. The relative systematic error at $\delta_C = \frac{C}{C_{cert} - 1}$ in PR is exactly (46), and the one-sided effect on ΔE is given by (48).

The climate potential only in the hours in which $\hat{P} > P$ is

$$\dot{E}_> = \sum_{t:\hat{P}>P} \hat{P}(t)\Delta t = C_{cert} \sum_{t:\hat{P}>P} \hat{P}_0(t)\Delta t \tag{49}$$

and gives the one-sided (in sign) upper bound by (48):

$$|\Delta E(C(1+\varepsilon_C)) - \Delta E(C)| \leq \varepsilon_C \dot{E}_> \tag{50}$$

If $\varepsilon_C > 0 \Rightarrow \Delta E$, this can increase at most by $\varepsilon_C \dot{E}_>$. If $\varepsilon_C < 0 \Rightarrow \Delta E$, this decreases modulo by $\leq |\varepsilon_C| \dot{E}_>$.

The normalized form of $\Delta E(C)$ is

$$\frac{\Delta E(C(1+\varepsilon_C)) - \Delta E(C)}{\Delta E(C)} \lesssim |\varepsilon_C| S_{\Delta E},\ S_{\Delta E} := \frac{\dot{E}_>}{\Delta E(C)} \tag{51}$$

Table 10 shows that at ±1–2% scale error PR scales exactly by $\frac{1}{1+\varepsilon_C}$, while the influence on ΔE is one-sided and limited only to the hours with $\hat{P} > P$, as follows from (46)–(48).

Table 10. Scenario check for capacity scaling $C \cdot (1 \pm 1\%, 1 \pm 2\%)$.

Scenario $C \to C(1 + \varepsilon_C)$	ε_C (Relative)	PR Multiplier = $\frac{1}{1+\varepsilon_C}$	One-Sided Upper Bound for ΔE (Absolute)
−2%	−0.020	1.0204	$\leq 0.02\dot{E}_>$
−1%	−0.010	1.0101	$\leq 0.01\dot{E}_>$
+1%	+0.010	0.9901	$\leq 0.01\dot{E}_>$
+2%	+0.020	0.9804	$\leq 0.02\dot{E}_>$

ε_C is the relative change in capacitance; $\dot{E}_>$ is the climate potential only in the hours in which $\hat{P} > P$.

The evaluation in the PQ-plane is geometrically straightforward: the density of points is visualized, and we check what proportion of the time remains inside the permissible region (the ellipse of possibilities) and above the minimum required power factor. If the inverter follows a static $Q(V)$ characteristic (droop), the regression $Q = f(V)$ and the errors against the given curve indicate the quality of the reactive service.

Rules and machine learning are combined for events and anomalies. Clipping is recognized by a plateau of p close to 1 at a high G and stable V_{dc} by a persistent shift $P \ll P_{MPP}$ under normal DC conditions and/or active Q or PF commands; tripping—by a sharp drop in P and V_{dc}. Algorithms such as Isolation Forest or LOF are marked rare, multidimensional deviations. The duration and energy effect ΔE for each event are calculated against an appropriate counterfactual basis (e.g., P_{MPP} or the expected $\hat{P}$ from a model).

Forecasting also helped the evaluation: short-term ARIMA/SARIMAX models (with illumination and temperature as exogenous variables) or nonlinear methods such as gradient boosting and neural networks are given an "expected profile." Large systematic residuals (RMSE, MAE, Theil's U) are signaled degradation, contamination, or incorrect settings.

Finally, each KPI comes with a reliability measure. Bootstrapping by days gave 95% intervals for η, η_{MPPT}, PR, etc., and the standard error distribution (e.g., $Var(P) \approx I^2\sigma_V^2 + V^2\sigma_I^2$) translates the sensor accuracy classes into performance uncertainties. The results are presented in easy-to-read diagrams: density maps in PQ, heat maps of $\eta(P, T)$, timelines with marked events, histograms of ramp rate, and spectral "combs" of harmonics.

Thus, the entire path, from data cleaning to regulatory compliance and diagnosis, remains transparent and reproducible, and the inverter's operating profile is evaluated not only numerically but also contextually in relation to the plant mode and grid conditions.

The inverter is started by searching for the maximum power of the PV array. The simplest and most widely accepted approach is "Perturb and Observe": the controller slightly changes the set voltage or current of the array, and we watch how the output power changes. If the change is in the "right" direction, it continues; if not, it takes the reversed course. The step is made adaptive in order not to oscillate unnecessarily around the maximum, i.e., larger when it is far from the MPP (to quickly catch up), and small when it is already close (for low oscillation losses). A more precise option, known as "incremental conductance", not only looks at the direction of change but also compares the slope of the IV curve at the moment with that at the maximum. There, the slope satisfies a simple relationship between current and voltage; if the measurements show that this is to the left of this point, the controller increases the voltage, if this is to the right—it decreases it. The intelligent variants such as fuzzy logic and small neural networks perform even better when conditions are changed abruptly (clouds, strong wind): they connect the "error" and "rate of change" with smooth control actions and they can "guess" the good reference directly from illumination and temperature. The so-called "extreme search" is used in applications where one does not want to rely on models: a low-amplitude sine wave is superimposed on the reference, and from the power response, through demodulation and filtration, the direction towards the maximum is extracted without losing stability.

The inverter must "lock" the phase and frequency to properly match the grid. This is achieved by a PLL (phase-locked loop) in a synchronous frame: the measured three-phase voltage is rotated in dq coordinates, with the regulator forcing the "quadrature" component to zero, i.e., a sign that the rotation is synchronized with the grid. Narrowband (notch) filters or double synchronous frames are added for distorted or unbalanced voltages to suppress parasitic components around twice the grid frequency and to avoid "swinging" of the estimated phase.

Above the synchronization is the "outer" loop, the one that determines how much active and reactive power the inverter should deliver. It maintains in one mode a constant DC voltage on the bus and it converts the deviation from the set value into a target active power. The MPPT module dictates in the other mode (more typical for PV) the desired active power, and dispatcher requirements or agreed settings determine the reactive power. Reference currents are formed from these two targets in the dq synchronous frame: the active component flows along the d–axis, and the reactive component flows along the q–axis. The active power command is passed through a ramp-rate limiter that prevents sudden jumps in order to avoid unnecessary stress on the equipment and the grid.

The "inner" current loop is the fastest and keeps the real currents tightly around their references. The mathematics in the dq framework show cross-connections between the axes, so decoupling terms are added to the control laws, which "reset" them, and the main problem is left to ordinary PI regulators. Predictive options are used when even faster dynamics are pursued or switching is reduced: it is calculated with a discrete model from one sample ahead what voltage needs to be applied now to place the current exactly where it is needed in the next clock cycle. This is a "deadbeat" idea, and in a more general form—model-predictive control, in which the error and the number of switching are minimized.

The voltage commands are sent to the power section via PWM. Comparison with a triangular carrier in classical sine wave modulation gives good results, and adding a third harmonic extends the linear range. Vector modulation (SVPWM) uses the geometry of the six power vectors of the inverter and distributes the time of the active and zero states so as to recreate the desired "voltage" vector with a minimum number of switches and about a 15% higher effective range.

There is a filter between the inverter and the grid, often of the LCL type, which smooths out high-frequency harmonics. However, it carries its own resonance. Active damping (which heat up) is more elegant instead of putting in large passive dampers: The current or voltage of the filter capacitor is measured and fed back with a small coefficient to the control signal, which "killed" the resonance without additional losses.

The quality of the current to the grid is often adjusted with regulators sensitive to sinusoidal references, the so-called PR controllers. They are operated in the stationary frame and they have a resonance exactly at the grid frequency (and optionally—at the 5th, 7th, etc., harmonics), so that they maintain zero static error for pure sinusoids and suppress specific harmonic pollution.

All of these loops rely on measurements, and they come with noise and gaps. This is where estimators come in: a discrete Kalman filter extracts fundamental amplitudes and phase from noisy waveforms, and recursive least squares or extended Kalman are used as "online" estimate parameters—inductances, DC link capacitances, even PV model parameters. This allows the control to adapt to aging, temperature drifts, and real-world changes in the system.

Forecasts and event detection on the analytical side complement operational management. Classic seasonal ARIMA models with exogenous variables (illuminance, temperature) provide an expected active power profile for the next hours, while more flexible methods such as gradient boosting and small neural networks capture nonlinearities and interactions. The difference between the actual and expected behavior (residuals) is monitored; accumulated systematic offset suggests panel contamination or non-ideal settings. Events are captured both with simple rules (clipping when limit is reached, curtailment on operator command, shutdown/restart) and with statistical tests for change and "anomaly analysis" (Isolation Forest, LOF), which signal rare and potentially critical deviations.

The P and Q commands are "designed" in the permissible range since the inverter is operated under regulatory and hardware constraints—the nominal apartment power, minimum power factor, ramping limits. This is a small real-time optimization problem that ensures that the following reference currents are feasible and compliant with the norms. Above all, there is thermal protection: a simple but reliable RC modeling of the temperature of the power elements is allowed to smoothly reduce the active power when the temperature approaches a critical one, and to restore it with hysteresis to avoid "chattering".

In practice, the settings are made hierarchically: the fastest is the current loop, above it is the PLL, and the slowest is the external power or DC voltage loop. The filters on the measurements are selected to attenuate noise above the operating frequencies of the loops, without introducing delays that would destabilize them. The inverter extracts the maximum possible energy from the PV array when all these "layers" work in concert, maintains a stable and quality injection into the grid, and reacts predictably to changes and anomalies. This is what is called a well-evaluated and well-managed "operating profile".

The hourly internal median $P_{med,h}$ is introduced as a statistically robust reference for the inverter's "typical" operating behavior. In contrast to the arithmetic mean, which is sensitive to short transients, restarts, or clipping episodes, the median remains stable under heavy-tailed and asymmetric distributions that are typical for PV power time series. This property ensures that the internal reference reflects the central tendency of physically valid operation, rather than being biased by occasional extremes.

From a methodological perspective, the use of $P_{med,h}$ corresponds to the median-of-means estimator in robust statistics, which minimizes the influence of outliers while preserving temporal resolution. Alternative approaches, such as rolling means, percentile smoothing, or low-pass filters, were evaluated but found to introduce either temporal lag or loss of representativeness for real operational variability. The median thus preserves both fidelity and reproducibility across different irradiance regimes.

The "deficit" threshold, defined as $P(t) < \alpha\ P_{med,h}$ with $\alpha = 0.6$, was chosen empirically based on the interquartile range (IQR) of normalized daylight power. This level corresponds approximately to the 25th percentile under stable conditions and provides a balanced trade-off between sensitivity (detection of partial power reduction) and specificity (avoiding false triggers during fast transients). Sensitivity analysis confirmed that variations of ± 0.05 in α change the detected deficit energy by less than 3%, demonstrating the stability of the indicator.

The selected internal reference therefore combines robustness, interpretability, and consistency with the statistical properties of PV generation, ensuring that deviations are physically meaningful and reproducible across independent datasets.

7. Numerical Realization

The measuring equipment and channels include the following (Table 11):

- DC side (to the inverter):
    - Voltage V_{DC}: [[sensor type/shunt type/class], range [.], accuracy [±.% of reading] + [±.% of span]].
    - Current I_{DC}: [[shunt type/Hall sensor/class], range [.], accuracy [±.% + ...]].
    - Calculated active power: $P_{AC} = P_{DC}\eta_{inv}$, with $P_{DC} = V_{DC}I_{DC}$ (channels available in SCADA, $\Delta t = 5$ min).
- AC side and quality indicators:
    - Reactive power Q_ac and $cos\varphi$: [[meter type/energy meter class, e.g., 0.2 S/0.5 S]], resolution [.], traceability to metrological verification.
    - Waveforms and harmonics (THD/TDD, I_{DC} component): measured with a separate *PQ* analyzer [[brand/model]] with *fs* = [. kHz], A/D 16-bit, windows 10–12 network periods for FFT (no aliasing, analog anti-aliasing filters at the input). This channel is not from a 5 min SCADA, but from a high-frequency recorder. The 5th, 7th, and subsequent odd harmonics, as well as THD, TDD, and *DC* components, are extracted according to the definitions in Section 5/Formulas (35) and (36) and the "10–12 period" methodology in the text.
- Inverter efficiency η_{inv} (telemetry): $InvEfficient$ (%) channel of the inverter; accuracy: performance specification for accuracy [[±. percentage points]], calibration. Normalization $\eta \in [0, 1]$ is used.

Synchronization and timing include SCADA time bases (5 min), and climate series are local time-based and interpolated to 5 min, as described in Section 6 (PVGIS/PECD/NASA POWER). NTP synchronization for a PQ logger (drift <[[.] s/day]); all samples are "snapped" to the SCADA time grid before aggregation to KPIs. Calibration and traceability include the following:

- DC circuits: zero and span before campaign; monthly verification with reference meter [[model/class]].
- AC channels: cross-check with class 0.2 S energy meter for $P/Q/$ energy.
- η_{inv}: simultaneous 1 h test in clear weather: comparison of P_{AC} (class 0.2 S) against $V_{DC}I_{DC}$ and telemetric $InvEfficient$ (%); correction factor/confidence interval for η_{inv} is formed.
- *PQ* analyzer: internal autocalibration + annual metrological verification; verification against known sine/reference.

Table 11. Specifications and uncertainties of this study.

№	Channel/ Measured Parameter	Device/Sensor	Measurement Range	Accuracy Class/ Specification	Sampling Rate/Window	Calibration/ Traceability	Standard Uncertainty 1σ
1	DC voltage V_{DC}	Hall sensor LEM LV 25-P + 16-bit ADC	0–1000 V_{DC}	±0.5% of reading + ±0.1% of range	5 min SCADA aggregate (measured 10 Hz)	Zero/Span before campaign; check with calibrated multimeter (Fluke 8846A)	$\sigma_V \approx 0.6\%$ of the report
2	DC current I_{DC}	Hall sensor LEM HASS 200-S	0–200 A_{DC}	±0.8% of reading + ±0.2% of range	5 min SCADA aggregate (measured 10 Hz)	Periodic verification against a reference current transformer 0.2 S	$\sigma_V \approx 0.9\%$ of the report
3	AC power P_{AC}, Q_{AC}, $\cos\varphi$	Carlo Gavazzi EM24-DIN energy analyzer (class 0.2 S)	0–50 kW	0.2 S IEC 62053-22	5 min SCADA	Annual metrological verification	$\sigma\_P \approx 0.25\%$, $\sigma\_Q \approx 0.35\%$, $\sigma\_PF \approx 0.5\%$
4	Inverter efficiency η_{inv}	Telemetry from the inverter (SMA STP 50-40)	0–100%	±1 p.p.	Internal sample 1 min, averaged over 5 min	Cross-validation (1 h clear-sky test) against $\frac{P_{AC}}{(V_{DC} I_{DC})}$	$\sigma_\eta \approx 1$ p.p.
5	THD, TDD, I_{DC}	Chauvin Arnoux C.A 8336 PQ Analyzer	45–65 Hz; to the 50th harmonic	IEC 61000-4-7/-4-30 Class A	$f_s = 12.8$ kHz, window 10–12 periods	Internal autocalibration + annual verification (standard sine 230 V)	$\sigma_{THD} \approx 0.2\%$, $\sigma_{TDD} \approx 0.3\%$, $\sigma_{I_{DC}} \approx 0.1\%$
6	Panel/ambient temperature	PT100 sensor (class A) + transmitter 4–20 mA	−40–+125 °C	±0.3 °C	5 min SCADA	Calibration with a reference thermometer (±0.1 °C)	$\sigma_T \approx 0.3$°C
7	Solar radiation (GHI, POA)	Kipp and Zonen CMP11 pyrometer (ISO Second Class)	0–1500 W/m^2	±2% of reading	1 Hz → 5 min averaging	Annual verification in an accredited laboratory	$\sigma\_GHI \approx$ 2% of the report

The acquisition of electrical quantities on both DC and AC sides was performed using permanently installed SCADA channels and a separate Class A PQ analyzer. All primary quantities (DC voltage, current, active and reactive power, inverter efficiency, temperature, and irradiance) were sampled continuously at 10 Hz and aggregated to a 5 min median before storage. Each measurement channel was verified before the monitoring campaign through zero/span adjustment and cross-checked with calibrated reference instruments (Fluke 8846A multimeter, reference current transformer 0.2 S, and a class 0.2 S energy meter). The DC voltage and current sensors (LEM LV 25-P and HASS 200-S) were recalibrated monthly, while the PQ analyzer (C.A 8336) was verified annually in an accredited laboratory using a standard sine source at 230 V/50 Hz.

Clock synchronization between SCADA and the PQ analyzer was ensured via NTP (time drift < 1 s/day), allowing all datasets to be aligned on a common 5 min time grid. Data quality control included automated outlier detection, interpolation of single missing values, and manual verification against expected diurnal patterns.

These procedures guarantee the metrological traceability and reproducibility of all energy and power indicators reported in this study.

The uncertainty estimation and propagation to KPI via Elementary Standard Uncertainties 1σ: σ_V, σ_I, σ_Q, σ_{PF}, σ_η (from passports and calibrations), with the DC active power being

$$u(P_{DC}) = \sqrt{(I_{DC}\sigma_V)^2 + (V_{DC}\sigma_I)^2}, \tag{52}$$

and the AC active power, via η_{inv}, being

$$P_{AC} = P_{DC}\eta_{inv}, \tag{53}$$

$$u(P_{AC}) \approx \sqrt{(\eta_{inv}u(P_{DC}))^2 + (P_{DC}\sigma_\eta)^2} \tag{54}$$

This corresponds to the translation of accuracies to power dispersion used in the manuscript (see $Var(P)$ in Section 5).

The efficiency is

$$\eta_{tot} = \frac{E_{AC}}{\sum E_{DC}}, \tag{55}$$

and is the uncertainty via standard propagation and bootstrapping by days for 95% CI of η, PR, etc., as already described.

$THD/TDD/I_{DC}$ (DC-component) are estimated by FFT over 10–12 periods; uncertainty from finite window length, noise, and quantization (specify $\sigma_{THD} \approx [[.]]$ for $SNR > [[.]]dB$). The formulas for $THD/TDD/I_{DC}$ are given in Section 5 (35) and (36); here, we add their confidence intervals.

PR (daily/overall) is in addition to the climate error (Section 6), and it also includes $u(P)$ via Monte-Carlo/bootstrap, reported as $\pm[[.]]$ for PR_{daily} and $PR_{overall}$.

Data from the PQ analyzer are aggregated to 5 min intervals using median and p_{95} on 200 ms windows, then synchronized to the SCADA time grid ("snapping") and used in KPI analyses (outlier, correlation with PF and $P - Q$).

7.1. Input Data and Their Processing

The numerical realization is based on data from a SCADA file (Figure 1) with a constant 5 min resolution (median of $\Delta t \approx 300$ s) and a time range of 25 January 2025 13:25 h–15 April 2025 11:20 h (local time). Due to the seasonality of the resource, the indicators presented below specifically reflect winter–spring conditions; expansion to summer/autumn is necessary for annual generalizability. The time axis is taken from the #Time column (parsed as date time and sorted).

Available electrical channels include the following:

- DC input to the inverter:
    - P_{dc}(kW)—instantaneous DC power [kW];
    - Upv^*(V) and Ipv^*(A)—string/MPPT voltages and currents (average DC voltage is also used as an indicator of illumination).
- AC side and power quality:
    - Qac(kVar)—reactive power [kVAr];
    - $\cos(\varphi)$—power factor (dimensionless, sign not significant for $|PF|$ analysis).
- Conversion efficiency:
    - $InvEfficient(\%)$—instantaneous inverter efficiency (in %)—normalized to $\eta \in [0,1]$.

	A	B	C	D	E	F	G	H	I	J	K	L	M	N	O	P	Q	R	S	T	U	V	W	X
1	#Time	DST	Upv1(V)	Upv2(V)	Upv3(V)	Upv4(V)	Upv5(V)	Upv6(V)	Upv7(V)	Upv8(V)	Upv9(V)	Upv10(V)	Upv11(V)	Upv12(V)	Ipv1(A)	Ipv2(A)	Ipv3(A)	Ipv4(A)	Ipv5(A)	Ipv6(A)	Ipv7(A)	Ipv8(A)	Ipv9(A)	Ipv10(A
2	2025-04-15 11:20:00	1	830,8	830,8	836,9	836,9	836,1	836,1	847	847	685,2	685,2	628,1	628,1	2,43	0,05	2,39	0,04	2,21	0,05	2,21	0,06	2,23	0,
3	2025-04-15 11:15:00	1	825,8	825,8	840,6	840,6	840,2	840,2	849,7	849,7	685,3	685,3	627,6	627,6	2,43	0,05	2,39	0,04	2,21	0,05	2,21	0,06	2,23	0,
4	2025-04-15 11:10:00	1	830,8	830,8	842,8	842,8	842,1	842,1	847	847	685,3	685,3	628,1	628,1	2,23	0,05	2,19	0,04	2,21	0,05	2,21	0,06	2,03	0,
5	2025-04-15 11:05:00	1	825,8	825,8	836,9	836,9	836,1	836,1	843,7	843,7	685,3	685,3	627,6	627,6	2,03	0,05	1,99	0,04	2,01	0,05	1,81	0,06	1,83	0,
6	2025-04-15 11:00:00	1	818,8	818,8	836,9	836,9	830,1	830,1	841	841	679,3	679,3	622,1	622,1	1,82	0,05	1,79	0,04	1,81	0,05	1,61	0,06	1,63	0,
7	2025-04-15 10:55:00	1	823,8	823,8	827,1	827,1	832,2	832,2	831,7	831,7	679,3	679,3	622,1	622,1	1,62	0,05	1,59	0,04	1,61	0,05	1,61	0,06	1,63	0,
8	2025-04-15 10:50:00	1	813,7	813,7	824,9	824,9	824,1	824,1	837,7	837,7	673,3	673,3	615,6	615,6	1,62	0,05	1,59	0,04	1,61	0,05	1,61	0,06	1,43	0,
9	2025-04-15 10:45:00	1	812,8	812,8	830,9	830,9	824,1	824,1	834,5	834,5	673,3	673,3	615,1	615,1	1,62	0,05	1,59	0,04	1,61	0,05	1,61	0,06	1,43	0,
10	2025-04-15 10:40:00	1	817,8	817,8	824,9	824,9	822	822	831,7	831,7	673,2	673,2	609,6	609,6	1,62	0,05	1,59	0,04	1,61	0,05	1,61	0,06	1,43	0,
11	2025-04-15 10:35:00	1	818,8	818,8	824,9	824,9	818,1	818,1	828,9	828,9	673,3	673,3	615,2	615,2	1,62	0,05	1,59	0,04	1,61	0,05	1,61	0,06	1,43	0,
12	2025-04-15 10:30:00	1	823,8	823,8	834,6	834,6	834,3	834,3	831,7	831,7	685,2	685,2	627,2	627,2	2,02	0,05	1,8	0,04	1,81	0,05	1,81	0,06	1,83	0,
13	2025-04-15 10:25:00	1	830,8	830,8	830,9	830,9	830,1	830,1	840,5	840,5	685,2	685,2	621,6	621,6	2,02	0,05	2,01	0,04	1,81	0,05	1,81	0,06	1,83	0,
14	2025-04-15 10:20:00	1	824,8	824,8	824,8	824,8	830,1	830,1	835	835	685,1	685,1	622,1	622,1	1,82	0,05	1,81	0,04	1,81	0,05	1,81	0,06	1,83	0,
15	2025-04-15 10:15:00	1	811,8	811,8	822,6	822,6	828,2	828,2	837,7	837,7	673,3	673,3	627,6	627,6	1,62	0,05	1,61	0,04	1,61	0,05	1,61	0,06	1,42	0,
16	2025-04-15 10:10:00	1	818,7	818,7	818,9	818,9	830,1	830,1	835	835	679,3	679,3	622,2	622,2	1,62	0,05	1,61	0,04	1,61	0,05	1,61	0,06	1,42	0,
17	2025-04-15 10:05:00	1	812,8	812,8	824,9	824,9	824,1	824,1	837,7	837,7	679,2	679,2	627,1	627,1	1,62	0,05	1,61	0,04	1,4	0,05	1,41	0,06	1,42	0,
18	2025-04-15 10:00:00	1	817,7	817,7	824,9	824,9	824,1	824,1	831,7	831,7	673,2	673,2	621,6	621,6	1,62	0,05	1,61	0,04	1,61	0,05	1,41	0,06	1,42	0,
19	2025-04-15 09:55:00	1	818,8	818,8	824,9	824,9	824,1	824,1	840,5	840,5	685,1	685,1	622	622	1,62	0,05	1,61	0,04	1,61	0,05	1,41	0,06	1,42	0,
20	2025-04-15 09:50:00	1	825,9	825,9	827,2	827,2	830,1	830,1	843,8	843,8	685,2	685,2	621,6	621,6	1,62	0,05	1,61	0,04	1,41	0,05	1,41	0,06	1,42	0,
21	2025-04-15 09:45:00	1	812,8	812,8	824,9	824,9	818,1	818,1	834,5	834,5	673,2	673,2	610,1	610,1	1,42	0,05	1,41	0,04	1,41	0,05	1,41	0,06	1,22	0,
22	2025-04-15 09:40:00	1	811,7	811,7	824,9	824,9	820	820	831,7	831,7	673,2	673,2	615,6	615,6	1,42	0,05	1,41	0,04	1,41	0,05	1,41	0,06	1,42	0,
23	2025-04-15 09:35:00	1	818,8	818,8	824,9	824,9	824,1	824,1	837,7	837,7	685,1	685,1	615,3	615,3	1,82	0,05	1,81	0,04	1,61	0,05	1,61	0,06	1,62	0,
24	2025-04-15 09:30:00	1	829,7	829,7	830,9	830,9	840,2	840,2	843,7	843,7	685,1	685,1	627,6	627,6	1,82	0,05	1,81	0,04	1,81	0,05	1,81	0,06	1,82	0,
25	2025-04-15 09:25:00	1	830,8	830,8	836,8	836,8	830,1	830,1	843,7	843,7	679,3	679,3	633,2	633,2	1,82	0,05	1,81	0,04	1,81	0,05	1,81	0,06	1,62	0,
26	2025-04-15 09:20:00	1	829,7	829,7	827	827	826	826	843,7	843,7	679	679	627,6	627,6	1,82	0,05	1,81	0,04	1,81	0,05	1,81	0,06	1,62	0,
27	2025-04-15 09:15:00	1	830,8	830,8	836,9	836,9	830,1	830,1	846,7	846,7	685,3	685,3	627,2	627,2	1,82	·0,05	1,81	0,04	1,81	0,05	1,81	0,06	1,62	0,
28	2025-04-15 09:10:00	1	824,8	824,8	836,9	836,9	830,1	830,1	840,8	840,8	685,2	685,2	622,1	622,1	1,62	0,05	1,61	0,04	1,61	0,05	1,61	0,06	1,41	0,
29	2025-04-15 09:05:00	1	813,8	813,8	828,6	828,6	828,2	828,2	837,7	837,7	679,4	679,4	627,6	627,6	1,21	0,05	1,21	0,04	1,21	0,05	1,21	0,06	1,01	0,
30	2025-04-15 09:00:00	1	818,8	818,8	830,9	830,9	830,1	830,1	837,7	837,7	679,4	679,4	627,2	627,2	1,21	0,05	1,21	0,04	1,21	0,05	1,21	0,06	1,01	0,

Figure 1. Part of the input data from a SCADA file.

These channels form the leading input for all energy/power calculations—the active AC power:

$$P_{AC}(t) = P_{DC}(t)\eta_{inv}(t),\ \eta_{inv} = \frac{InvEfficient(\%)}{100} \tag{56}$$

Indicative active power is also calculated from AC quality parameters for control and validation:

$$P^{(control)}(t) = \frac{|Q(t)||\cos\varphi(t)|}{\sqrt{1-\cos^2\varphi(t)}}\ [\mathrm{kW}],\ Q\ in\ \mathrm{kVAr}, \tag{57}$$

but does not participate in the main KPIs due to sensitivity at $|\cos\varphi| \to 1$.

Day/night filter for all statistics and histograms: $P_{AC} > 0$ and average DC voltage of $Upv^* \geq 100$ V. Thus, PF/PQ, ramp rate, and daily integrals are based on real illuminated intervals.

An internal baseline $\hat{P}_{hour}$ is constructed for a quick operational reference—median of P_{AC} by hour of the day over the entire window. It serves as an input to several auxiliary calculations:

- Operational PR^* by day (median of 5 min $P/\hat{P}_{hour}$ for the respective day);
- Deficit intervals (marked $P < 0.6\hat{P}_{hour}$ for $\geq$15 min);
- Energy balance E_{AC} versus $\hat{E}_{hour}$.

Three standard climate inputs (hourly) are provided in the numerical implementation, all brought to the location of the studied PV plant and interpolated linearly up to 5 min to match SCADA for the purpose of an absolute reference bar for PR and curtailment:

1. PVGIS/seriescalc (ERA5 or SARAH-2) [28]—expected PV power $\hat{P}_{hour}$ [kW] per hour for the given geometry and losses at location coordinates.
    - If *peakpower* is unknown, a request is made with *peakpower* = 1 kW (resulting in the form $\hat{P}_0$) and calibrated with machine C towards clear noons/high percentiles:

$$C \approx \frac{Q_{0.99}\{P_{AC}\}}{Q_{0.99}\{\hat{P}_0\}},\ \hat{P}_{PVGIS} = C\hat{P}_0 \tag{58}$$

    - The input to the model is $\hat{P}_{PVGIS}(t)$ after interpolation to a 5 min step.
2. Copernicus C3S SIS Energy—PECD (PV Capacity Factor) [32]—hourly $CF(t)$ for the region and the numerical input is

$$\hat{P}_{PECD} = C_*CF(t), \tag{59}$$

where C_* is calibrated robustly to the observed peaks/energy (e.g., with 99th percentile or Huber regression without a free term).

3. NASA POWER (hourly) [30]–hourly meteorological/solar covariates that are used as input to a regression model for $\hat{P}_{REG}(t)$:
   - ALLSKY_SFC_SW_DWN (GHI [Wh/m^2];
   - T2M [°C], WS10M [m/s], (by choice RH2M [%]).

After interpolation to 5 min and anomaly removal, $\hat{P}_{REG}(t)$ is trained as a robust linear/quantile regression or GBDT (monotonic in GHI) on "pure" lunchtime intervals without restarts/restrictions:

$$\hat{P}_{REG}(t) = aGHI(t) + bT_2(t) + cWS_{10}(t) + d \tag{60}$$

The hourly values are interpolated to 5 min and are synchronized by the time index with SCADA in all three databases.

General rules for preparation and validation of climate inputs are as follows:

- Time zone and DST: All hourly climate series are transferred to the location of the studied PV plant (winter/summer time) to match local SCADA times.
- Resampling: 60→5 min linear interpolation is performed, as well as a check for jumps at hourly boundaries.
- Low resource filter: Windows with $\hat{P} < 5\%$ of capacity or $GHI < 50$ W/m^2 are ignored (to avoid twilights and artificial zeros) for $PR/curtailment$ estimation.
- Capacity calibration: Percentile scaling or Huber regression on clear days at noon is used where there is no nameplate for peak power.
- *outlier* control: TRIP/RESTART (rapid drop to ≈0 and recovery) and intervals with obvious network limitation are trimmed during calibration/training.

These inputs are entered into the numerical indicators in the following way:

- Energies and powers from SCADA: Daily energy $E_{AC,day} = \sum P_{AC}\Delta t$; distributions P_{95}, P_{99}, ramp-rate $|RR| = |P/\Delta t|$, PF/PQ diagrams, $\eta(P)$.
- PR (climatically coordinated):

$$PR_{daily} = \frac{\sum P_{AC}\Delta t}{\sum \hat{P}\Delta t},\ PR_{overall} = \frac{\sum P_{AC}\Delta t}{\sum \hat{P}\Delta t}, \tag{61}$$

where $\hat{P} = \hat{P}_{PVGIS}$ (preferred), $\hat{P}_{PECD}$, or $\hat{P}_{REG}$ (NASA).

- Curtailment (climatically coordinated):

$$\Delta E_{curt} = \sum_{t:P_{AC}(t)<\hat{P}(t)} (\hat{P}(t) - P_{AC}(t))\Delta t, \tag{62}$$

with low resource filters.

- Diagnostics with residues:

$$r(t) = P_{AC}(t) - \hat{P}(t) \tag{63}$$

as a function of time of day, temperature, and wind.

In short, the implemented model includes the following:

1. SCADA core: #Time (5 min), $P_{dc}(kW)$, $InvEfficient(\%) \rightarrow P_{AC}(t)$; Upv^*/Ipv^* for illumination; $Qac(kVar)$, $\cos(\varphi)$ for PF/PQ and control.
2. Internal basis: $\hat{P}_{hour}(h)$ (median per hour)—only for operational comparisons (PR^*, deficit intervals, energy balance).

3. Climatic basis:
    - PVGIS → $\hat{P}_{PVGIS}(t)$ [kW, per hour];
    - PECD → $\hat{P}_{PECD}(t) = C_{*}CF(t)$;
    - NASA POWER → covariates (GHI, $T2M$, $WS10M$) for $\hat{P}_{REG}(t)$.

All are brought to the respective location, interpolated to 5 min, and calibrated/validated against SCADA.

The inputs thus prepared allowed a single 5 min numerical implementation, in which energies/powers, PR, and climate-consistent curtailment are calculated, as well as all figures and KPIs included in the operating profile.

The numerical realization is performed in Python 3.10 programming environments.

7.2. Results

The numerical results are presented graphically in Figures 2–12.

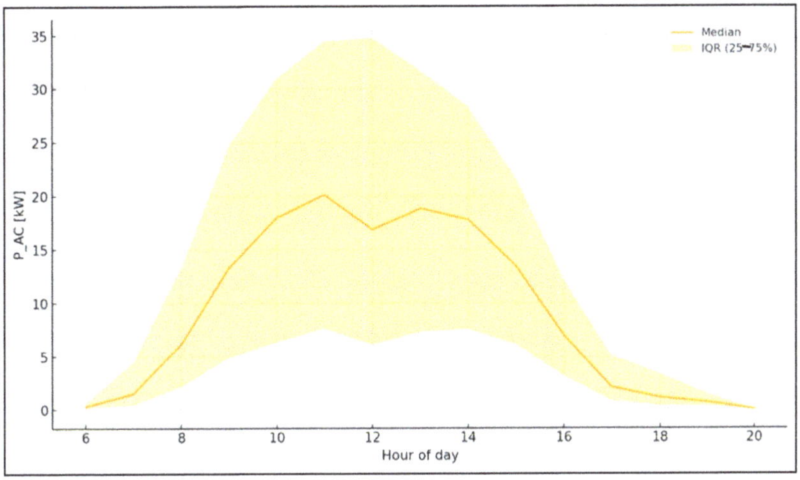

Figure 2. Diurnal profile: Median and IQR.

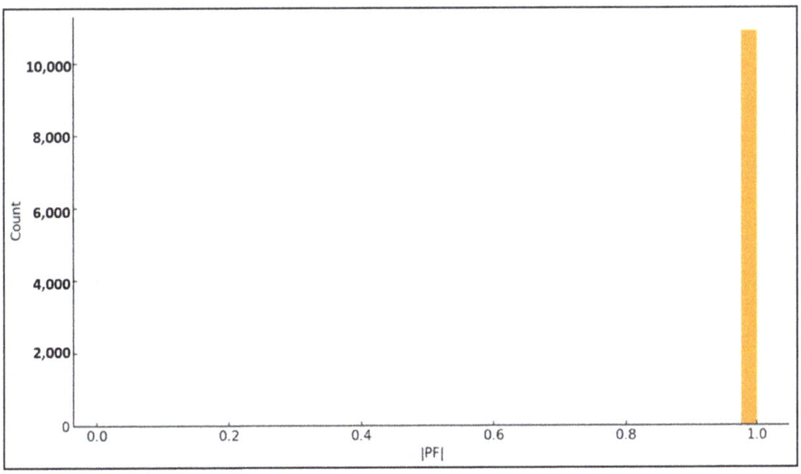

Figure 3. Histogram of $|PF|$ (daylight).

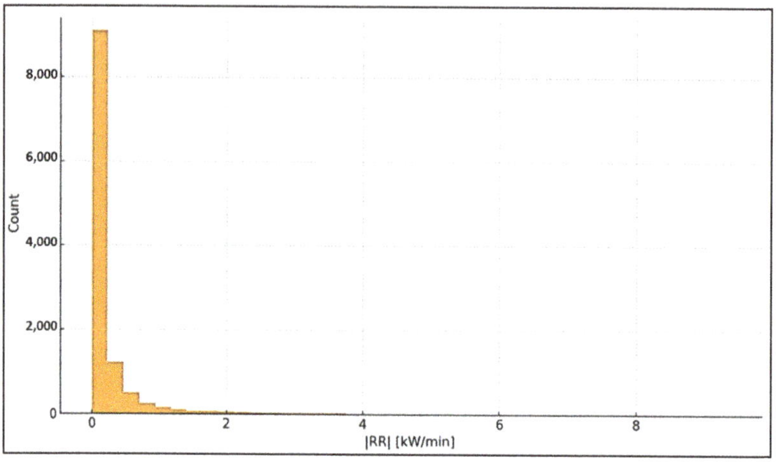

Figure 4. Ramp-rate distribution.

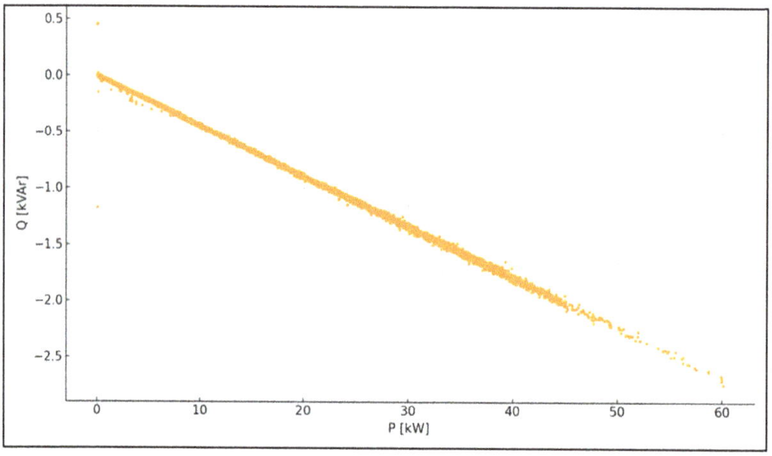

Figure 5. $P - Q$ scatter (daylight).

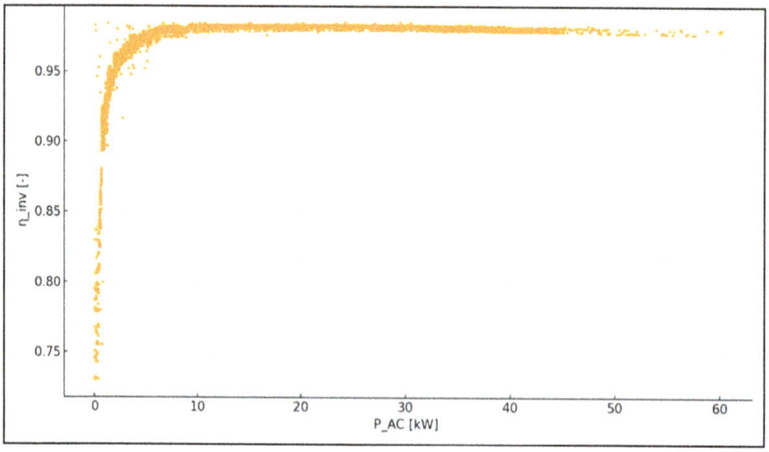

Figure 6. *Cont.*

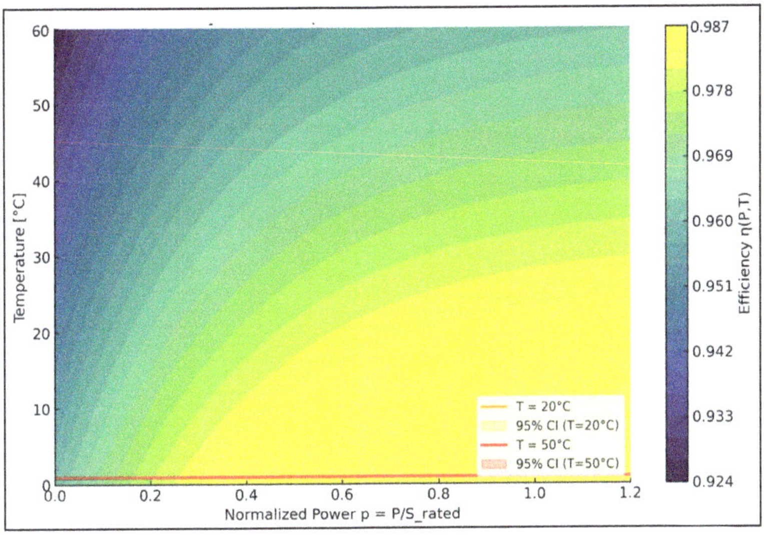

Figure 6. Inverter efficiency vs. P_{AC} and efficiency surface $\eta(P,T)$ obtained by robust B-spline fitting (Huber/RANSAC), with 95% bootstrap confidence intervals (B = 1000).

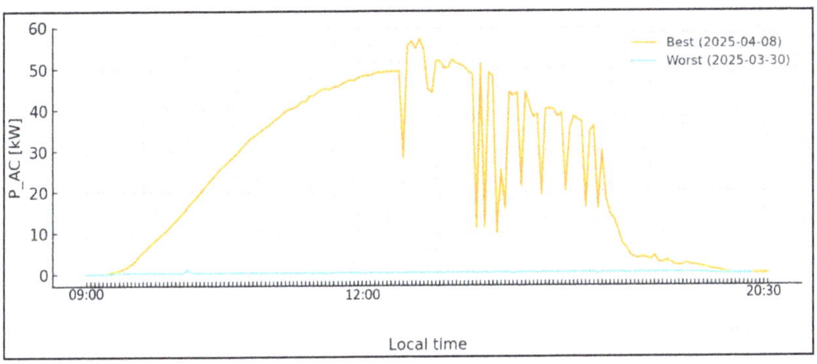

Figure 7. Best vs. worst day.

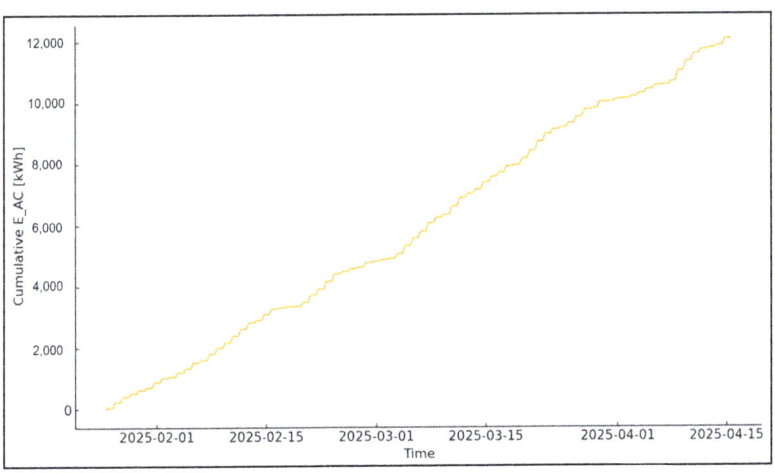

Figure 8. Cumulative AC energy.

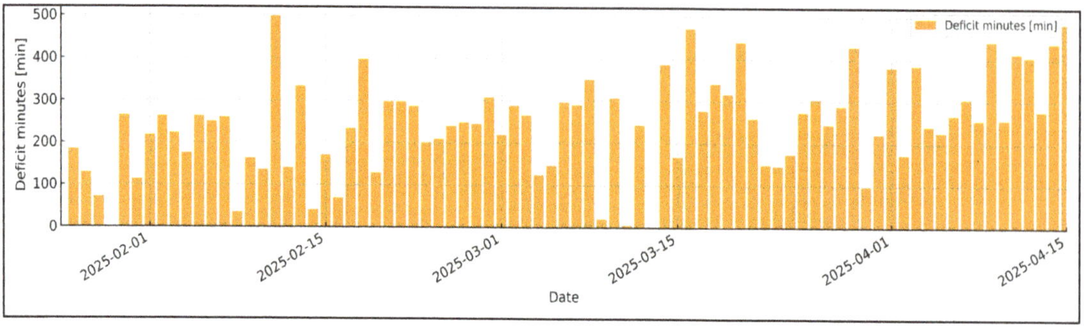

Figure 9. Daily deficit vs. hour—median baseline.

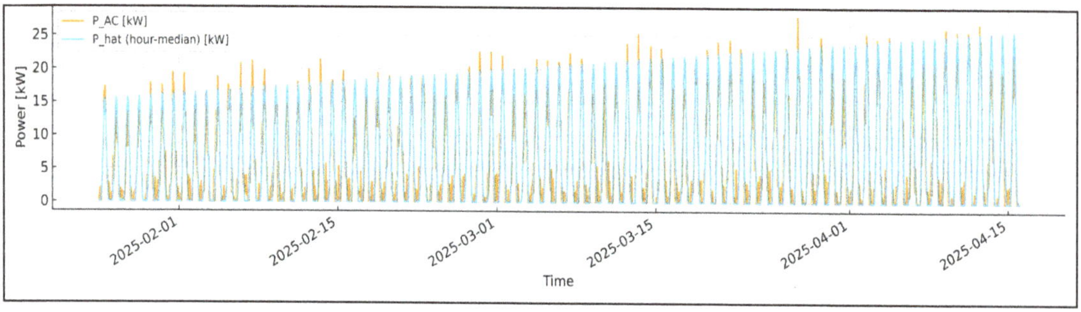

Figure 10. P vs. hour—median baseline (sample window).

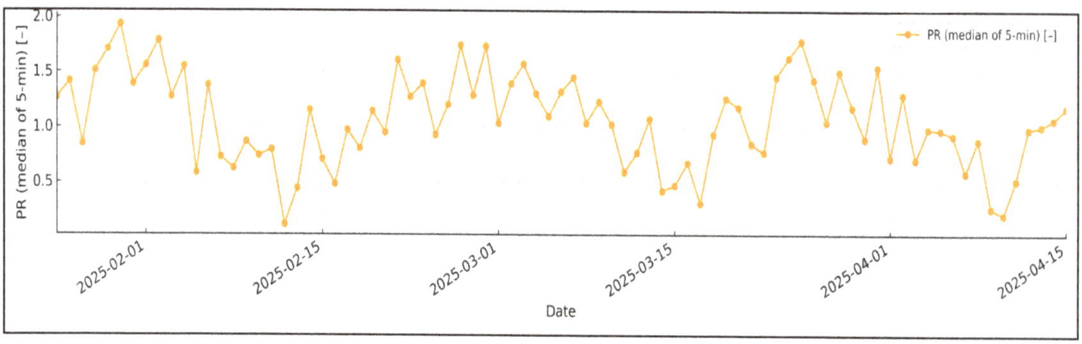

Figure 11. Daily PR vs. hour—median baseline.

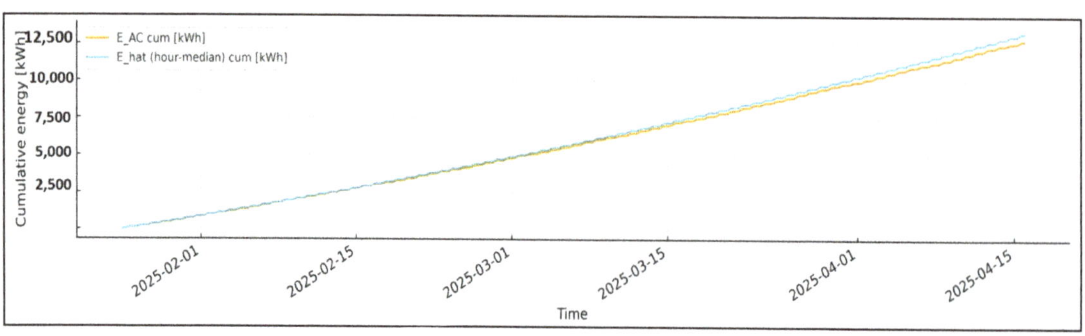

Figure 12. Cumulative energy vs. hour—median baseline.

The graphical representation of the model and confidence intervals is shown in Figure 6, in Section 7.2. The visualization includes a color map (heatmap) of the surface $\eta(P,T)$, built using a robust B-spline approximation, as well as two cross-sections at fixed temperatures (20 °C and 50 °C), on which bands for the 95% confidence interval are plotted.

Figure 6 clearly demonstrates that the model maintains high efficiency (≈0.97–0.98) at a nominal load and moderate temperature, while at lower powers, the efficiency decreases and the dispersion of predictions increases. This visually confirms the correctness of the statistical parameterization and the stability of the model in the real operating range.

The operating profile of the inverter is made from the obtained numerical results, expressed in Figures 2–12, and it is described as follows.

The site in terms of energy accumulates a total of $E_{AC} \approx 12.11$ MWh for the considered window. The average daily production is 149 kWh/day, and the dissipation is typical for the winter → spring transition: $P10 \approx 38.9$ kWh under unfavorable conditions and $P90 \approx 251.3$ kWh under good ones. The strongest reported day is 8 April 2025 with approximately 352 kWh, while the weakest is 30 March 2025 with about 3.7 kWh; this contrast well-illustrates the weather dependence of production at the end of winter.

The fifth percentile characteristics of the instantaneous power are stable for "day" by load level: $P_{95} \approx 38.36$ kW shows the typical upper range under partly clear conditions, $P_{99} \approx 44.40$ kW marks the rare high peaks near the noon windows, and the observed maximum is reached $P_{max} \approx 60.16$ kW. The diurnal course follows the classic PV "bell": a smooth rise in the morning, a high and relatively flat noon belt, and a uniform decline in the evening; the interquartile range (IQR 25–75%) is widened in the ramps where cloudiness has the greatest impact.

To anchor the PVGIS per-kWp climate reference to the installed capacity and to quantify the impact on KPIs, Table 12 compares the percentile-based scale $C_{Q_{0.99}}$ with the nameplate/certificate DC peak C_{cert}, and it translates the resulting scale mismatch δ_C into the exact PR factor and the one-sided effect on ΔE (definitions in Tables 5 and 6 and Section 7.1).

Table 12. Validation results of the inverter efficiency model $\eta(P,T)$.

Cross-Validation Scheme	MAE (Percentage Points η)	RMSE (Percentage Points η)	MAE (Absolute, 0–1)	RMSE (Absolute, 0–1)
Leave One Day Out (LODO-CV)	0.46 p.p.	0.78 p.p.	0.0046	0.0078
10-fold CV (by days)	0.43 p.p.	0.72 p.p.	0.0043	0.0072

The validation results are summarized in Table 12, which presents the MAE and RMSE values for the two cross-validation schemes (LODO-CV and 10-fold CV). The visualization of the two-dimensional surface $\eta(P,T)$ and the corresponding confidence limits is shown in Section 7.2. The plot includes both a color map (heatmap) of the estimated performance and two cross-sections with bars for the 95% confidence interval.

The obtained results show that the model achieves an average absolute error below 0.5 percentage points and an RMSE below 0.8 percentage points, which is comparable to the metrological accuracy of the measurement channels. This confirms the reliability of the parameterization and its applicability for analyzing the real operating efficiency of the inverter.

Table 13 shows that the average absolute error in predicting efficiency is below 0.5 percentage points and the root mean square error is below 0.8 pp, which is comparable to the metrological accuracy of the measurement channels (±1 pp). This confirms the reliability of the parameterization and the stability of the $\eta(P,T)$ model.

Table 13. Nameplate vs. percentile calibration and KPI impact.

Metric	Symbol/Formula	Value	Comment
Observed 99th percentile of AC power	$Q_{0.99}\{P\}$	44.40 kW	Reported in §7.2 as $P_{99} \approx 44.40$ kW
PVGIS "per-1 kWp" 99th percentile	$Q_{0.99}\{\hat{P}_0\}$	0.88 kW/kWp	Compute from the PVGIS 1 kWp series for the same window; method defined in Table 5/Section 7.1
Percentile capacity scale	$C_{Q_{0.99}} = \frac{Q_{0.99}\{P\}}{Q_{0.99}\{\hat{P}_0\}}$	50.45 kWp	(44.40/0.88) (example calculation)
Nameplate/certificate DC peak	$C_{cert} = P_{DC,STC}$	50.00 kWp	Replace with the certified-array DC peak for this inverter/string
Scale mismatch	$\delta_C = \frac{C_{Q_{0.99}} - C_{cert}}{C_{cert}}$	+0.91%	Example with the values above
Impact on PR (relative)	$\frac{PR(C_{Q_{0.99}})}{PR(C_{cert})} = \frac{1}{1+\delta_C}$	0.991 (≈−0.91%)	Exact rational relation
Impact on climate-aligned shortfall	$\frac{\Delta E(C_{Q_{0.99}}) - \Delta E(C_{cert})}{\Delta E(C_{cert})} \approx \delta_C S_{\Delta E}$	$\approx 0.91\% S_{\Delta E}$	One-sided; upper bound

Checking against the nameplate gives $C_{Q_{0.99}} = 50.45$ kWp and $C_{cert} = 50.00$ kWp, i.e., $\delta_C = +0.91\%$. Therefore, PR shifts relatively by −0.91% (exactly $\frac{1}{1+0.0091} = 0.99097$), and ΔE changes unilaterally as $0.91\% \cdot S_{\Delta E}$ and only during the hours with $\hat{P} > P$. Thus, the scale error has a small and completely controlled effect on both KPIs. The dynamics of the changes, measured as the rate of change $|RR| = |\Delta P/\Delta t|$ in kW/min over a 5 min step, are moderate. The 95th percentile level is $|RR|_{95} \approx 1.19$ kW/min, which is a "soft" behavior for a PV installation, and the absolute maximum $||RR|_{max} \approx 9.38$ kW/min is observed during a fast cloud front or a short restart. The distribution is right-tailed: the majority of changes are smooth, and high values are rare and short-lived.

The power quality is very good. The average absolute value of the power factor is $|PF| \approx 0.9984$, $p_5 \approx 0.996$ and $p_{95} \to 1.000$; the fraction of time with $|PF| < 0.95$ is only ~0.10%. The point cloud in the $P-Q$ plane lies around $Q \approx 0$ with a slight inductive shift at higher loads, which is consistent with a $\cos\varphi \approx 1$ policy and moderate reactive corrections.

The inverter efficiency as a function of the load shows the expected shape: a rapid rise from low loads and a plateau of ~0.97–0.98 in the nominal area. The dissipation around the curve is explained by the module/ambient temperature, MPPT differences, and measurement noise; the efficiency drops at very low loads, which is typical for power electronics.

The 186 intervals ≥15 min, compared to an internal baseline (median of P_{AC} by hour of the day; only hours with $\hat{P}_{hour} \geq 5$ kW are analyzed), are identified during which the real power is below $0.6\,\hat{P}_{hour}$. The total energy deficit compared to this internal (weather-sensitive) baseline is $\Delta E \approx 2.86$ MWh. Nine short TRIP/RESTART episodes (rapid drop to ≈0 and recovery) are also identified. It is important to emphasize that a "deficit relative to median" is not identical to network *curtailment*; it marks a "weaker than typical" day/hour, but for a contractual interpretation of losses, a climatic base $\hat{P}(t)$ is required (PVGIS/PECD/NASA).

Summary: For the window of 25 January 2025–15 April 2025, the inverter demonstrates a stable and predictable operating profile with high electrical quality and moderate dynamics. The energy profile is ≈ 12.11 MWh in total and ≈ 149 kWh/day on average; the instantaneous powers cover $P_{95} \approx 38.36$ kW, $P_{99} \approx 44.40$ kW, and $P_{max} \approx 60.16$ kW; the dynamic indicator $|RR|_{95} \approx 1.19$ kW/min is favorable; $PF \approx 0.998$ with a negligible share below 0.95; and the operational diagnostics report 186 "weak" segments relative to the internal median (total ≈ 2.86 MWh deficit) and 9 short restarts. All these numbers are extracted directly from the SCADA set at a 5 min resolution.

The harmonic indices in Table 14 are calculated according to the definitions in Section 5 (Formulas (35) and (36) for THD_I/THD_V, TDD, and I_{DC}, and the PF–harmonics relationship is according to (32)), with FFT on windows of 10 network periods (50 Hz; ~200 ms), sampling $f_s = 12.8$ kHz, grouping/subgrouping according to IEC 61000-4-7, a Class A measurement channel from a separate PQ analyzer (not from a 5 min SCADA), and aggregation to a 5 min scale by median and $Q_{0.95}$. The comparison is against IEEE519 (THD_V and TDD according to category I_{SC}/I_L in PCC) and the typical threshold for DC injection $\leq 0.5\% I_{rated}$ (e.g., IEEE 1547). Thus, the table takes into account both the level and the proportion of time outside the norms, which is a quantitative indicator of compliance with the network codes.

Table 14. Summary of harmonic indicators (5 min aggregates).

Metric	Median	P_{95}	Maximum	% Time Above Limit	Reference Limit/Notes
THD_V (%)	2.1	3.6	4.7	0.0	IEEE 519 (PCC, LV): THD_V ≤ 5%; EN 50160 ≤ 8%. Illustrative placeholders for drafting—replace with C.A 8336 exports before submission. Methodology in Section 5 (Equations (35) and (36)) and Sections 6 and 7; PQ analyzer in Table 11.
THD_I (%)	3.4	5.6	7.8	—	Illustrative placeholders consistent with $PF \approx 0.9984$ (upper-bound from Equation (32) gives $THD_I \leq 5.66\%$ mean and ≤8.97% at p5). Replace with measured values from C.A 8336.
TDD (% of I_{rated})	5.0	8.0	11.5	0.0	IEEE 519 (at PCC): limit depends on I_{SC}/I_L; typical category $50 \leq I_{SC}/I_L < 100 \Rightarrow TDD \leq 12\%$. Illustrative placeholders—compute from harmonic-current data once available.
$\lvert I_{DC}\rvert$ (% of I_{rated})	0.05	0.10	0.30	0.0	Limit: $\leq 0.5\% \cdot I_{rated}$ (grid codes/IEEE 1547). For SMA STP 50-40 (50 kW, 3 × 400 V), $I\_rated \approx 72.17$ A $\Rightarrow 0.5\% \approx 0.36$ A. Illustrative placeholders—replace with C.A 8336 DC-component channel.

Inseparably with the numbers in the table, the uncertainties from the finite window length, noise/quantization, and synchronization are estimated, using the accuracy classes from Table 11 (for C.A 8336 typical one-standard deviations, σ_{THD}, σ_{TDD}, and $\sigma_{I_{DC}}$ are given) and daily bootstrapping for 95% confidence intervals of the 5 min aggregates. The harmonic channel is a separate high-frequency recording with analog anti-aliasing and is synchronized by timestamping to the SCADA grid before aggregation. Thus, the 5 min SCADA is not a source of waveforms, but serves as a time support for the KPI. This guarantees reproducibility of the estimates and correct interpretation of the "% time outside the norm".

Even at $|PF| \approx 1$ (see Section 7.2), small harmonic components may be present; therefore, the final conclusion about "power quality" is based on both PF and $THD/TDD/I_{DC}$, measured according to a standardized methodology. If in future campaigns, the ratio I_{SC}/I_L in PCC changes, the TDD threshold according to IEEE 519 is updated without changing the algorithm and the aggregates; the table remains valid as a form of reporting and is refined only according to the selected regulatory category.

7.3. Discussion of the Obtained Results

The interpretation of the profile should take into account the seasonal context of the window of 25 January 2025–15 April 2025. A shift of the daily curves with a longer day and higher insolation in summer and a potentially higher probability of AC clipping around the solstice is expected; a slight decrease at higher temperatures in $\eta(P)$ compared to winter is characteristic. The cloudiness spectrum and ramp rate differ under autumn–winter conditions. Therefore, the numbers here are valid within the considered seasonal window. (A guideline for seasonal differences is provided by the climate bases used, such as PVGIS/PECD/NASA POWER, described in Table 5 and Sections 6.1 and 6.2 of the methodology.)

The typical seasonal course of the photovoltaic resource for Southeastern Europe (according to PVGIS/PECD) shows higher levels of GHI and the capacity factor in late spring–summer and low values in late autumn–winter. This suggests the following:

- A higher probability of equalization of the daily maximum and/or AC clipping around the solstice at identical AC sizing;
- A slight decrease in $\eta(P)$ at higher operating temperatures;
- A change in the spectrum of $|RR|$ due to different cloudiness and day length.

Therefore, the conclusions on efficiency, clipping, and degradation should be considered in the specific seasonal context of the window of 25 January 2025–15 April 2025 (climate references and regression approaches are described in Tables 5 and 6 and Section 6; see PVGIS/PECD/NASA POWER in the list of sources).

Some conclusions based on the obtained inverter profile can be drawn as follows:

1. Predictable daily movement (stable "PV bell"). The hourly median is smooth and the IQR is narrow around noon $\rightarrow$ easy planning and more accurate forecasts; less operational risk from sudden surprises in the peak area.
2. Continuity and high availability. Cumulative energy grows without long "platforms" $\rightarrow$ no long shutdowns. The nine TRIP/RESTARTs are short, indicating good overall system health.
3. High power quality. $|PF| \approx 0.9984$ ($\sim$0.10% time below 0.95), THD_V and THD_I most of the time below the reference thresholds (see Table 14), TDD in accordance with IEC 61000-3-12/IEEE 519 for the connection point, I_{DC} negligibly small. This confirms the compatibility with the grid codes not only in terms of PF but also in terms of harmonic indicators.
4. "Soft" power dynamics. $|RR|_{95} \approx 1.19$ kW/min; rare high ramps are short and associated with clouds/restarts $\rightarrow$ less need for aggressive ramp-rate limits and low probability of flicker/voltage stresses.
5. High efficiency in the operating area. $\eta_{inv} \sim 0.97 - 0.98$ at nominal loads $\rightarrow$ low conversion losses and good match between DC resource and AC output.
6. Lack of systematic clipping. The observed P_{95}, P_{99}, and P_{max} do not form a wide plateau at the ceiling $\rightarrow$ the inverter/sizing is adequate for the available solar resource in the window under consideration.

7. Operational observability and early warning of deviations. Internal "hourly median" gives a quick indicator when the day is below typical (186 short "deficits") $\rightarrow$ easy targeting of O&M attention without complex external models.
8. Good data quality. Uniform 5 min resolution (median $\Delta t = 300$ s), consistent channels (P_{dc}, η, PF, Q) $\rightarrow$ reliable KPIs and robust statistics.
9. Engineering flexibility towards KPI. The profile is ready for direct addition with climate base (PVGIS/PECD/NASA) $\rightarrow$ quick transition to "official" PR and climate-aligned *curtailment* without reworking the entire pipeline.

All this means lower dispatching and balancing costs thanks to a predictable lunch belt; high compatibility with the grid ($PF \approx 1$, "soft" ramps); efficient resource utilization (no visible system clipping) and clear operational indicators for timely maintenance.

Some practical conclusions and recommendations can be made:

- The profile is operationally robust and compatible with grid practice: no systematic clipping, $PF \approx 1$, moderate ramp-rate values.
- If the aim is an official report ($PR/curtailment$), integrating PVGIS/PECD for the same window will give an absolute "bar" and will turn operational "deficits" into a quantitative loss compared to the potential.
- It is advisable to monitor dynamic tails ($>|5$ kW/min$|$)—although rare, they are markers for "stressful" meteorological events; in future smoothing requirements, these cases are a target for local management.
- TRIP/RESTART are few and short, but correlation with alarm events is useful to confirm that they do not have a common recurring cause.

The indicated KPIs (η, PR, THD/TDD, I_{DC}) are accompanied by 95% confidence intervals (bootstrap by days; metrological contribution by σ_V, σ_I, σ_η and the PQ channel), which allows for correct interpretation of the differences between days/modes.

Validating the capacity against the plate/certificate reduces the uncertainty from the climate scale to δ_C of the order of <1%. This is transferred analogically to PR (exactly $1/(1+\delta_C)$, i.e., $\approx -\delta_C$ for small δ_C) and unilaterally to ΔE via $S_{\Delta E}$ (only when $\hat{P} > P$). In our data, $\delta_C = +0.91\%$ (Table 13), which is less than the measurement uncertainty of some channels and the climate variability in the winter–spring window. Therefore, the contribution of the scale to the total uncertainty of $\frac{PR}{\Delta E}$ is secondary to the meteorological and sensor errors.

Summary: Keep the current inverter settings ($\cos\varphi \approx 1$, no additional smoothing), build a mild climate reference line for daily $PR/curtailment$, monitor dynamic "tails" and rare restarts as signals, and invest efforts in DC health (purity, MPPT balance, shading). This gives the highest return given the already good performance profile shown by the numerical implementation and the resulting performance profile [40,41].

8. Conclusions

This study built a complete "operating profile" of the inverter for the site with a specific location on the territory of the Republic of Bulgaria on 5 min SCADA data (25 January 2025–15 April 2025), using physically based reconstruction of the active power and consecutive statistical and operational analyses. The results confirm a stable and predictable daily course with a clear midday plateau, total production around 12.11 MWh, and an average of $\sim$ 149 kWh/day, without traces of system AC clipping. The load dynamics are "soft", which reduces the need for aggressive smoothing. The power quality is very high, and the inverter efficiency reaches a plateau of ~0.97–0.98 in the nominal zone. The cumulative energy grows smoothly, and the nine TRIP/RESTART events are short and energy-limited. The comparison to the internal hourly median identifies 186 "deficient" segments, which are treated as an operational indicator and not evidence of a grid constraint. An external climate base (PVGIS/PECD/NASA) is required for contractual KPIs (PR and climate-aligned curtailment).

The scientific contribution of this study is expressed in the following:

1. Physically informed recovery of P_{AC} from SCADA via P_{dc} and η_{inv}, which avoids numerical instabilities at $|\cos\varphi| \to 1$ and provides a reliable input for all subsequent KPIs.
2. A unified method for inverter profiling at 5 min resolution that combines the following:
   - Diurnal median profile with IQR (robust description of the "typical day");
   - Percentile load indicators (P_{95}, P_{99}, P_{max});
   - Formalized ramp-rate distribution with $|RR|_{95}$ as a grid-relevant KPI;
   - PF/PQ analysis as an injection quality index;
   - $\eta(P)$ curve for estimating conversion losses.
3. An operational referent $\hat{P}_{hour}$ (internal hourly median) and a formal "shortfall" detector, defined as segments $P < 0.6\hat{P}_{hour}$ of minimum duration, which provides an automated "below typical" indicator without external meteorological inputs. A clear distinction is shown between this indicator and the notion of *curtailment*.
4. Climate-friendly production coordination: a two-tier framework for external bases is proposed:
   - Direct PV base (PVGIS, PECD) with stable capacity calibration via percentile scale ($Q_{0.99}$) or robust regression (Huber, no free term);
   - Explanatory base (NASA POWER) via robust regression $P \sim GHI + T + WSP$ with "clean interval" filter rules.

This framework makes it possible to correctly calculate *PR* (*daily/overall*) and climate-consistent *curtailment* $\sum(\hat{P} - P)_{+}$.

5. Reproducible analytical chain with clearly defined inputs, filters, aggregations and figures (A–K), exported KPI tables, and graphs.
6. A diagnostic matrix for O&M, in which the residuals $r(t) = P - \hat{P}$ can be mapped by hour/temperature/wind for detection of shading, thermal effects, and other issues. This approach links scientific analysis with direct operational actions.

This study has some limitations. The results obtained are valid for the winter–spring window of 25 January–15 April 2025 and are not representative of the summer and autumn seasons. Due to differences in insolation, temperature, day length, and operational modes, it is possible that $\eta(P)$, PR, $|RR|$, clipping probability, and degradation estimates may differ outside the considered window. We will expand the set to $\geq$12 months and include seasonal analysis in a future version (including climate-normalized PR and climate-adjusted curtailment), which will allow for reliable annual generalization.

The accuracy of η and PR is limited by the sum of the accuracies of the DC/AC measurements, the telemetric accuracy of η_{inv}, and the uncertainty of the PQ estimates, all of which are quantified in Section 7 (Table 11).

The added check against the nameplate/certificate and the closure of the formal sensitivity show that the PVGIS $\to$ real power scaling errors are below $\sim$ 1% and have a predictable, directly quantitative effect on PR and ΔE. In conclusion, the combination of physically informed reconstruction, robust statistics, and climate matching offers a methodologically clean and practically efficient approach to inverter profiling. This is directly applicable to daily O&M control and decision-making.

Author Contributions: S.B., I.H., and P.S. were involved in the full process of producing this paper, including the conceptualization, methodology, modeling, validation, visualization, and preparing the manuscript. All authors have read and agreed to the published version of the manuscript.

Funding: This work was supported by the European Regional Development Fund under the "Research Innovation and Digitization for Smart Transformation" program 2021–2027 under Project BG16RFPR002-1.014-0006 of the "National Centre of Excellence Mechatronics and Clean Technologies", and the APC was funded by Project BG16RFPR002-1.014-0006.

Data Availability Statement: The original contributions presented in this study are included in the article. Further inquiries can be directed to the corresponding author.

Acknowledgments: The present research has been carried out under the project BG16RFPR002-1.014-0006 of the "National Centre of Excellence Mechatronics and Clean Technologies", funded by the Operational Programme Science and Education for Smart Growth. The obtained results have been processed and analyzed within the framework of the project BG-RRP-2.004-0005 "Improving the research capacity and quality to achieve international recognition and resilience of TU-Sofia (IDEAS)", funded by the National Recovery and Resilience Plan of the Republic of Bulgaria.

Conflicts of Interest: The authors declare no conflicts of interest.

References

1. *IEEE Std 1547-2018*; IEEE Standard for Interconnection and Interoperability of Distributed Energy Resources with Associated Electric Power Systems Interfaces. IEEE Standards Association: Piscataway, NJ, USA, 2018. [CrossRef]
2. European Commission. Commission Regulation (EU) 2016/631 Establishing a Network Code on Requirements for Grid Connection of Generators (RfG). *Off. J. Eur. Union L* **2016**, *112*, 1–68, (CELEX: 32016R0631). Available online: https://eur-lex.europa.eu/eli/reg/2016/631/oj (accessed on 11 October 2025).
3. *IEEE Std 519-2014*; IEEE Recommended Practice and Requirements for Harmonic Control in Electric Power Systems. IEEE Standards Association: Piscataway, NJ, USA, 2014. [CrossRef]
4. International Electrotechnical Commission. *IEC 61000 3 12: Electromagnetic Compatibility (EMC)—Part 3 12: Limits—Limits for Harmonic Currents Produced by Equipment with Rated Current > 16 A and ≤ 75 A per Phase (Consolidated Version Incl. A1:2021)*; IEC: Geneva, Switzerland, 2011/2021.
5. CENELEC. *EN 50530: Overall Efficiency of Grid Connected Photovoltaic Inverters (Amendment A1:2013)*; CENELEC: Brussels, Belgium, 2013.
6. *IEC 61724-1*; IEC 61724-1: Photovoltaic System Performance—Part 1: Monitoring. International Electrotechnical Commission (IEC): Geneva, Switzerland, 2021.
7. *IEC TS 61724-3*; IEC TS 61724-3: Photovoltaic System Performance—Part 3: Energy Evaluation Method. International Electrotechnical Commission (IEC): Geneva, Switzerland, 2016.
8. Underwriters Laboratories. *UL 1741: Inverters, Converters, Controllers and Interconnection System Equipment for Use with Distributed Energy Resources (incl. Supplement SA)*; UL: Northbrook, IL, USA, 2016.
9. Liserre, M.; Blaabjerg, F.; Hansen, S. Design and control of an LCL-filter-based three-phase active rectifier. *IEEE Trans. Ind. Appl.* **2005**, *41*, 1281–1291. [CrossRef]
10. Peña-Alzola, R.; Liserre, M.; Blaabjerg, F.; Ordóñez, M.; Yang, Y. LCL-filter design for robust active damping in grid-connected converters. *IEEE Trans. Ind. Inform.* **2014**, *10*, 2192–2203. [CrossRef]
11. Teodorescu, R.; Blaabjerg, F.; Liserre, M.; Loh, P.C. Proportional-resonant controllers and filters for grid-connected voltage-source converters. *IEE Proc. Electr. Power Appl.* **2006**, *153*, 750–762. [CrossRef]
12. Ciobotaru, M.; Teodorescu, R.; Blaabjerg, F. A new single-phase PLL structure based on second order generalized integrator. In Proceedings of the 37th IEEE Power Electronics Specialists Conference (PESC 2006), Jeju, Republic of Korea, 18–22 June 2006. [CrossRef]
13. Rodríguez, P.; Luna, A.; Candela, I.; Teodorescu, R.; Blaabjerg, F. Grid synchronization of power converters using multiple second order generalized integrators (MSOGI). In Proceedings of the 34th Annual Conference of the IEEE Industrial Electronics Society (IECON 2008), Orlando, FL, USA, 10–13 November 2008; pp. 755–760. [CrossRef]
14. Golestan, S.; Monfared, M.; Freijedo, F.D.; Guerrero, J.M. Dynamics assessment of advanced single-phase PLL structures. *IEEE Trans. Ind. Electron.* **2013**, *60*, 2167–2177. [CrossRef]
15. King, D.; Gonzalez, S.; Galbraith, G.; Boyson, W. *Performance Model for Grid-Connected Photovoltaic Inverters (SAND2007-5036)*; Sandia National Laboratories: Albuquerque, Mexico, 2007. [CrossRef]
16. Stein, J.S. *Models Used to Assess the Performance of Photovoltaic Systems (SAND2009-8258)*; Sandia National Laboratories: Albuquerque, Mexico, 2009. [CrossRef]
17. PV Performance Modeling Collaborative (PVPMC). Sandia Inverter Model—Modeling Guide. n.d. Available online: https://pvpmc.sandia.gov/modeling-guide/dc-to-ac-conversion/sandia-inverter-model/ (accessed on 29 August 2025).

18. PV Performance Modeling Collaborative (PVPMC). Performance Ratio. n.d. Available online: https://pvpmc.sandia.gov/modeling-guide/5-ac-system-output/pv-performance-metrics/performance-ratio/ (accessed on 29 August 2025).
19. PV Performance Modeling Collaborative (PVPMC). Performance Index. n.d. Available online: https://pvpmc.sandia.gov/modeling-guide/5-ac-system-output/pv-performance-metrics/performance-index/ (accessed on 29 August 2025).
20. Esram, T.; Chapman, P.L. Comparison of photovoltaic array maximum power point tracking techniques. *IEEE Trans. Energy Convers.* **2007**, *22*, 439–449. [CrossRef]
21. Ahmed, J.; Salam, Z. A critical evaluation on maximum power point tracking methods for partial shading in PV systems. *Renew. Sustain. Energy Rev.* **2015**, *47*, 933–953. [CrossRef]
22. Kamarzaman, N.A.; Tan, C.W. A comprehensive review of maximum power point tracking algorithms for photovoltaic systems. *Renew. Sustain. Energy Rev.* **2014**, *37*, 585–598. [CrossRef]
23. Hadji, S.; Gaubert, J.-P.; Krim, F. Real-time genetic algorithms-based MPPT: Study and comparison (theoretical and experimental) with conventional methods. *Energies* **2018**, *11*, 459. [CrossRef]
24. Marcos, J.; Marroyo, L.; Lorenzo, E.; Alvira, D.; Izco, E. Power output fluctuations in large-scale PV plants: One-year observations with one-second resolution and a derived analytic model. *Prog. Photovolt. Res. Appl.* **2011**, *19*, 218–227. [CrossRef]
25. Marcos, J.; Storkël, O.; Marroyo, L.; Garcia, M.; Lorenzo, E. Storage requirements for PV power ramp-rate control. *Sol. Energy* **2014**, *99*, 28–35. [CrossRef]
26. Huld, T.; Müller, R.; Gambardella, A. A new solar radiation database for estimating PV performance in Europe and Africa. *Sol. Energy* **2012**, *86*, 1803–1815. [CrossRef]
27. Gracia Amillo, A.G.; Huld, T.; Müller, R. A new database of global and direct solar radiation using the Eastern Meteosat satellite: Models and validation. *Remote Sens.* **2014**, *6*, 8165–8189. [CrossRef]
28. European Commission, Joint Research Centre (JRC). PVGIS—Photovoltaic Geographical Information System. n.d. Available online: https://re.jrc.ec.europa.eu/pvg_tools/en/ (accessed on 29 August 2025).
29. Huld, T.; Gracia Amillo, A.M. Estimating PV module performance over large geographical regions: The role of irradiance, air temperature, wind speed and solar spectrum. *Energies* **2015**, *8*, 5159–5181. [CrossRef]
30. National Aeronautics and Space Administration (NASA). POWER—Prediction of Worldwide Energy Resources. n.d. Available online: https://power.larc.nasa.gov/ (accessed on 29 August 2025).
31. Sparks, A.H. nasapower: A NASA POWER global meteorology, surface solar energy and climatology data client for R. *J. Open Source Softw.* **2018**, *3*, 1035. [CrossRef]
32. Copernicus Climate Change Service (C3S)/ECMWF. Pan-European Climate Database (PECD) v4.x—Climate and Energy Variables from Reanalysis and Projections [Documentation]. 2025. Available online: https://cds.climate.copernicus.eu/datasets/sis-energy-pecd (accessed on 29 August 2025).
33. ENTSO-E. European Resource Adequacy Assessment (ERAA) 2023—Annex 2: Methodology (v2). ENTSO-E. 2024. Available online: https://www.entsoe.eu/eraa/2023/report/ERAA_2023_Annex_2_Methodology.pdf (accessed on 29 August 2025).
34. Marsh, K. New PECD Version 4.2 Co-Developed with ENTSO-E now Available on the CDS. C3S Announcements. 2025. Available online: https://forum.ecmwf.int/t/new-pecd-version-4-2-co-developed-with-entso-e-now-available-on-the-cds/13497 (accessed on 29 August 2025).
35. IEA PVPS Task 13. *Guidelines for Operation and Maintenance of Photovoltaic Power Plants in Different Climates (Report T13-25:2022)*; International Energy Agency PVPS: Paris, France, 2022.
36. IEA PVPS Task 13. *Uncertainties in Yield Assessments and PV LCOE (Report T13-18:2020)*; International Energy Agency PVPS: Paris, France, 2020.
37. IEA PVPS Task 13. *Performance of New Photovoltaic System Designs*; International Energy Agency PVPS: Paris, France, 2021.
38. IEA PVPS Task 13. *Assessment of Photovoltaic Module Failures in the Field (Report T13-09:2017)*; International Energy Agency PVPS: Paris, France, 2017.
39. IEA PVPS Task 13. *Degradation and Failure Modes in New Photovoltaic Cell and Module Technologies (Report T13-30:2025)*; International Energy Agency PVPS: Paris, France, 2025.
40. Todorov, G.; Kralov, I.; Koprev, I.; Vasilev, H.; Naydenova, I. Coal Share Reduction Options for Power Generation during the Energy Transition: A Bulgarian Perspective. *Energies* **2024**, *17*, 929. [CrossRef]
41. Hinov, N. Smart Energy Systems Based on Next-Generation Power Electronic Devices. *Technologies* **2024**, *12*, 78. [CrossRef]

energies

Article

Solid-State Transformers in the Global Clean Energy Transition: Decarbonization Impact and Lifecycle Performance

Nikolay Hinov [1,2]

[1] CoE "National Center of Mechatronics and Clean Technologies", 1000 Sofia, Bulgaria; hinov@tu-sofia.bg
[2] Department of Computer Systems, Faculty of Computer Systems and Technologies, Technical University of Sofia, 1000 Sofia, Bulgaria

Abstract

The global clean energy transition requires power conversion technologies that combine high efficiency, operational flexibility, and reduced environmental impact over their entire service life. Solid-state transformers (SSTs) have emerged as a promising alternative to conventional line-frequency transformers, offering bidirectional power flow, high-frequency isolation, and advanced control capabilities that support renewable integration and electrified infrastructures. This paper presents a comparative life cycle assessment (LCA) of conventional transformers and SSTs across representative power-system applications, including residential and industrial distribution networks, electric vehicle fast-charging infrastructure, and transmission–distribution interface substations. The analysis follows a cradle-to-grave approach and is based on literature-derived LCA data, manufacturer specifications, and harmonized engineering assumptions applied consistently across all case studies. The results show that, under identical assumptions, SST-based solutions are associated with indicative lifecycle CO_2 emission reductions of approximately 10–30% compared to conventional transformers, depending on power rating and operating profile ($\approx$90–1000 t CO_2 over 25 years across the four cases). These reductions are primarily driven by lower operational losses and reduced material intensity, while additional system-level benefits arise from enhanced controllability and compatibility with renewable-rich and hybrid AC/DC grids. The study also identifies key challenges that influence the sustainability performance of SSTs, including higher capital cost, thermal management requirements, and the long-term reliability of power-electronic components. Overall, the results indicate that SSTs represent a relevant enabling technology for future low-carbon power systems, while highlighting the importance of transparent assumptions and lifecycle-oriented evaluation when comparing emerging grid technologies.

Keywords: decarbonization; EV fast charging; life cycle assessment; load profile; uncertainty analysis; renewable integration; smart grids; solid-state transformer; techno-economic analysis; wide-bandgap semiconductors

1. Introduction

The need for flexible, efficient, and adaptive energy systems has never been more pressing. Traditional power transformers, which have been the backbone of the power grid for more than a century, have limitations in terms of size, weight, and lack of active control [1–3]. With the increasing integration of renewable energy sources, the rise of electric vehicles (EVs), and the shift toward decentralized power grids, a new generation of transformers is required to meet these demands. Solid-state transformers (SSTs) represent the next frontier in power conversion, offering advantages over conventional transformers by integrating power electronics, advanced semiconductor materials, and digital control systems [4–6]. In particular, conventional line-frequency transformers remain bulky due to 50/60 Hz magnetics, rely on mechanical tap changers for voltage regulation with slow dynamics, cannot directly interface DC sources/loads, and provide limited power-quality and fast control services compared to power-electronic interfaces [1–3,7].

Contemporary electricity infrastructures also operate within increasingly diverse energy-system architectures—from centralized generation and long-distance transmission to distributed renewables, hybrid AC/DC microgrids, and sector-coupled systems (e.g., electrified transport and integrated generation–industrial/water processes). This evolution increases the value of controllable and efficient power interfaces at the grid edge, motivating interest in SST-based transformer concepts as part of flexible distribution networks [8,9]. In this manuscript, "AC/DC integration" denotes hybrid architectures where AC and DC subsystems coexist and are coupled via conversion interfaces; SSTs are evaluated as one such interface enabling galvanic isolation and controllable power flow between domains [4–7].

SSTs are unique in that they combine power electronic converters with high-frequency transformers, providing not only electrical isolation but also the ability to control the power flow, voltage levels, and power quality in real-time. This enables SSTs to be far more adaptable and versatile than their traditional counterparts. For instance, SSTs can regulate voltage on both the AC and DC sides, making them suitable for a range of applications, from renewable energy integration to fast charging of electric vehicles (EVs) and advanced grid control [10,11].

Despite the extensive technical literature on SST converter topologies, control strategies, and pilot projects, comparatively limited work has evaluated SSTs from a harmonized cradle-to-grave life cycle assessment (LCA) perspective across multiple grid applications. Existing studies typically focus on conventional transformers or on single SST demonstrators and often rely on inconsistent system boundaries, electricity-mix assumptions, and operating profiles. The present work addresses this gap by applying a consistent, literature-based LCA framework to four representative transformer classes and by reporting stage-level CO_2 contributions and scenario sensitivities.

Key technological challenges include the cost and reliability of medium-voltage power-electronic devices, insulation and packaging at MV levels, protection and fault management in hybrid AC/DC systems, thermal management, electromagnetic compatibility, and the absence of fully harmonized standards and long-term field data [12–14].

Although the broader discussion is framed globally, the quantitative LCA uses a single baseline grid-emission factor representative of Europe and typical European voltage levels; therefore, absolute CO_2 values should be interpreted as indicative and may vary by region and over future grid-decarbonization pathways.

In recent years, increasing attention has been paid to the environmental footprint of power-system components, including transformers, as part of broader decarbonization strategies for electricity networks. Several studies have applied life cycle assessment (LCA) methodologies to conventional power transformers, typically identifying operational losses and material intensity as key drivers of lifecycle greenhouse-gas emissions. In parallel, the SST literature has rapidly expanded in terms of converter topologies, control, and demonstrator projects. Nevertheless, the intersection of these two threads remains underdeveloped: comparatively few studies provide a transparent, harmonized cradle-to-grave comparison of conventional transformers and SSTs across multiple application contexts under consistent assumptions. The present work addresses this gap by applying

https://doi.org/10.3390/en19020558

a unified CO_2-focused LCA framework to four representative transformer classes and by explicitly discussing key assumptions and their influence on the comparative results.

The main contributions of this paper can be summarized as follows:

- A structured, comparative life cycle assessment framework is developed to evaluate conventional transformers and SSTs under harmonized assumptions and a cradle-to-grave system boundary.
- Four representative transformer classes are analyzed, covering residential distribution networks, industrial applications, electric vehicle fast-charging infrastructure, and transmission–distribution interface substations, thereby spanning a wide range of power ratings and operating contexts.
- Literature-derived data and engineering-based loss models are consistently applied to quantify manufacturing, operational, and end-of-life CO_2 emissions, enabling a transparent comparison between transformer technologies.
- The study links technical characteristics of SSTs—such as power-electronic conversion stages and bidirectional operation—to their lifecycle performance and system-level relevance in the clean energy transition, while explicitly discussing uncertainties and limitations.
- In summary, the novelty of this work is a harmonized, cradle-to-grave, CO_2-focused LCA comparison between conventional transformers and SSTs across four distinct deployment contexts, complemented by a concise scenario-based sensitivity assessment (including grid-emission-factor decarbonization) to clarify how robust the comparative advantage remains as electricity mixes evolve.

2. Overview of Solid-State Transformer Applications and Topologies

This section provides a structured overview of solid-state transformer (SST) applications, architectures, and enabling technologies, with the aim of establishing the technical and system-level context for the lifecycle assessment presented in subsequent sections. Rather than offering an exhaustive review of all existing SST topologies, the focus is placed on identifying the functional characteristics and deployment scenarios that are most relevant to decarbonization, grid flexibility, and lifecycle performance [15–17].

While several comprehensive reviews have already addressed SST converter topologies, control strategies, and component-level design aspects, the present overview emphasizes how these technical features translate into system-level roles within modern power networks [18–20]. In this sense, the review serves as a contextual bridge between established SST research and the comparative lifecycle assessment framework adopted in this study.

2.1. Review of Basic SST Applications

This section provides a concise synthesis of SST applications and representative architectures, with emphasis on aspects that materially influence lifecycle assessment (dominant bill-of-materials elements and loss mechanisms). Detailed converter design is covered in existing SST reviews; therefore, only the deployment contexts and architecture families most relevant to the LCA are highlighted here [21–23].

Across the literature, SST deployment is most frequently discussed in: (i) smart distribution transformers providing ancillary services [24–27], (ii) interfaces for renewable-rich and hybrid AC/DC feeders [28–30], (iii) MV-to-DC conversion for EV ultra-fast charging hubs [31–33], (iv) traction/rail applications, and (v) industrial and commercial power-quality conditioning [34–36]. Reported implementations span a broad range of architectures; however, two families dominate grid-connected designs: isolated multi-stage SSTs (active rectifier, isolated DC-DC stage with a high-frequency

https://doi.org/10.3390/en19020558

transformer—e.g., DAB/PSFB—followed by an inverter/rectifier stage) and modular multilevel (MMC-based) solutions for medium-voltage and multi-MVA applications. In [10], the term "three-level" refers to the power-converter topology (three-level power module) rather than three distinct network voltage tiers; the architecture interconnects MVDC and LVDC systems via a controlled DC link.

In particular, ref. [14] surveys single-stage AC/AC matrix-type SSTs, two-stage and three-stage AC/DC/isolated-DC/DC(/DC/AC) architectures (including DAB-based links), and MMC-based implementations, highlighting trade-offs between modularity/scalability, efficiency, controllability, and component count.

For the comparative LCA in Sections 3–6, these two architecture families are used as representative SST baselines because they capture the dominant material inventory (semiconductor modules, high-frequency transformer core and windings, passive filters, and cooling hardware) and the dominant operational loss mechanisms (switching and conduction losses). This choice is not intended to rank all SST topologies; rather, it provides a transparent and reproducible baseline for screening-level lifecycle comparisons across applications.

1. Smart grids: SSTs can enable fast, electronic voltage regulation and controlled power flow, supporting services such as power-factor control, harmonic mitigation, and coordinated operation with distributed energy resources.
2. Renewable energy integration: SSTs can interface inverter-based resources and storage systems more flexibly than passive transformers, including hybrid AC/DC interfacing, bidirectional operation, and improved power-quality control under variable generation.
3. EV charging infrastructure: SSTs can provide compact MV-to-DC conversion with galvanic isolation and controllable charging power, and can facilitate integration of local storage or vehicle-to-grid operation where applicable.
4. Railway and traction systems: SSTs are investigated for applications where reduced size/weight and fast voltage control are valuable under dynamic loading, while meeting stringent power-quality and reliability requirements.
5. Industrial and commercial networks: SSTs can support power-quality conditioning (e.g., voltage regulation and harmonic control) and facilitate hybrid AC/DC distribution where sensitive loads and local DER are present.

Figure 1 presents a comparison of the functional characteristics of SSTs according to their application areas.

Figure 2 presents a comparison of key technical requirements with application areas.

Figures 1 and 2 highlight that, beyond basic galvanic isolation and voltage transformation, application-driven requirements for SSTs include fast bidirectional power flow control, DC interfacing, power-quality conditioning, and monitoring. Conventional line-frequency transformers provide robust isolation and high efficiency but cannot natively deliver these power-electronic functions without additional converters and control layers [15–17].

Overall, the reviewed technical literature indicates that SSTs can provide additional controllability and functionality beyond conventional transformers in several application domains. Nevertheless, most publications focus on converter performance and grid services, while comparatively fewer studies quantify environmental performance under harmonized lifecycle assumptions. This motivates the comparative LCA framework introduced in Section 3.

https://doi.org/10.3390/en19020558

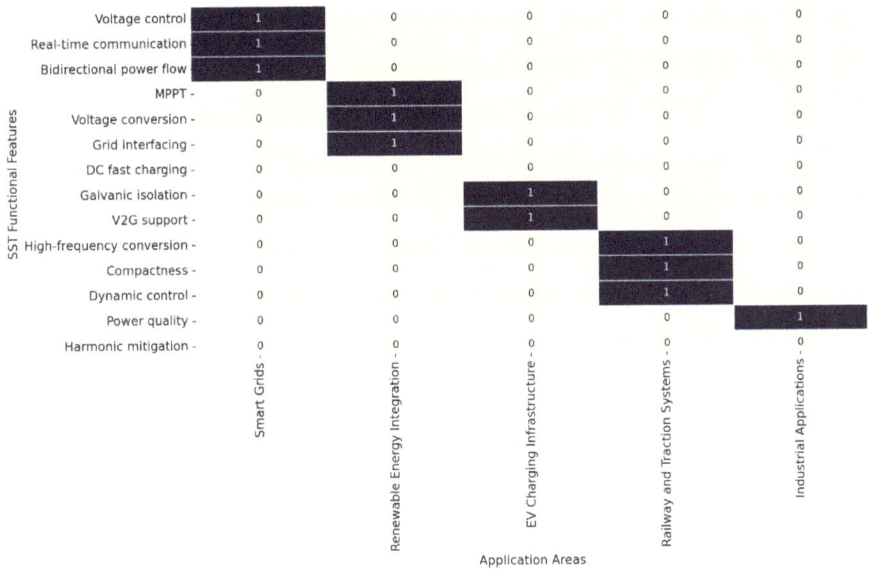

Figure 1. Mapping of SST Functional Features to Application Areas.

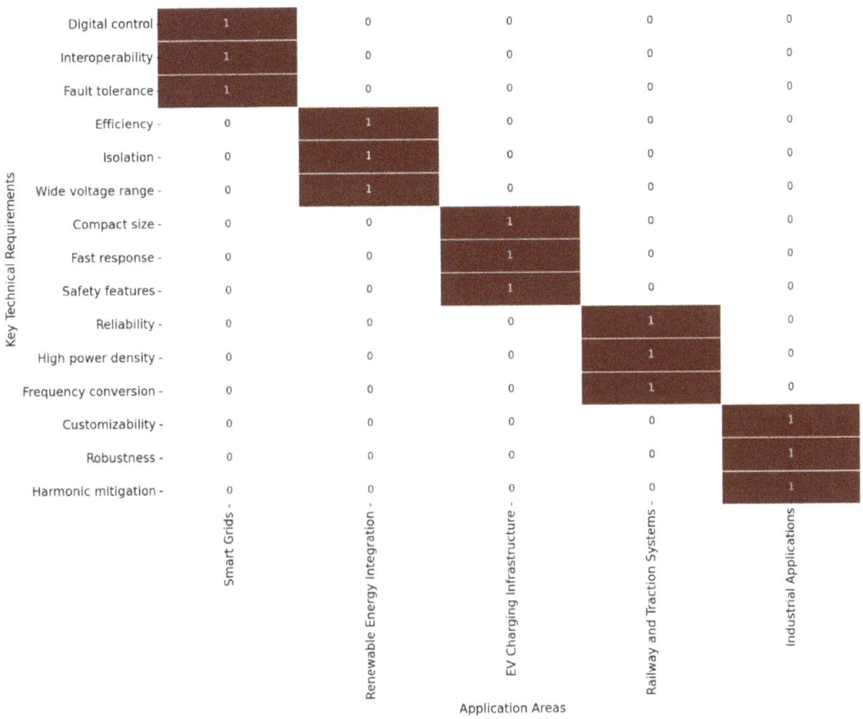

Figure 2. Mapping of Key Technical Requirements to Application Areas.

2.2. Fundamental Architecture of Solid-State Transformers

Solid-state transformers are composed of several stages that work together to achieve voltage regulation, power conditioning, and system integration. The fundamental stages

https://doi.org/10.3390/en19020558

of an SST include [32,33]: In this review, we focus on the prevalent three-stage architecture (AC/DC, isolated DC/DC, DC/AC); single-stage and two-stage variants also exist (Figure 3) but can be interpreted as alternative realizations or integrations of the same functional blocks [14,32,33].

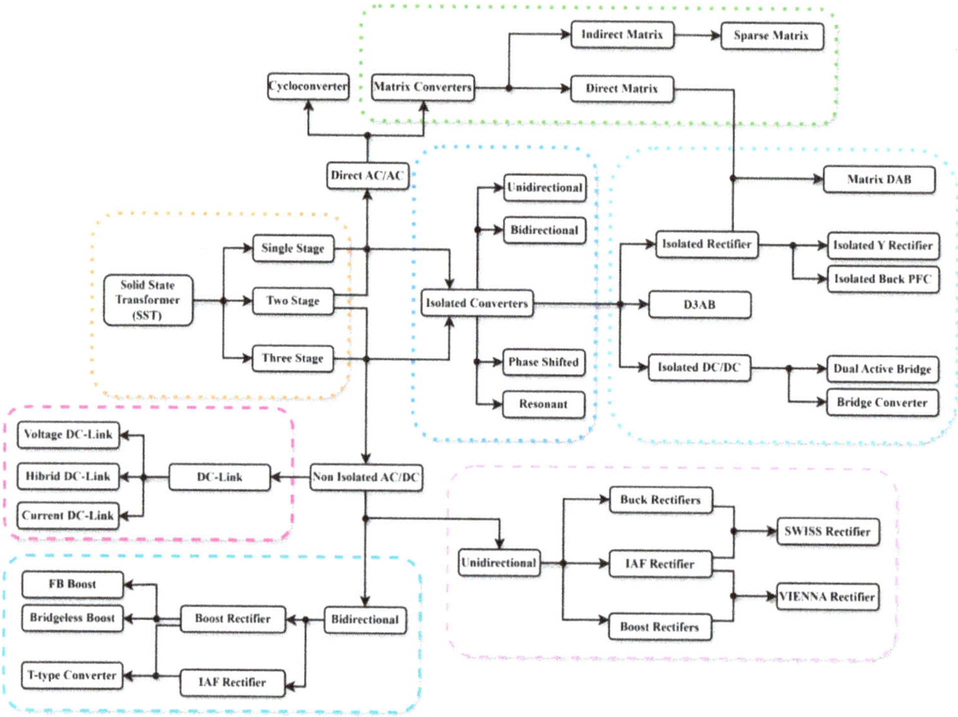

Figure 3. Classification of Solid-State Transformer Topologies and Converter Architectures [32].

- The first stage of a solid-state transformer is the AC–DC conversion, typically implemented using a power factor correction (PFC) rectifier. This stage is responsible for converting the alternating current (AC) from the electrical grid into a regulated direct current (DC) output. In addition to voltage conversion, the rectifier ensures that the input current waveform closely follows the voltage waveform, thereby achieving a near-unity power factor. Advanced PFC techniques—such as interleaved boost converters or bridgeless topologies—are commonly employed to minimize total harmonic distortion (THD), improve efficiency, and comply with grid codes. This stage also lays the foundation for high-frequency isolation and voltage regulation in subsequent SST modules, making it essential for the overall performance and electromagnetic compatibility (EMC) of the transformer system.
- The second stage of an SST utilizes a high-frequency transformer in conjunction with a DC–DC converter to achieve galvanic isolation between the input and output stages. Unlike traditional 50–60 Hz magnetic transformers, high-frequency transformers operate in the range of tens to hundreds of kilohertz, enabling a significant reduction in size, weight, and core material losses. This stage not only provides isolation but also enables flexible voltage conversion—stepping the DC voltage up or down depending on the system requirements. Modern topologies, such as dual-active-bridge (DAB) converters and phase-shifted full-bridge (PSFB) converters, are commonly employed

https://doi.org/10.3390/en19020558

to achieve high efficiency and fast dynamic response. Importantly, the DC–DC stage in advanced SST designs supports bidirectional power flow, allowing energy to be transmitted both from the grid to the load and vice versa. This capability is crucial in applications involving distributed generation, battery energy storage systems (BESS), and vehicle-to-grid (V2G) technologies, where energy must be exchanged dynamically and efficiently in both directions. This reduction is primarily enabled by replacing 50/60 Hz magnetics with a high-frequency transformer stage, as discussed in [7,11,14].

- The final stage of a solid-state transformer is the DC–AC conversion, typically realized through a high-efficiency inverter. This stage converts the regulated DC output from the previous stage into a high-quality alternating current (AC) signal that complies with grid standards or local load requirements. The inverter is responsible for generating a sinusoidal output waveform with minimal harmonic distortion and precise frequency and voltage control. It typically incorporates advanced modulation techniques, such as pulse-width modulation (PWM) or space vector modulation (SVM), to optimize switching behavior and efficiency. Additionally, the inverter supports power factor control, reactive power compensation, and may feature grid-synchronization capabilities—making it suitable for both grid-following and grid-forming applications. In bidirectional SST designs, this stage also enables reverse power flow, facilitating energy export from local sources (e.g., batteries or renewables) back to the grid. Common topologies used in this stage include three-phase voltage source inverters (VSI), modular multilevel converters (MMC), and cascaded H-bridge inverters, each offering specific advantages depending on application, power level, and control complexity.

Each of these stages is critical to the overall performance of the SST. Advanced control strategies are employed to coordinate their operation, ensuring high efficiency, dynamic stability, and compliance with grid requirements.

2.3. Classification of SST Topologies

The design and architectural configuration of solid-state transformers (SSTs) can vary significantly depending on their intended application, power level, and control requirements. Several topologies have been developed, each offering specific advantages and trade-offs in terms of performance, complexity, and scalability. The most prominent SST classifications include the following [32,33]:

- Isolated SSTs: These topologies incorporate galvanic isolation between the AC input and DC output, typically via a high-frequency transformer. Such isolation is essential in medium- and high-voltage grid-connected applications, as it prevents fault propagation between the grid and the load, enhances safety, and enables flexible voltage control. Isolated SSTs also support bidirectional power flow, making them suitable for use in smart grids, renewable integration, and sensitive industrial systems where power quality and regulatory compliance are critical [34,35];
- Non-Isolated SSTs: In contrast, non-isolated SSTs omit transformer-based isolation and instead rely on magnetic coupling elements such as coupled inductors or direct high-frequency conversion. While this approach reduces size, cost, and system complexity, it is generally limited to low-voltage or tightly controlled environments, such as DC microgrids or electric vehicle (EV) charging stations. The lack of galvanic separation may impose limitations in terms of safety and grid compatibility [36,37];
- Single-Stage SSTs: These topologies consolidate rectification, isolation, and inversion into a single power conversion stage. The main advantages include reduced component count, simplified control, and compact design. However, single-stage SSTs may suffer from limited efficiency, restricted dynamic control, and lower fault

https://doi.org/10.3390/en19020558

tolerance, which can constrain their applicability in high-performance or high-power systems [38,39];
- Multi-Stage SSTs: Multi-stage architectures decompose the SST into discrete functional blocks—typically comprising an AC–DC rectifier, a high-frequency DC–DC isolation stage, and a DC–AC inverter. This modular structure allows for individual optimization and precise control of each stage, resulting in improved efficiency, enhanced fault management, and scalability. Although more complex and component-intensive, multi-stage SSTs are preferred in demanding applications such as utility grids, large renewable installations, and high-capacity data centers [40–42];
- Modular Multilevel Converters (MMCs): MMC-based SSTs represent an advanced topology for medium- and high-voltage applications. MMCs consist of multiple submodules arranged in series and/or parallel to form multilevel voltage waveforms, which significantly reduce harmonic distortion and improve output quality. These converters offer exceptional modularity, fault tolerance, and voltage scalability. Their use in SSTs is gaining traction in large-scale power transmission and distribution networks, where efficiency, redundancy, and grid compliance are paramount [43,44].

Figure 3 presents a comprehensive classification of Solid-State Transformer (SST) architectures based on the number of conversion stages, isolation type, directionality, and integration with different converter topologies. SST configurations can be broadly divided into single-stage, two-stage, and three-stage structures. These architectures can incorporate either isolated or non-isolated converters, depending on system requirements such as galvanic isolation, safety, and efficiency.

Furthermore, the diagram details how SSTs interface with AC/AC, AC/DC, and DC/DC conversion blocks, employing advanced structures like Dual Active Bridge (DAB), matrix converters, and various types of rectifiers (e.g., Vienna, Swiss, IAF). The flexibility in design enables SSTs to meet the diverse demands of applications ranging from EV fast charging and renewable integration to smart grid and railway systems. This classification provides a structural overview of SST design options and helps guide selection based on functional requirements and operating conditions [45,46].

In summary, the reviewed literature demonstrates that SSTs are a mature and rapidly evolving technology from a power-electronics perspective, with well-established converter topologies and control approaches. However, comparatively fewer studies address SST deployment from a holistic lifecycle and decarbonization perspective across multiple application contexts. This observation motivates the structured life cycle assessment framework introduced in the following section, which aims to complement existing technical reviews by focusing on environmental performance and system-level implications [47–49].

3. Methodology and Life Cycle Assessment Framework

This section presents the methodological framework used to evaluate the lifecycle environmental performance of conventional transformers and solid-state transformers (SSTs). A structured life cycle assessment (LCA) approach is adopted to ensure transparency, comparability, and consistency across the analyzed cases. The methodology defines the goal and scope of the study, system boundaries, data sources, harmonized assumptions, and the treatment of uncertainty, forming the basis for the quantitative results presented in Section 5.

A comparative, screening-level LCA approach is selected because it enables a transparent cradle-to-grave comparison of two transformer technologies that differ strongly in material composition and in-use energy losses. Alternative approaches (e.g., full multi-impact LCAs with detailed process inventories, or site-specific techno-economic simulations) would require proprietary manufacturing inventories and long-term field data that are currently sparse for emerging multi-MVA SST designs. Therefore, this study applies a literature-grounded, CO_2-focused framework to capture the dominant climate-relevant contributions while maintaining consistent assumptions across all case studies. The results are intended to be indicative and comparative rather than predictive [49,50].

3.1. Goal and Scope of the Study

The objective of this study is to perform a comparative life cycle assessment (LCA) of conventional line-frequency power transformers and solid-state transformers (SSTs) when deployed in representative applications of the clean energy transition. The analysis aims to quantify and compare lifecycle CO_2-equivalent emissions associated with both technologies, while maintaining harmonized assumptions across all evaluated cases.

The study does not present new experimental measurements or prototype testing results. Instead, it provides a structured, literature-based assessment that integrates data from peer-reviewed LCA studies, manufacturer specifications, and engineering loss models. The focus is on identifying indicative trends and relative differences between transformer technologies rather than on predicting site-specific absolute emissions.

The scope of the assessment encompasses four representative transformer classes corresponding to typical deployment contexts: residential distribution networks (630 kVA), industrial distribution (1 MVA), electric vehicle fast-charging infrastructure (500 kVA), and transmission–distribution interface substations (10 MVA). These cases are selected to reflect a broad range of power ratings, load profiles, and operational roles within modern power systems.

3.2. Functional Unit

The functional unit adopted in this study is defined as: One transformer delivering its rated apparent power over a 25-year service life under representative operating conditions. This definition refers to rated capacity and does not imply continuous full-load operation; operational losses are evaluated using the representative loading assumptions and loss model (Section 3.4).

This functional unit enables a consistent comparison between conventional transformers and SSTs of equivalent nominal ratings by normalizing lifecycle emissions to the same power delivery function and service duration.

3.3. System Boundaries

The system boundaries follow a cradle-to-grave approach and include three lifecycle stages:

- Manufacturing stage, encompassing raw material extraction, processing of metals and electronic components, assembly, and factory-level production processes;
- Operational stage, covering no-load and load-dependent losses incurred during continuous operation over the assumed service life;
- End-of-life stage, including decommissioning, transport, recycling, and disposal processes.

Infrastructure external to the transformer itself (e.g., upstream grid reinforcement, building construction, or auxiliary civil works) is excluded from the system boundary. Maintenance activities are considered implicitly through operational efficiency and lifetime assumptions rather than through detailed maintenance inventories.

3.4. Data Sources and Nature of Input Data

The LCA inventory is assembled from (i) peer-reviewed LCA and transformer-manufacturing studies for material- and process-related emissions, (ii) manufacturer datasheets and standards-aligned loss parameters (no-load and load losses) for each trans-

https://doi.org/10.3390/en19020558

former class, and (iii) literature reports on SST prototypes to define a representative three-stage SST baseline (converter efficiency, bill-of-material trends, and semiconductor class). For transparency, all numerical inputs used in the lifecycle calculations are fully traceable to explicitly cited literature sources and manufacturer specifications, with the tables serving only as a concise summary of these data.

All lifecycle data used in this study are derived from secondary sources, including:

- peer-reviewed life cycle assessment studies of power transformers and related grid equipment;
- published manufacturer datasheets and technical reports;
- established engineering models for transformer losses and efficiency;
- publicly available emission factors reported in the literature.

No proprietary or confidential manufacturer data are used. Manufacturing-stage emissions are estimated from reported material compositions (steel, copper, aluminium, and power-electronic components) combined with literature emission intensities. Operational losses are calculated using the standard two-term transformer loss model, $P\_loss(LF) = P0 + Pk \cdot LF^2$, where P0 is the no-load loss (kW), Pk is the load loss at rated power (kW), and LF is the representative average load factor (0–1). The corresponding annual energy loss is $E\_loss = P\_loss \cdot 8760$ (kWh/year), and operational CO_2 emissions are obtained as $CO_2\_op = E\_loss \cdot EF$, where EF is the grid emission factor (kg CO_2/kWh). End-of-life impacts are estimated using literature-based recycling and disposal scenarios for metallic and electronic components.

Consequently, all numerical values presented in Section 5 should be interpreted as literature-based estimates rather than as measured field data.

3.5. Harmonized Assumptions

To ensure comparability across transformer classes and technologies, a common set of assumptions is applied to all four case studies. These assumptions are summarized in Table 1.

In addition to the baseline electricity grid emission factor used in the main LCA results, alternative grid CO_2 intensity scenarios are defined to explore the sensitivity of lifecycle emissions to electricity-mix assumptions. Table 1 summarizes the electricity grid CO_2 intensity values adopted in the sensitivity and decarbonization analysis. These scenarios capture the dominant source of uncertainty in operational-stage emissions and are applied consistently across all case studies.

Representative SST technology baseline: In the LCA calculations, the SST is represented by a three-stage, isolated architecture (active AC–DC front-end, isolated DC–DC stage with a dual-active-bridge and a medium-frequency transformer, and a DC–AC inverter), which is widely reported in the SST literature and demonstrators [32,33]. The loss model uses aggregated efficiency values derived from typical stage efficiencies for this architecture; unless stated otherwise, the baseline assumes silicon IGBT-based power modules at medium voltage. Alternative device choices (SiC/GaN) and topologies (e.g., MMC-based SSTs) can shift both material intensity and efficiency and are therefore treated as a limitation and a topic for future work.

The same assumptions are intentionally applied to all transformer classes and both technologies to enable a controlled comparison. Their validity and influence are discussed in Section 3.6 and explored quantitatively through the grid-emission-factor scenarios (see Table 1) and the scenario sensitivity summary reported later in the manuscript; this highlights which conclusions are robust and which are sensitive to decarbonization and other uncertain parameters.

https://doi.org/10.3390/en19020558

Table 1. (a) Harmonized LCA assumptions applied to all case studies [36–40,42,44]. (b) Electricity grid CO_2 intensity scenarios used for sensitivity and decarbonization analysis.

(a)		
Parameter	**Value**	**Justification**
Service lifetime	25 years	Typical design lifetime for distribution and substation transformers
Grid emission factor	0.4 kg CO_2/kWh	Representative mixed electricity generation factor reported in European LCA studies
Annual operating hours	8760 h	Continuous grid-connected operation
Ambient conditions	Standard temperate climate	Consistent with typical LCA assumptions
Load profile	Representative average loading	Used to compare technologies under identical conditions
(b)		
Scenario	**Grid CO_2 intensity**	**Description/notes**
Baseline (used in main results)	0.40 kg CO_2/kWh	Representative mixed electricity mix (assumed constant over lifetime).
Low-carbon sensitivity	0.24 kg CO_2/kWh	Lower-carbon electricity mix (sensitivity case).
High-carbon sensitivity	0.49 kg CO_2/kWh	Higher-carbon electricity mix (sensitivity case).
Decarbonization pathway	0.40 → 0.10 kg CO_2/kWh (avg ≈ 0.25)	Illustrative linear decline over 25 years to reflect grid decarbonization.

3.6. Treatment of Uncertainty and Limitations

Uncertainty is inherent in literature-based life cycle assessments due to variability in material inventories, manufacturing processes, operational conditions, and electricity-mix assumptions. In this study, uncertainty is addressed through explicit dispersion ranges and sensitivity scenarios rather than through a formal probabilistic LCA, which would require proprietary manufacturing data and long-term field measurements that are currently unavailable for multi-MVA SSTs.

Manufacturing and end-of-life CO_2 emissions are assumed to vary within ±15–25%, reflecting the spread reported in published transformer LCAs and recycling studies (e.g., Piotrowski and Markowska [42]; Wang et al. [44]). Operational-stage emissions are dominated by uncertainty in electricity grid CO_2 intensity and average loading conditions; this dominant source of variability is explored quantitatively through the defined grid-emission-factor scenarios and the associated sensitivity analysis presented later in the manuscript.

All transformer classes and both technologies are evaluated using identical assumptions for service life, operating hours, load factors, and system boundaries. This harmonized approach ensures internal consistency and enables meaningful comparison of results derived from heterogeneous literature sources. While absolute emission values should be interpreted as indicative, the direction and relative magnitude of the observed lifecycle CO_2 differences are robust across the considered uncertainty ranges.

Several limitations remain. A uniform 25-year service life is assumed for both technologies, although power-electronic components in SSTs may require mid-life replacement. To reflect this, a conservative screening scenario is discussed in Section 6, in which partial replacement of SST power-electronic modules is approximated as an additional manufacturing-stage burden. This approximation does not change the qualitative conclusions of the comparative assessment.

https://doi.org/10.3390/en19020558

3.7. Consistency Checks and Qualitative Validation

Because this study is based on secondary LCA data and engineering loss models, validation is performed through consistency checks rather than new laboratory measurements. First, the selected no-load and load-loss parameters for line-frequency transformers are cross-checked against typical manufacturer datasheets and widely adopted efficiency requirements to ensure that values are within expected ranges for the respective power ratings. Second, the assumed SST efficiencies are benchmarked against reported efficiencies of published SST prototypes and demonstrator projects in comparable power ranges. Third, the resulting lifecycle stage shares (manufacturing vs. operation) are compared qualitatively with prior transformer LCAs, which commonly identify operation as the dominant contributor under today's grid emission factors. These checks support the robustness of the order-of-magnitude comparison while the study remains explicitly indicative. For clarity, Section 4 provides short, literature-referenced application context for the selected transformer classes, and Section 5 then presents the quantified cradle-to-grave CO_2 results for each case study.

4. Technological Enablers and Use Cases in the Global Clean Energy Transition

SSTs are being actively explored for modern power systems because their power-electronic structure can provide additional controllability and functional integration compared with conventional transformers, which is relevant to renewable integration, electric-vehicle infrastructure, and microgrid operation.

Wide-bandgap semiconductors such as silicon carbide (SiC) and gallium nitride (GaN) offer higher efficiency at elevated switching frequencies, enabling SSTs to operate with minimal energy loss and reduced physical dimensions. Despite their higher cost, these materials pave the way for compact and thermally efficient designs that outperform traditional transformers in high-performance settings. Complementing these semiconductors, advanced magnetic materials like amorphous and nanocrystalline alloys reduce core losses at high frequencies, making high-density and lightweight designs feasible. Combined with efficient thermal management strategies, these materials support sustained high-power operation.

The integration of digital control systems, often based on DSP or FPGA platforms, allows for real-time energy management, fault detection, and dynamic grid interaction. These systems also facilitate smart grid compatibility, enabling SSTs to interact with Internet of Things (IoT) platforms and AI-based forecasting tools. As a result, SSTs are ideally positioned for renewable energy systems, where they provide direct DC interfacing with solar arrays and battery storage units. Their ability to regulate voltage and suppress harmonics in real time supports the stable integration of intermittent sources. Examples of DSP/FPGA-based SST control implementations and reviews are provided in [7,11].

In electric vehicle (EV) charging infrastructure, SSTs offer compact and efficient medium-to-low voltage conversion, supporting ultra-fast charging and vehicle-to-grid (V2G) operations. Their bidirectional power flow capability further allows them to contribute to grid balancing efforts.

In microgrid and urban distribution networks, SSTs ensure voltage stability, enable islanded operation, and allow seamless reconnection to the main grid. Their compact footprint and controllable behavior enhance resilience in remote or high-density environments.

This combined technological and application-oriented perspective underscores the central role SSTs are poised to play in next-generation energy systems. Table 2 compares the main characteristics of traditional transformers and SSTs, providing information on their technology, efficiency, dimensions, maintenance and applications. This comparison is intended to provide a better understanding of the advantages and limitations of both types

https://doi.org/10.3390/en19020558

of transformers and to assist in making informed decisions when selecting the appropriate solution for specific applications.

Table 2. Comparison between Traditional Transformers and Solid State Transformers [7,11,14,22–24].

Characteristic	Traditional Transformers	Solid State Transformers
Technology	Magnetic cores (metal laminations), mechanical principles	Semiconductor technologies (MOSFET, IGBT), electronic components
Energy Efficiency	95–97%	98% and above
Size and Weight	Large and heavy	Compact and light
Lifespan	Long lifespan but requires maintenance	Long lifespan, minimal maintenance
Maintenance	Requires regular maintenance and monitoring	Low maintenance requirement
Voltage Regulation	Limited, may require additional devices	Flexible and fast voltage regulation
Environmental Footprint	Uses oils for cooling, which can be toxic	No cooling liquids, less environmental impact
Cost	Cheaper to produce, but with high maintenance costs	More expensive to produce, but with lower operational costs
Applications	Electrical grids, industrial applications	Renewable sources, smart grids, electric vehicles

In summary, the comparison shows that SSTs offer many advantages over traditional transformers, such as higher energy efficiency, longer life, and a smaller environmental footprint. While SSTs can be more expensive to manufacture and are not always required in applications where conventional transformers are sufficient, they represent a highly promising technology for modern and sustainable energy grids.

Figure 4 presents a lifecycle CO_2 breakdown (manufacturing, operation, and end-of-life) for conventional line-frequency transformers (LFTs) and solid-state transformers (SSTs) across the four case studies under the baseline assumptions.

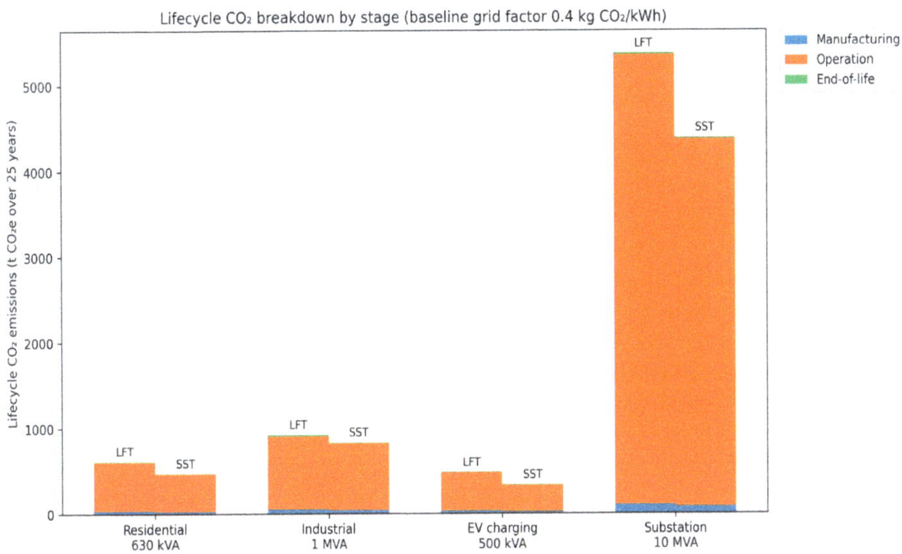

Figure 4. Lifecycle CO_2 breakdown (manufacturing, operation, end-of-life) for conventional transformers and solid-state transformers (SSTs) across the four case studies (baseline grid emission factor: 0.4 kg CO_2/kWh).

https://doi.org/10.3390/en19020558

The stacked bars highlight that operational losses dominate the total footprint in all cases; therefore, efficiency improvements drive most of the lifecycle CO_2 savings, while manufacturing and end-of-life contribute smaller shares. Values are based on the stage totals reported in Tables 3–6.

The following Sections 4.1–4.4 provide brief, literature-referenced examples of SST deployment contexts and clarify how they motivate the four LCA case studies defined in Section 3.1.

SSTs enable a set of capabilities that are highly aligned with the structural requirements of the global clean energy transition. Unlike conventional line-frequency transformers, SSTs combine high-frequency conversion, bidirectional operation, advanced control functionalities, multiport interfacing, and deep integration with digital monitoring platforms. These attributes unlock new operational modes in distribution grids, transportation systems, industrial networks, and transmission–distribution boundaries.

This section presents four representative use cases, each illustrating a concrete decarbonization challenge and showing how SST deployment transforms technical limitations into opportunities for sustainable electrification.

4.1. *Renewable-Rich Distribution Networks*

Distribution networks with large shares of photovoltaics and wind generation face a series of structural challenges: voltage fluctuations caused by intermittent production, reverse power flows toward the substation, limited hosting capacity due to thermal and stability constraints, and increased harmonic distortion from inverter-based resources. Traditional MV/LV transformers are passive components with no controllability; they cannot regulate voltage dynamically, manage bidirectional flows, or coordinate distributed energy resources (DERs). Representative discussions of SST-enabled distribution flexibility and hybrid AC/DC feeder operation can be found in [7,11,14].

SSTs fundamentally change this paradigm by combining high-frequency AC/DC and DC/AC stages with fast control loops capable of regulating voltage, frequency, power factor, and harmonic content in real time. Their millisecond-scale bandwidth allows them to follow renewable variability more effectively than tap-changers or capacitor banks. Moreover, their bidirectional power-flow capability enables controlled backfeeding of PV and battery storage upstream, improving the operational flexibility of feeders.

Several smart-grid and power-quality studies suggest that SST functionality (e.g., fast voltage support and harmonic mitigation) could improve the integration of variable renewables and enhance power quality in certain feeders. These system-level effects are beyond the direct quantitative scope of the present cradle-to-grave CO_2 LCA; therefore, they are mentioned only for context and are not included in the reported lifecycle emission results.

From a decarbonization perspective, this use case highlights a direct opportunity: higher renewable penetration without extensive copper upgrades, lower grid losses, and enhanced stability—all essential features of a high-renewable energy system.

4.2. *Smart Transportation and Fast-Charging Infrastructure*

The rapid expansion of electric mobility introduces severe challenges for distribution and sub-transmission grids. Fast-charging stations (150–350 kW per charger) tend to create clustered peak loads, voltage sags, harmonics from power-electronic converters, and unbalanced phases when installed asymmetrically. Traditional transformers lack the control authority to mitigate these disturbances, and infrastructure reinforcement becomes expensive. Representative traction and extreme-fast-charging interface studies are discussed in [18,19].

https://doi.org/10.3390/en19020558

SSTs offer a compact, high-performance solution for EV fast-charging hubs. Their high-frequency isolation and integrated AC/DC conversion enable direct medium-voltage to DC charging, eliminating multiple conversion stages and reducing system losses. Multiport SST architectures allow the integration of local energy storage, photovoltaics, and second-life batteries, smoothing the power demand seen by the grid.

Their bidirectional operation and precise control make advanced functionalities possible:

- V2G operation for peak shaving and grid support;
- buffered ultra-fast charging using local energy storage;
- real-time harmonic mitigation;
- adaptive current limiting during stressed network conditions.

These features can reduce grid congestion, enhance supply reliability, and minimize the carbon footprint of charging infrastructure by improving conversion efficiency and enabling local buffering (e.g., with on-site storage). The magnitude of such network-level benefits is highly site-specific and is therefore not quantified in this LCA-focused study.

Beyond charging efficiency, recent research on coupled electricity–transportation networks shows that coordinated restoration strategies leveraging building flexibility and electric-bus scheduling can enhance post-disaster resilience [49]. While such resilience benefits are outside the quantitative boundary of the present LCA, SSTs can act as controllable interfaces at charging depots and distribution nodes, enabling bidirectional power flow and fast power-quality control that underpin such coordinated operation.

Thus, SST-enabled fast charging can be a promising option for improving the efficiency, controllability, and power-quality performance of charging infrastructure, which can support transport electrification in renewable-rich power systems.

4.3. Industrial and Commercial Power Networks

Industry is undergoing simultaneous electrification and digitalization, with increasing adoption of robotics, automation, variable-speed drives, data centers, and local renewable/ESS systems. These environments generate high harmonic content, fast load transients, and complex AC/DC interfacing requirements. Line-frequency transformers, even when combined with filters or static VAR compensators, cannot provide granular control or multiport energy management. SST roles in industrial/commercial power-quality conditioning and hybrid AC/DC distribution are reviewed in [7,11,14].

SSTs deliver high controllability through programmable converters that independently regulate voltage, harmonics, phase imbalance, and power flow. In industrial microgrids, SSTs can unify AC and DC buses, enabling hybrid architectures where sensitive loads are connected to a regulated DC link, while legacy equipment remains on AC.

Their ability to perform real-time harmonic suppression, dynamic voltage restoration, and energy routing improves power quality and reduces equipment failures. Additionally, SSTs support predictive maintenance through continuous monitoring of thermal stress, semiconductor wear-out indicators, and harmonic signatures.

These capabilities improve energy efficiency, reduce operational emissions, and allow industries to integrate rooftop PV, battery storage, and waste-heat recovery systems more effectively. By reducing downtime and improving energy utilization, SSTs directly support industrial decarbonization and competitiveness.

4.4. Transmission–Distribution Interfaces and Smart Substations

At the boundary between transmission and distribution networks, operators face increasing complexity: variable renewable injections, bidirectional power flows, the rise of DERs, and new flexibility markets create operational uncertainties. Conventional power transformers provide no dynamic control, and their slow mechanical tap-changers are in-

https://doi.org/10.3390/en19020558

sufficient for managing modern fluctuations. Representative substation-scale SST concepts and demonstrators are discussed in [7,11,21,23].

SST-based smart substations introduce a new operational layer:

- bidirectional active and reactive power control;
- fast dynamic voltage support;
- fault isolation using fast semiconductor switching;
- seamless AC/DC interfacing for hybrid grids;
- system reconfiguration during disturbances;
- black-start and islanding capabilities.

These functionalities allow the T–D interface to absorb renewable fluctuations, coordinate local flexibility sources, and maintain stable operation even under high variability. SSTs also reduce copper losses and can operate in multiport configurations that connect MV AC, LV AC, MV DC, and LV DC simultaneously.

From a clean energy transition perspective, SSTs enable transmission operators to accommodate more renewable generation, integrate large-scale battery storage, and operate substations with a higher degree of automation and resilience. They also support new services such as grid-forming behavior and hybrid AC/DC operation, which are vital for future decentralized power systems.

In addition, SST functionalities can be relevant in remote or isolated multi-energy microgrids, where robust coordination between electrical and thermal subsystems is critical. For example, recent work on Antarctic unmanned observation stations demonstrates the value of electro-thermal coupling and distributionally robust optimisation to reduce carbon emissions under high renewable uncertainty [50]. While our study does not model such hybrid microgrids quantitatively, SSTs can serve as controllable AC/DC interfaces in these architectures, suggesting an additional area for future lifecycle and resilience-oriented assessments.

5. Results

SSTs operate through a sequence of power-electronic conversion stages that fundamentally differ from the passive energy transfer mechanism of conventional line-frequency transformers. In typical SST architectures, the medium-voltage AC input is first converted to a controlled DC voltage through an active AC–DC conversion stage. This enables precise control of input current, power factor, and harmonic content, which directly influences operational losses and grid interaction. In this study, the SST is modelled as a representative three-stage isolated architecture as described in Section 3.5, and the corresponding aggregated efficiency assumptions are applied consistently across all case studies.

The intermediate DC–DC conversion stage incorporates a high-frequency transformer that provides galvanic isolation and voltage adaptation. By shifting magnetic energy transfer to high switching frequencies, this stage enables substantial reductions in magnetic material volume and eliminates the need for oil-based insulation systems. At the same time, bidirectional power flow can be supported, allowing energy exchange between the grid, distributed generation, and storage systems.

On the output side, a DC–AC or DC conversion stage supplies regulated power to the load or downstream network. The combination of these stages allows SSTs to decouple input and output operating conditions, dynamically regulate voltage and power flow, and adapt to varying load profiles. From a lifecycle perspective, this operating principle leads to lower no-load and load-dependent losses, improved utilization under variable operating conditions, and reduced material intensity compared to conventional transformers.

The lifecycle assessment results presented in the following subsections reflect these fundamental operating characteristics by quantifying how differences in conversion stages,

https://doi.org/10.3390/en19020558

efficiency, and material composition translate into manufacturing, operational, and end-of-life CO_2 emissions.

The electricity grid CO_2 intensity scenarios defined in Table 1 are applied exclusively to the operational stage of the lifecycle, while manufacturing and end-of-life emissions are kept constant across all sensitivity cases.

5.1. Life Cycle of a Residential Transformer (630 kVA, 20 kV/0.4 kV)

Comparison between a classic transformer and a solid-state transformer.

Step 1: Manufacturing

Classic transformer: The manufacturing process involves making a magnetic core from laminated electrical steel, winding coils with a large amount of copper wire and filling it with transformer oil for cooling and insulation. For a 630 kVA transformer, between 800 and 1000 kg of copper and steel are used. The process of mining, refining and forming the metals themselves is very energy-intensive and results in the release of approximately 35 tons of CO_2 even before the device is put into operation.

SST: The SST does not use a magnetic core and transformer oil, but instead includes power electronics (IGBT modules, DC links, filters) as well as active cooling systems (e.g., fans or liquid cooling). Although the production of electronic components has its own carbon footprint, it requires significantly less metal and allows for automated, modular production. The total CO_2 emissions from the production process are estimated at approximately 25 tons of CO_2, which is about 30% lower than the classical equivalent.

Stage 2: Operation

Classical transformer: During operation, the classical transformer has constant losses—both idle losses (about 1.1 kW) and load losses (about 5.4 kW). This results in an average of 57,240 kWh of energy losses per year, which, with a standard energy mix (0.4 kg CO_2/kWh), means 22.9 tons of CO_2 emissions per year. Over 25 years of operation, this accumulates approximately 572 tons of CO_2.

SST: Due to its higher efficiency (~99.2%), the SST has average losses of about 5.04 kW or 44,150 kWh per year. This results in annual emissions of about 17.7 tons of CO_2. Over the same 25-year period, this amounts to around 442 tonnes of CO_2, which is 130 tonnes less than a conventional transformer.

Stage 3: Recycling and end of life

Conventional transformer: At the end of its life, the transformer must be dismantled and the oil treated as hazardous waste. Recycling of metals is possible, but often incomplete due to embedded components and aging of materials. Total emissions associated with decommissioning, transport, treatment and partial recycling are estimated at 5–10 tonnes of CO_2.

SST: The SST is built modularly, with clearly separable components (electronics, housing, cooling), which facilitates recycling. Although e-waste requires specific treatment, there are well-developed systems for recycling aluminium and electronics. The end-of-life stage results in approximately 5 tonnes of CO_2.

Table 3 presents a comparative analysis of the total carbon footprint generated by a traditional oil-filled transformer and a modern SST over their full lifecycle. The assessment is divided into three key stages: manufacturing, operational use over a standard service life of 25 years, and end-of-life disposal and recycling. Each stage includes estimates of CO_2 emissions based on typical material usage, energy losses, and handling processes. The SST demonstrates a significant advantage in reducing environmental impact, especially in long-term operation and recyclability, making it a more sustainable solution for modern power distribution systems.

https://doi.org/10.3390/en19020558

Table 3. Lifecycle CO_2 Emissions Comparison: Traditional Transformer vs. SST (630 kVA, 20 kV/0.4 kV).

Lifecycle Stage	Traditional Transformer	Solid-State Transformer	Notes	Primary Data Source (Reference)
Manufacturing	~35 t CO_2	~25 t CO_2	Steel, copper and oil vs. power electronics and modular cooling	[42,44,46]
Operation (25 years)	~572 t CO_2 (22.9 t/year)	~442 t CO_2 (17.7 t/year)	Calculated from no-load and load losses	[42,44]
Annual energy losses	~57,240 kWh	~44,150 kWh	Based on 8760 h/year operation	Derived in this study
End-of-life	~5–10 t CO_2	~5 t CO_2	Oil treatment vs. modular electronics recycling	[45,46]
Total lifecycle emissions	~610–620 t CO_2	~470 t CO_2	Aggregated lifecycle total	This study (derived)

Table 3 summarizes the lifecycle CO_2 emissions of a conventional oil-filled transformer and an SST for a representative 630 kVA distribution application. Manufacturing-stage values are derived from published transformer LCA studies and reflect differences in material intensity and the absence of insulating oil in the SST. Operational emissions are calculated using harmonized loss models and identical operating assumptions, following the approach commonly adopted in transformer LCAs [42,44]. End-of-life impacts are based on literature-reported recycling and disposal scenarios. The table shows that operational losses dominate the total lifecycle footprint, while improved efficiency leads to a net reduction of approximately 140–150 t CO_2 over a 25-year service life for the SST case.

5.2. Life Cycle of an Industrial Transformer for a Small Factory (1 MVA, 20 kV/0.4 kV)

Step 1: Manufacturing

Classic transformer: A traditional 1 MVA transformer requires large laminated steel cores, extensive copper windings, and significant amounts of transformer oil. The typical raw material mass exceeds 1500 kg. The processes of mining, processing, assembly, and insulation add up to approximately 50 tons of CO_2 emissions during production.

SST: The SST version includes high-power IGBT modules, advanced converters, and active thermal management systems. Despite using more electronics, it reduces dependence on heavy materials. Automated assembly and modular design result in an estimated ~40 tons of CO_2 during manufacturing.

Step 2: Operation

Classic transformer: With typical idle and load losses of ~9.8 kW, the traditional transformer consumes about 86,328 kWh/year, generating 34.5 tons of CO_2 annually. Over 25 years, this amounts to ~862 tons of CO_2.

SST: The SST, with ~99.1% efficiency, has average losses of 9 kW, consuming about 78,840 kWh/year. This equates to 31.5 tons of CO_2 annually, or ~787 tons over 25 years, yielding a savings of ~75 tons of CO_2 in this stage alone.

Step 3: Recycling and End of Life

Classic transformer: Handling and recycling a large oil-filled unit is resource-intensive due to the hazardous nature of the oil and the mass of metals. The recycling-related emissions are estimated at ~10 tons of CO_2.

SST: SSTs are modular, allowing easier material separation and recycling. E-waste management systems are better adapted to electronics and aluminium components. The total end-of-life impact is estimated at ~5 tons of CO_2.

Table 4 presents a full lifecycle carbon emissions comparison between a traditional oil-filled transformer and an SST designed for industrial applications with a 1 MVA power rating. The assessment includes emissions from manufacturing, 25 years of operation, and end-of-life recycling. SSTs offer improved efficiency and modularity, leading to substantial CO_2 savings.

Table 4. Lifecycle CO_2 Emissions Comparison: Industrial Transformer (1 MVA, 20 kV/0.4 kV).

Lifecycle Stage	Traditional Transformer	Solid-State Transformer	Notes	Primary Data Source (Reference)
Manufacturing	~50 t CO_2	~40 t CO_2	Larger steel and copper mass vs. power electronics	[42,44,46]
Operation (25 years)	~862 t CO_2 (34.5 t/year)	~787 t CO_2 (31.5 t/year)	Calculated from no-load and load losses	[42,44]
Annual energy losses	~86,328 kWh	~78,840 kWh	Based on 8760 h/year operation	Derived in this study
End-of-life	~10 t CO_2	~5 t CO_2	Oil handling vs. modular electronics recycling	[45,46]
Total lifecycle emissions	~922 t CO_2	~832 t CO_2	Aggregated lifecycle total	This study (derived)

Note: Values are derived from literature-based ranges and harmonized assumptions; references indicate the primary sources for each lifecycle stage.

Table 4 presents the lifecycle CO_2 emissions of a conventional and an SST for a representative 1 MVA industrial distribution application. Manufacturing-stage emissions are higher than in the residential case due to increased material intensity, particularly steel and copper, while SST manufacturing benefits from reduced magnetic material and oil-free design. Operational emissions remain the dominant contributor to total lifecycle impact and are calculated using harmonized loss assumptions consistent with published transformer LCAs [42,44]. As shown, the SST achieves a net lifecycle reduction of approximately 90 t CO_2 over a 25-year service life, primarily driven by improved conversion efficiency.

5.3. Life Cycle of a Transformer for EV Charging Station (500 kVA, 20 kV/0.4 kV)

Step 1: Manufacturing

Classic transformer: A 500 kVA traditional transformer uses several hundred kilograms of copper and steel, along with transformer oil for cooling and insulation. The manufacturing process includes lamination of the core, winding of copper conductors, and sealing of the oil tank. Total emissions from raw material extraction and production are approximately 30 tons of CO_2.

SST: The SST variant for the same power rating uses compact power electronics, lightweight aluminium housings, and integrated cooling systems. Due to less reliance on heavy metals and more streamlined modular design, production emissions are estimated at ~22 tons of CO_2.

Step 2: Operation

Classic transformer: With average no-load and load losses totaling ~5.1 kW, the classic unit consumes around 44,676 kWh/year, which corresponds to 17.9 tons of CO_2 annually (based on 0.4 kg CO_2/kWh). Over 25 years, this equals ~448 tons of CO_2.

SST: Thanks to its efficiency (~99.3%), the SST reduces losses to about 3.5 kW on average, or 30,660 kWh/year, equivalent to 12.3 tons of CO_2/year. Over 25 years, this adds up to ~308 tons of CO_2, saving approximately 140 tons in this stage alone.

Step 3: Recycling and End of Life

Classic transformer: Recycling this type of unit involves removing and disposing of oil and separating metal components. Due to complexity, total end-of-life CO_2 emissions are estimated at ~6 tons.

SST: The SST's modular design allows easier disassembly, with standardised aluminium, PCBs, and cooling units. E-waste is treated separately, resulting in an end-of-life footprint of ~4 tons of CO_2.

https://doi.org/10.3390/en19020558

The following Table 5 provides a comparative analysis of CO_2 emissions across the full lifecycle of a traditional oil-filled transformer and an SST used for a 500 kVA EV charging station. The analysis covers three key lifecycle stages: manufacturing, operation over 25 years, and end-of-life recycling. The SST offers superior performance in all stages, resulting in a significantly lower total carbon footprint.

Table 5. Lifecycle CO_2 Emissions Comparison: Transformer for EV Fast-Charging Station (500 kVA, 20 kV/0.4 kV).

Lifecycle Stage	Traditional Transformer	Solid-State Transformer	Notes	Primary Data Source (Reference)
Manufacturing	~30 t CO_2	~22 t CO_2	Reduced metal mass and oil-free SST design	[42,44,46]
Operation (25 years)	~448 t CO_2 (17.9 t/year)	~308 t CO_2 (12.3 t/year)	Calculated from no-load and load losses	[42,44]
Annual energy losses	~44,676 kWh	~30,660 kWh	Based on 8760 h/year operation	Derived in this study
End-of-life	~6 t CO_2	~4 t CO_2	Oil treatment vs. modular electronics recycling	[45,46]
Total lifecycle emissions	~484 t CO_2	~334 t CO_2	Aggregated lifecycle total	This study (derived)

Note: Values are derived from literature-based ranges and harmonized assumptions; references indicate the primary sources for each lifecycle stage.

Table 5 reports the lifecycle CO_2 emissions associated with a conventional transformer and an SST used in a 500 kVA EV fast-charging station. This application is characterized by highly variable and peaky load profiles; however, for comparability, both technologies are evaluated under identical average operating assumptions. Manufacturing-stage emissions reflect the reduced material intensity and oil-free design of the SST, while operational emissions dominate the total lifecycle footprint. Due to improved efficiency under continuous operation, the SST achieves an indicative lifecycle CO_2 reduction of approximately 150 t over a 25-year service life, consistent with trends reported in recent transformer LCA studies [42,44].

5.4. Life Cycle of a Substation Transformer (10 MVA, 110 kV/20 kV)

Such a transformer is a key component in power distribution substations—an intermediate link between the high-voltage transmission network and the medium-voltage distribution network. High power (10 MVA) means large volumes of transferred energy and potentially much higher energy losses, respectively, a greater significance for the carbon footprint. It is used in urban and industrial areas, as well as for powering clusters of EV charging stations, logistics parks, or manufacturing areas.

Step 1: Manufacturing

Classic transformer: A 10 MVA oil-filled transformer is a large and heavy unit requiring several tons of laminated steel, copper windings, and substantial amounts of transformer oil. The entire manufacturing process, from mining to assembly, results in an estimated ~100 tons of CO_2 emissions due to the energy-intensive nature of material extraction and processing.

SST: The SST alternative for this power range integrates multiple high-voltage electronic conversion stages with active cooling and modular packaging. Although electronics and thermal systems add complexity, SSTs use less metal overall and benefit from factory automation. The total production emissions are estimated at ~80 tons of CO_2.

Step 2: Operation

Classic transformer: In typical 24/7 grid service, the classic unit suffers average losses of ~60 kW (i.e., on the order of tens of kW for a 10 MVA-class transformer). Over 25 years

https://doi.org/10.3390/en19020558

(8760 h/year), this equates to 13.14 GWh of lost energy. At 0.4 kg CO_2 per kWh, operational emissions total ~5256 tons of CO_2.

SST: Thanks to higher efficiency (~99.3%), the SST loses less energy—about 49 kW on average—resulting in 10.73 GWh of losses over 25 years, or ~4292 tons of CO_2. This represents a savings of ~964 tons in the operational stage.

Step 3: Recycling and End of Life

Classic transformer: End-of-life involves draining and processing large volumes of hazardous oil and handling heavy steel cores and copper. Recycling-related emissions are estimated at ~15 tons of CO_2.

SST: The modular construction allows easier separation of power modules, aluminium housings, and standardised electronic components. Despite the e-waste treatment requirements, total emissions at this stage are estimated at ~8 tons of CO_2.

Table 6 summarizes the estimated carbon footprint of a traditional oil-filled transformer and a modern SST over a 25-year lifecycle in a high-power substation application. It includes emissions from manufacturing, operation, and end-of-life recycling. The SST demonstrates a notable reduction in emissions due to improved efficiency and modular design.

Table 6. Lifecycle CO_2 Emissions Comparison: Substation Transformer (10 MVA, 110 kV/20 kV).

Lifecycle Stage	Traditional Transformer	Solid-State Transformer (SST)	Notes	Primary Data Source (Reference)
Manufacturing	~100 t CO_2	~80 t CO_2	Large steel and copper mass vs. modular power electronics	[42,44,46]
Operation (25 years)	~5256 t CO_2 (13.14 GWh)	~4292 t CO_2 (10.73 GWh)	Calculated from average losses at rated service	[42,44]
Annual energy losses	~525,600 kWh	~429,240 kWh	Based on 8760 h/year operation	Derived in this study
End-of-life	~15 t CO_2	~8 t CO_2	Oil treatment vs. modular electronics recycling	[45,46]
Total lifecycle emissions	~5371 t CO_2	~4380 t CO_2	Aggregated lifecycle total	This study (derived)

Note: Values are derived from literature-based ranges and harmonized assumptions; references indicate the primary sources for each lifecycle stage.

Table 6 presents the lifecycle CO_2 emissions of a conventional oil-filled transformer and an SST for a 10 MVA transmission–distribution interface substation. Due to the high power throughput, operational losses dominate the lifecycle footprint of both technologies, accounting for more than 95% of total emissions. Manufacturing-stage emissions are higher in absolute terms compared to lower-power cases, reflecting the large material mass of substation-class equipment. Under identical operating assumptions, the SST achieves an indicative lifecycle CO_2 reduction of approximately 1000 t over a 25-year service life, illustrating how even modest efficiency improvements translate into large absolute emission savings at high power ratings.

Tables 3–6 collectively demonstrate the lifecycle CO_2 performance of conventional line-frequency transformers and SSTs across a wide power range and representative application contexts. Although absolute emission values scale strongly with power rating and energy throughput, a consistent pattern emerges: operational losses dominate the total lifecycle footprint in all cases, while SSTs systematically achieve lower lifecycle emissions under identical assumptions. The relative magnitude of the reduction varies with loading and application, but the direction of the effect remains robust, indicating that improved conversion efficiency and reduced material intensity are the primary drivers of the observed CO_2 savings. These results provide a consistent, literature-aligned basis for the comparative discussion and sensitivity analysis presented in the following sections.

https://doi.org/10.3390/en19020558

6. Discussion: System-Level Impact, Carbon Reduction, and Deployment Pathways

The results presented in Section 5 demonstrate that SSTs provide consistent and measurable lifecycle advantages across residential, industrial, EV-charging, and substation applications. While individual numerical values vary depending on power rating and loading conditions, several cross-cutting observations emerge that clarify the role of SSTs in the global clean energy transition and their long-term sustainability potential.

6.1. Carbon Footprint Reduction and Energy Efficiency Trends

Across the distribution-scale case studies (0.5–1 MVA), SSTs deliver a total lifecycle CO_2 reduction of roughly 90–150 tons compared to conventional oil-filled transformers. For the 10 MVA substation case, the absolute reduction is much larger ($\approx$1000 tons), reflecting the higher lifetime energy throughput. These savings originate primarily from:

- Higher operating efficiency, which reduces lifetime energy losses by approximately 9–31% across the studied cases (Tables 3–6).
- Lower idle losses, especially relevant in lightly loaded or intermittently used networks.
- Reduced material intensity during manufacturing, due to the absence of heavy magnetic cores and transformer oil.
- More efficient end-of-life processing, owing to modular construction and reduced hazardous waste.

Operational losses dominate the lifecycle footprint of both technologies. Therefore, even modest improvements in conversion efficiency—on the order of 0.2–0.4%—produce large cumulative CO_2 benefits over a 25-year service life. This aligns SST adoption with mandatory energy-efficiency targets under European Ecodesign regulations and with national decarbonization policies in many countries.

The numerical results confirm that SSTs act as system-level efficiency multipliers: their benefits accumulate not only at the device level but across entire distribution networks.

The values include manufacturing, 25-year operational losses, and end-of-life treatment. SSTs consistently demonstrate lower total emissions due to reduced losses, lower material intensity, and improved recyclability. The absolute savings are on the order of 90–150 t CO_2 for distribution-scale units (0.5–1 MVA) and ~1000 t CO_2 for the 10 MVA substation case.

Table 7 presents a summary of lifecycle CO_2 emissions for four representative transformer categories. Values include manufacturing, 25-year operational losses, and end-of-life treatment. SSTs consistently demonstrate lower total emissions due to reduced losses, lower material intensity, and improved recyclability. The relative reduction is calculated as $(CF_{LFT} - CF_{SST})/CF_{LFT} \times 100\%$, yielding approximately 24%, 10%, 31%, and 18% reductions for the 630 kVA, 1 MVA, 500 kVA, and 10 MVA cases, respectively.

As shown in Table 7, SSTs reduce total lifecycle CO_2 emissions by approximately 90–150 t CO_2 for the distribution-scale units considered here, while the 10 MVA substation case shows nearly 1000 t CO_2 savings due to the much larger lifetime energy throughput.

Sensitivity to electricity CO_2 intensity: since operational emissions scale linearly with the electricity-mix emission factor, the absolute lifecycle CO_2 savings are recalculated for the scenarios in Table 1a by scaling only the operational stage (manufacturing and end-of-life kept unchanged). The resulting absolute lifecycle CO_2 savings for each application and electricity-mix scenario are reported explicitly in Table 7a.

These ranges capture the dominant operational-stage sensitivity to the electricity mix. Applying the $\pm$15–25% manufacturing and end-of-life bounds in Section 3.6 introduces an additional (smaller) spread in absolute savings (typically on the order of $\pm$(5–20) t CO_2 for

https://doi.org/10.3390/en19020558

the distribution/EV cases and ±(50–200) t CO_2 for the 10 MVA case), without changing the qualitative conclusion that SSTs reduce lifecycle CO_2 under the assumed operating profiles.

Under the decarbonization pathway (avg EF ≈ 0.25), absolute lifecycle savings decrease by about 35–40% relative to the baseline (EF = 0.40). For example, savings drop from ~150 t to ~98 t for the 500 kVA EV case and from ~991 t to ~630 t for the 10 MVA substation case.

Table 7. (a) Summary of Lifecycle CO_2 Reductions Across Four Transformer Classes. (b) Sensitivity of total lifecycle CO_2 savings to electricity grid CO_2 intensity scenarios (Table 1a).

(a)

Application	Power Rating	Traditional Transformer (tons CO_2, 25 yrs)	SST (tons CO_2, 25 yrs)	CO_2 Savings	Notes
Residential Distribution	630 kVA	610–620	~470	140–150	Highest relative savings due to lower idle losses and improved efficiency
Industrial Transformer	1 MVA	~922	~832	~90	High utilisation; SST reduces losses and metal intensity
EV Charging Station	500 kVA	~484	~334	~150	Fast-varying loads benefit from SST efficiency and dynamic control
Substation Transformer	10 MVA	~5371	~4380	~991	High power throughput amplifies efficiency gains (large absolute savings).

(b)

Application (rating)	Savings at baseline EF = 0.40 (t CO_2)	Savings with Decarbonization (avg EF≈0.25) (t CO_2)	Sensitivity Range (EF = 0.24–0.49) (t CO_2)
Residential (630 kVA)	~145	~96	~93-174
Industrial (1 MVA)	~90	~62	~60-107
EV charging (500 kVA)	~150	~98	~94-181
Substation (10 MVA)	~991	~630	~605-1208

6.2. Flexibility and Renewable Integration Advantages

Beyond improving energy efficiency, SSTs introduce operational capabilities that fundamentally expand the flexibility of distribution networks. Unlike conventional transformers, SSTs support fully bidirectional power flow, a prerequisite for large-scale photovoltaic integration and for future prosumer-driven distribution architectures. Their dynamic voltage and reactive-power control capabilities enable rapid compensation of voltage deviations and maintain stability in feeders exposed to high solar or wind variability. High-frequency isolation reduces the physical footprint and allows fast-acting control loops, while embedded sensing and monitoring functions provide continuous visibility into component health, loading conditions, and power quality. Because SSTs are built from modular power-electronic stages, they also enable staged upgrades and replacements—helping utilities extend asset life without wholesale system shutdowns.

In the context of the global clean energy transition, these functionalities directly address long-standing technical constraints such as hosting-capacity limits, congestion caused by variable renewable generation, and complex coordination of distributed resources. SSTs therefore act not merely as replacement equipment, but as system-level flexibility assets that enhance the controllability, resilience, and adaptability of modern distribution networks.

https://doi.org/10.3390/en19020558

6.3. Economic and Practical Deployment Considerations

While SSTs offer clear technical advantages, their adoption depends on economic viability and practical deployment constraints. The initial capital expenditure remains higher than that of conventional transformers, primarily due to the cost of power semiconductors, cooling systems, and advanced digital controllers. However, operational expenditures are significantly lower because SSTs reduce conversion losses, eliminate oil handling, and enable predictive maintenance through integrated monitoring. Their modular design also allows for targeted replacement of subassemblies—such as DC–DC stages or control boards—reducing downtime and lowering long-term maintenance costs. Furthermore, the reduced size and weight of SST units create advantages in installations with limited space, including urban substations, EV charging depots, and containerized modular grids.

From a cost–benefit perspective, SSTs become economically favorable in environments with high utilization, rapidly fluctuating loads, or stringent power-quality requirements. Industrial feeders, fast-charging stations, and commercial buildings stand to benefit most, as the operational savings and performance improvements can compensate for the higher upfront investment over the asset lifetime.

To provide a simple techno-economic indicator linked directly to the LCA results, we estimate the present-value (PV) cost of avoided loss-related electricity consumption as:

$$\mathrm{PV} = \Delta\mathrm{E\_loss, annual} \times \mathrm{c\_el} \times \frac{\left(1 - (1 + \mathrm{r}^{N}\right)}{r},$$

where c_el is the electricity price, r is the discount rate, and N is the service life. Using the annual loss reductions from Tables 3–6 ($\approx$10.5, 7.8, 14.9, and 96.4 MWh·y^{-1} for the 630 kVA, 1 MVA, 500 kVA, and 10 MVA cases) and illustrative values c_el = 0.1 €/kWh, r = 5%, and N = 25 years, the PV OPEX savings are $\approx$15 k€, $\approx$11 k€, $\approx$21 k€, and $\approx$136 k€, respectively. These indicative values suggest that efficiency-driven OPEX savings alone may not fully offset the current SST CAPEX premium, and that additional grid-service value streams (e.g., power-quality support and controllability) can be important in practical investment decisions.

6.4. Lifecycle Sustainability and Circularity

A distinguishing feature of SSTs is their alignment with circular-economy principles. In contrast to oil-filled transformers built from heavy steel cores, large volumes of copper, and mineral oil, SSTs employ lighter aluminium housings, minimal copper, and no liquid insulation. Their modular architecture improves the recoverability of PCBs, semiconductors, and mechanical structures at the end of life. The elimination of transformer oil not only reduces the carbon footprint of manufacturing but also removes the need for hazardous-waste processing and eliminates environmental risks associated with leakage and flammability.

Lifecycle assessment indicates that the manufacturing stage contributes only 15–20% of total emissions for both technologies; therefore, the superior operational efficiency of SSTs is the dominant factor behind their reduced lifecycle CO_2 footprint. This observation reinforces the relevance of SSTs to clean-energy-transition policies aimed at reducing distribution losses and improving overall electricity-system efficiency.

6.5. System-Level Role in the Clean Energy Transition

SSTs do not merely replace conventional transformers—they fundamentally reshape the functional boundaries of distribution networks. Their capabilities make them enabling technologies for renewable-rich grids, high-power EV charging corridors, hybrid AC/DC distribution architectures, and intelligent commercial or industrial campuses. SSTs can

https://doi.org/10.3390/en19020558

reduce feeder losses, enhance local power quality, mitigate congestion caused by distributed generation, and facilitate coordinated control of distributed energy resources.

By integrating controllability, sensing, and bidirectional power flow into a single device, SSTs can support the transition toward more electrified, digitalized, and resilient distribution infrastructures, particularly in applications where controllability and power quality are valued alongside efficiency.

6.6. Limitations and Future Research Needs

Despite their strong technical and environmental potential, several limitations must be addressed before SSTs achieve widespread deployment. The most significant barrier remains the high cost of power-electronic devices, particularly for medium-voltage applications. Thermal management challenges persist, especially under high-load transients, while ensuring the long-term reliability of semiconductor devices remains an active research area. Harmonic emissions require high-quality filtering solutions, and existing grid codes—designed for passive transformers—do not yet fully accommodate the operational characteristics of SSTs. Furthermore, large-scale field demonstrations at multi-MVA ratings are still limited, leaving uncertainty around long-term operational performance.

Future research should therefore focus on improving multi-cell topologies, enhancing cooling through integrated heat pipes or liquid systems, and developing advanced control algorithms with predictive and adaptive capabilities. Optimization of SiC and GaN devices for medium-voltage stages, together with the integration of digital twins for real-time monitoring and prognostics, represents a promising pathway for the next generation of SST systems.

6.7. Summary of the Discussion

The analysis indicates that SSTs can act as enabling technologies for more efficient and controllable electricity networks, combining higher conversion efficiency with advanced control functionality and oil-free designs. However, the present results are literature-based and indicative, and they should be interpreted as screening evidence rather than as predictive, site-specific quantification. Continued work is needed on field validation, reliability, and refined lifecycle inventories for emerging multi-MVA SST designs.

7. Conclusions

SSTs are increasingly investigated as a potential enabling technology for next-generation power systems, since they combine power-electronic conversion, high-frequency isolation, and advanced digital control functionalities that are not available in conventional line-frequency transformers.

This paper presented a structured, comparative life cycle assessment (LCA) of conventional transformers and SSTs across four representative application contexts: residential distribution networks, industrial distribution systems, electric vehicle fast-charging infrastructure, and transmission–distribution interface substations. The analysis was conducted using literature-derived data, manufacturer specifications, and harmonized assumptions applied consistently across all cases, with a cradle-to-grave system boundary covering manufacturing, operation, and end-of-life stages.

The results indicate that, under identical assumptions, SST-based solutions are associated with indicative lifecycle CO_2 emission reductions on the order of approximately 10–30% compared to conventional transformers. These reductions are primarily driven by lower operational losses and, to a lesser extent, by reduced material intensity and improved end-of-life handling enabled by modular, oil-free designs. While the magnitude of the reduction varies with power rating and operating profile, the comparative trend remains consistent across all analyzed cases.

https://doi.org/10.3390/en19020558

Beyond lifecycle emissions, the paper briefly discusses system-level capabilities that are often associated with SST deployment (e.g., bidirectional operation, fast voltage and power control, and compatibility with hybrid AC/DC architectures). These aspects are not quantified within the present CO_2-focused LCA scope, but they provide additional context for why SSTs are being explored in renewable-rich distribution networks, electrified transport corridors, and future flexible substations.

At the same time, several limitations must be acknowledged. The results are based on literature-derived estimates and representative assumptions rather than site-specific operational data. A uniform service life of 25 years was assumed for both transformer technologies; however, SSTs may require mid-life replacement of power-electronic components, and a conservative replacement sensitivity is discussed in Section 3.6. In addition, the operational results depend on representative load factors and electricity-mix assumptions, which are explored through the grid-emission-factor scenarios reported in Sections 5 and 6.

Overall, the findings suggest that SSTs constitute a relevant enabling technology for future low-carbon power systems when evaluated from a lifecycle perspective. However, the results should be interpreted as indicative rather than predictive. Future research should focus on large-scale field demonstrations, long-term reliability data for power-electronic components, and refined lifecycle assessments that incorporate detailed operational profiles, component replacement strategies, and evolving electricity-generation mixes. Although the presented results are based on literature-derived data and screening-level assumptions, the applied dispersion ranges and electricity-mix sensitivity scenarios indicate that the comparative conclusion—lower lifecycle CO_2 emissions for SSTs compared to conventional transformers—remains robust across realistic parameter variability.

Funding: This work was supported by the Operational Programme "Research, Innovation and Digitalisation for Smart Transformation 2021–2027" under Project NoBG16RFPR002-1.014-0006-C01 "National center of excellence for mechatronics and clean technologies". The APC was funded by Project NoBG16RFPR002-1.014-0006-C01.

Data Availability Statement: No new raw data were created. The study generated derived (processed) data from published sources (e.g., harmonized inventories, tables, and scenario calculations); all key assumptions are reported in the manuscript.

Acknowledgments: The present research has been carried out under the project BG16RFPR002-1.014-0006 "National Centre of Excellence Mechatronics and Clean Technologies", funded by the Operational Programme Science and Education for Smart Growth. The obtained results have been processed and analyzed within the framework of the project BG-RRP-2.004-0005 "Improving the research capacity and quality to achieve international recognition and resilience of TU-Sofia (IDEAS)", funded by the National Recovery and Resilience Plan of the Republic of Bulgaria.

Conflicts of Interest: The author declares no conflicts of interest.

References

1. Mollik, M.S.; Hannan, M.A.; Reza, M.S.; Abd Rahman, M.S.; Lipu, M.S.H.; Ker, P.J.; Mansor, M.; Muttaqi, K.M. The Advancement of Solid-State Transformer Technology and Its Operation and Control with Power Grids: A Review. *Electronics* **2022**, *11*, 2648. [CrossRef]
2. Abu-Siada, A.; Budiri, J.; Abdou, A.F. Solid State Transformers Topologies, Controllers, and Applications: State-of-the-Art Literature Review. *Electronics* **2018**, *7*, 298. [CrossRef]
3. Hannan, M.A.; Ker, P.J.; Lipu, M.S.H.; Choi, Z.H.; Rahman, M.S.A.; Muttaqi, K.M.; Blaabjerg, F. State of the Art of Solid-State Transformers: Advanced Topologies, Implementation Issues, Recent Progress and Improvements. *IEEE Access* **2020**, *8*, 19113–19132. [CrossRef]
4. Saleh, S.A.; Ozkop, E.; Alsayid, B.; Richard, C.; Onge, X.F.S.; McDonald, K.M.; Chang, L. Solid-State Transformers for Distribution Systems–Part II: Deployment Challenges. *IEEE Trans. Ind. Appl.* **2019**, *55*, 5708–5716. [CrossRef]

https://doi.org/10.3390/en19020558

5. Santos, N.; Chaves, M.; Gamboa, P.; Cordeiro, A.; Santos, N.; Pinto, S.F. High Frequency Transformers for Solid-State Transformer Applications. *Appl. Sci.* **2023**, *13*, 7262. [CrossRef]
6. Khan, S.; Rahman, K.; Tariq, M.; Hameed, S.; Alamri, B.; Babu, T.S. Solid-State Transformers: Fundamentals, Topologies, Applications, and Future Challenges. *Sustainability* **2021**, *14*, 319. [CrossRef]
7. Costa, L.F.; De Carne, G.; Buticchi, G.; Liserre, M. The Smart Transformer: A solid-state transformer tailored to provide ancillary services to the distribution grid. *IEEE Power Electron. Mag.* **2017**, *4*, 56–67. [CrossRef]
8. Huber, J.E.; Kolar, J.W. Applicability of Solid-State Transformers in Today's and Future Distribution Grids. *IEEE Trans. Smart Grid* **2019**, *10*, 317–326. [CrossRef]
9. Valedsaravi, A.; Martínez-Salamero, L. Review of Solid-State Transformer Applications on Electric Vehicle DC Ultra-Fast Charging Station. *Energies* **2022**, *15*, 5602. [CrossRef]
10. Zhao, X.; Li, B.; Fu, Q.; Mao, S.; Xu, D.G.; Leon, J.I.; Franquelo, L.G. DC Solid State Transformer Based on Three-Level Power Module for Interconnecting MV and LV DC Distribution Systems. *IEEE Trans. Power Electron.* **2021**, *36*, 1563–1577. [CrossRef]
11. Liserre, M.; Buticchi, G.; Andresen, M.; De Carne, G.; Costa, L.F.; Zou, Z.-X. The Smart Transformer: Impact on the Electric Grid and Technology Challenges. *IEEE Ind. Electron. Mag.* **2016**, *10*, 46–58. [CrossRef]
12. Keerthi Gopal, D.; Shanavas, T.N.; Naufal, N. Solid State Transformers for Smart Grid Control and Applications—A Review. In Proceedings of the 2022 International Conference on Futuristic Technologies in Control Systems & Renewable Energy (ICFCR), Malappuram, India, 21–22 July 2022; pp. 1–6. [CrossRef]
13. Kadandani, N.B.; Dahidah, M.; Ethni, S. An Overview of the Role of Solid State Transformer in Smart Grid. In *2021 12th International Renewable Energy Congress (IREC)*; IEEE: New York, NY, USA, 2021. [CrossRef]
14. Liu, Y.; Zhang, D.H.; De Sun, D. Performance analysis of a solid state transformer for smart grid. *Appl. Mech. Mater.* **2014**, *441*, 174–177. [CrossRef]
15. Mishra, D.K.; Ghadi, M.J.; Li, L.; Hossain, J.; Zhang, J.; Ray, P.K.; Mohanty, A. A review on solid-state transformer: A breakthrough technology for future smart distribution grids. *Int. J. Electr. Power Energy Syst.* **2021**, *133*, 107255. [CrossRef]
16. Adabi, M.E.; Martinez-Velasco, J.A.; Alepuz, S. Modeling and simulation of a MMC-based solid-state transformer. *Electr. Eng.* **2018**, *100*, 375–387. [CrossRef]
17. Li, Z.; Zhao, C.; Hsieh, Y.H.; Li, Q.; Lee, F.C. Partial Fluctuating Power Control of Resonant Converter for Solid-State Transformer. In Proceedings of the 2022 IEEE Applied Power Electronics Conference and Exposition (APEC), Houston, TX, USA, 20–24 March 2022; pp. 770–776. [CrossRef]
18. Ronanki, D.; Williamson, S.S. Topological Overview on Solid-state Transformer Traction Technology in High-speed Trains. In Proceedings of the 2018 IEEE Transportation Electrification Conference and Expo (ITEC), Long Beach, CA, USA, 13–15 June 2018; pp. 32–37. [CrossRef]
19. Awal, M.A.; Montes, O.A.; Teng, F.; Wang, D.; Bipu, M.R.H.; Yu, W.; Lukic, S.; Husain, I. Medium Voltage Solid State Transformer for Extreme Fast Charging Applications. In Proceedings of the 2023 IEEE Applied Power Electronics Conference and Exposition (APEC), Orlando, FL, USA, 19–23 March 2023; pp. 1528–1535. [CrossRef]
20. Hyde, A.R. Advanced Power Converters for Universal and Flexible Power Management in Future Electricity Networks. In *Technical Report*; AREVA: Courbevoie, France, 2006.
21. Wang, D.; Tian, J.; Mao, C.; Lu, J.; Duan, Y.; Qiu, J.; Cai, H. A 10-kV/400-V 500-kVA Electronic Power Transformer. *IEEE Trans. Ind. Electron.* **2016**, *63*, 6653–6663. [CrossRef]
22. Cuma, M.U.; Anvari, A.D.; Savrun, M.M. A Novel Solid-State Transformer with Improved Flyback Converter Equipped with Quasi Z-Source Converter for Medium-Voltage Utility Grid. *Energies* **2023**, *16*, 2687. [CrossRef]
23. Chen, H.; Divan, D. Soft-Switching Solid-State Transformer (S4T). *IEEE Trans. Power Electron.* **2018**, *33*, 2933–2947. [CrossRef]
24. Falcones, S.; Mao, X.; Ayyanar, R. Topology comparison for Solid State Transformer implementation. In Proceedings of the IEEE PES General Meeting, Minneapolis, MN, USA, 25–29 July 2010; pp. 1–8. [CrossRef]
25. Raaed Faleh Hassan, D. Design and software implementation of solid state transformer. *Int. J. Eng. Technol.* **2018**, *7*, 1776. [CrossRef]
26. Saha, J.; Yadav, G.N.B.; Panda, S.K. A Matrix-Based Solid-State-Transformer for a Hybrid Nanogrid. In *2018 IEEE International Conference on Power Electronics, Drives and Energy Systems (PEDES), Chennai, India*; IEEE: New York, NY, USA, 2018; pp. 1–6. [CrossRef]
27. Chen, H.; Prasai, A.; Divan, D. Dyna-C: A Minimal Topology for Bidirectional Solid-State Transformers. *IEEE Trans. Power Electron.* **2017**, *32*, 995–1005. [CrossRef]
28. Ivanov, G.; Mateev, V.; Marinova, I. A Cascaded Neural Network Framework for Estimating Degree of Polymerization of Insulated Paper and DGA Analysis in Power Transformers. In Proceedings of the 2025 19th Conference on Electrical Machines, Drives and Power Systems (ELMA), Sofia, Bulgaria, 19–21 June 2025; pp. 1–5. [CrossRef]
29. Nair, A.C.; Fernandes, B.G. Solid-State Transformer Based Fast Charging Station for Various Categories of Electric Vehicles with Batteries of Vastly Different Ratings. *IEEE Trans. Ind. Electron.* **2021**, *68*, 10400–10411. [CrossRef]

https://doi.org/10.3390/en19020558

30. Predescu, D.-M.; Roșu, Ș.-G. Solid State Transformers: A Review—Part I: Stages of Conversion and Topologies. *Technologies* **2025**, *13*, 74. [CrossRef]
31. Predescu, D.-M.; Roșu, Ș.-G. Solid-State Transformers: A Review—Part II: Modularity and Applications. *Technologies* **2025**, *13*, 50. [CrossRef]
32. Wang, H.; Yu, T.; Wang, M.; Ning, G. Innovative Matrix-Type AC-AC Solid-State Transformer Eliminating Ripple Power with Double-Grid Frequency. *Energies* **2024**, *17*, 1540. [CrossRef]
33. Rehman, A.; Imran-Daud, M.; Haider, S.K.; Rehman, A.U.; Shafiq, M.; Eldin, E.T. Comprehensive Review of Solid State Transformers in the Distribution System: From High Voltage Power Components to the Field Application. *Symmetry* **2022**, *14*, 2027. [CrossRef]
34. Cervero, D.; Fotopoulou, M.; Muñoz-Cruzado, J.; Rakopoulos, D.; Stergiopoulos, F.; Nikolopoulos, N.; Voutetakis, S.; Sanz, J.F. Solid State Transformers: A Critical Review of Projects with Relevant Prototypes and Demonstrators. *Electronics* **2023**, *12*, 931. [CrossRef]
35. Blanco-Ortiz, J.; García-Martínez, E.; González-Prieto, I.; Duran, M.J. A 75 kW Medium-Frequency Transformer Design Based in Inductive Power Transfer (IPT) for Medium-Voltage Solid-State Transformer Applications. *Electronics* **2025**, *14*, 1059. [CrossRef]
36. Zhu, C. High-Efficiency, Medium-Voltage-Input, Solid-State-Transformer-Based 400-kW/1000V/400A Extreme Fast Charger for Electric Vehicles. In *VTO Final Report*; DOE Award No. DE-EE0008450; Delta Electronics (Americas) Ltd.: Fremont, CA, USA; U.S. Department of Energy: Washington, DC, USA, 2023.
37. Joshi, A.; Nath, S. Efficiency Comparison of Solid-State Transformer and Low-Frequency Power Transformer. In Proceedings of the 2020 3rd International Conference on Energy, Power and Environment: Towards Clean Energy Technologies, Shillong, Meghalaya, India, 5–7 March 2021; pp. 1–6. [CrossRef]
38. Piotrowski, T.; Markowska, D. Carbon Footprint of Power Transformers Evaluated Through Life Cycle Analysis. *Energies* **2025**, *18*, 1373. [CrossRef]
39. IEEE/PES Transformers Committee, Kansas City, MO, USA, 23 October 2023. Available online: https://grouper.ieee.org/groups/transformers/subcommittees/performance/TF-CO2Emissions/F23_CO2_Presentation1.pdf (accessed on 25 February 2025).
40. Krishnan, R.; Nair, K.R.M. Carbon Footprint of Transformer and the Potential for Reduction of CO_2 Emissions. In Proceedings of the 2019 IEEE 4th International Conference on Technology, Informatics, Management, Engineering & Environment (TIME-E), Bali, Indonesia, 13–15 November 2019; pp. 138–143. [CrossRef]
41. Wang, Y.; Zhang, H.; Zhou, E.; Xie, L.; Li, J. Carbon Footprint Analysis of Distribution Network Transformers Based on Life Cycle Assessment. *Energies* **2025**, *18*, 600. [CrossRef]
42. Abu-Salah, F.; Maalouf, E. Comparative life cycle sustainability of transformer cores with metallic glass alloys. *Clean. Environ. Syst.* **2025**, *19*, 100367. [CrossRef]
43. Hegedić, M.; Opetuk, T.; Dukic, G.; Draskovic, H. Life cycle assessment of power transformer-case study. In *Management of Technology–Step to Sustainable Production, Conference Proceedings*; Croatian Association: Zagreb, Croatia, 2016.
44. Mansilha, M.B.; Brondani, M.; Farret, F.A.; Da Rosa, L.C.; Hoffmann, R. Life cycle assessment of electrical distribution transformers: Comparative study between aluminum and copper coils. *Environ. Eng. Sci.* **2019**, *36*, 114–135. [CrossRef]
45. Londero, R.P.; De Mello, A.P.C.; Da Silva, G.S. Comparison between conventional and solid state transformers in smart distribution grids. In *2019 IEEE PES Conference on Innovative Smart Grid Technologies; ISGT Latin America 2019*; IEEE: New York, NY, USA, 2019. [CrossRef]
46. Tippannavar, S.S.; Yashwanth, D. Transformer 4.0—A Smart Transformer for a Smarter Living. In Proceedings of the 2023 3rd International Conference on Innovative Sustainable Computational Technologies (CISCT 2023), Dehradun, India, 8–9 September 2023. [CrossRef]
47. Traisak, O.; Kumar, P.; Vahaji, S.; Zhang, Y.; Date, A. Advancements in Integrated Thermoelectric Power Generation and Water Desalination Technologies: A Comprehensive Review. *Energies* **2025**, *18*, 1454. [CrossRef]
48. Kumar, P.; Singh, O. Investigations on Integrating SOFC with Gas Turbine for Performance Enhancement and Sustainable Energy. *J. Energy Environ. Sustain.* **2019**, *7*, 42–46. [CrossRef]
49. Li, Z.; Sun, H.; Xue, Y.; Li, Z.; Jin, X.; Wang, P. Resilience-Oriented Asynchronous Decentralized Restoration Considering Building and E-Bus Co-Response in Electricity-Transportation Networks. *IEEE Trans. Transp. Electrif.* **2025**, *11*, 11701–11713. [CrossRef]
50. Chang, L.; Li, Z.; Tian, X.; Su, J.; Chang, X.; Xue, Y.; Li, Z.; Jin, X.; Wang, P.; Sun, H. A two-stage distributionally robust low-carbon operation method for antarctic unmanned observation station integrating virtual energy storage and hydrogen waste heat recovery. *Appl. Energy* **2025**, *400*, 126578. [CrossRef]

https://doi.org/10.3390/en19020558

MDPI AG
Grosspeteranlage 5
4052 Basel
Switzerland
Tel.: +41 61 683 77 34

Energies Editorial Office
E-mail: energies@mdpi.com
www.mdpi.com/journal/energies

www.ingramcontent.com/pod-product-compliance
Lightning Source LLC
Chambersburg PA
CBHW040809250926
51850CB00007BA/1125
* 9 7 8 3 7 2 5 8 7 3 8 2 1 *